A SCIENTIST
AT THE WHITE HOUSE

A Scientist
at the White House

*The Private Diary
of President Eisenhower's
Special Assistant
for Science and Technology*

GEORGE B. KISTIAKOWSKY

With an Introduction by
CHARLES S. MAIER

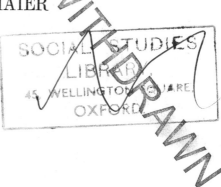
*Harvard University Press
Cambridge, Massachusetts, and
London, England   1976*

Library of Congress Cataloging in Publication Data

Kistiakowsky, George Bogdan.
    A scientist at the White House.

    Includes index.
    1. Science and state—United States.   2. Technology
and state—United States.   3. Scientists—Correspondence,
reminiscences, etc.   4. Kistiakowsky, George Bogdan.
I. Title.
Q127.U6K53   353.008′55′0924   [B]   76–19013
ISBN 0–674–79496–6

Preface

My APPOINTMENT as the special assistant to the President for science and technology grew out of my membership in the President's Science Advisory Committee (PSAC) after its establishment as a part of the White House Office, in the wake of public concern about our progress in science and technology after the Soviet launching of Sputnik I. At that time President Eisenhower appointed James R. Killian, Jr., as his first special assistant for science and technology. Killian expanded the committee that had existed since Truman's days on a lower level (in the Office of Defense Mobilization); I became one of the new members in November 1957.

Upon returning to Harvard from the Los Alamos project after the end of World War II, I had become involved, on a part-time basis, in a number of technical advisory bodies in the Department of Defense. Among these, I had been a member of the Ballistic Missiles Advisory Committee (the von Neumann committee) since its organization in 1953. It was probably my extensive knowledge of development projects related to the long-range ballistic missiles that led to my inclusion in PSAC, for I was immediately appointed by Killian as the chairman of the corresponding PSAC panel. In that capacity, I was taken more than once by Dr. Killian to brief the President on the progress and the troubles of our ICBM projects. I was also involved in several other panel activities of PSAC, so that my teaching and research suffered greatly after October 1957.

The early period of the existence of PSAC was an exciting time, with Dr. Killian playing an influential role in a number of major organizational changes: the transformation of NACA (National Advisory Committee on Aeronautics) into NASA (National Aeronautics and Space Administration) ; the creation of the high-level position of the Director of Defense Research and Engineering (DDRE) in the strengthened Office of the Secretary of Defense; and of the Advanced Research Projects Agency (ARPA) as another part of the same office. Killian was also instrumental in selecting appointees for several high-echelon positions and in greatly increasing the funding of the National Science Foundation as well as starting some specific scientific research projects. Members of PSAC then spent a goodly portion of their time in Washington, and a visitor to the PSAC-Killian offices was sure to encounter a few of them any day.

My own involvement in PSAC activities slackened somewhat at the end of 1958 and I was trying to get back into academic activities, but this process was drastically reversed once the President chose me in the spring of 1959 as Killian's successor.

On the day I was sworn in as special assistant, I started entries in a private journal concerning what was said in my presence by the people I came in contact with and the gist of my remarks to them. After holding this material in storage for more than a dozen years, I read the diary and showed it to a few people related to the publishing world to assess whether it was of historic interest. It was a pleasant surprise to discover that the Harvard University Press considered the material worthy of publication after only relatively minor additional work, and an especial pleasure that Charles Maier, formerly of the Harvard history department, agreed to participate in the preparation of the final manuscript and wrote the vitally needed overview that follows this preface. Maier was of inestimable help to me in editing the diary and in suggesting topics that called for explanatory notes. It was a pleasure to work with him; and I am especially grateful for his willingness to write the introductory remarks, which set this diary so clearly in the context of that epoch.

No substantive changes or additions have been made in the journal itself. In many places, however, the style was somewhat cryptic or telegraphic and needed editorial clarification. This has been done. Finally, for the sake of maintaining the reader's interest, I have removed entries concerning some ceremonial functions, personal matters, and office trivia. Maier and I have weighed these alterations so that no substantive event should be lost, deeming it wiser to sacrifice some streamlining for the sake of a more complete record.

I have added identifications of numerous individuals when their names are mentioned in the text, and here and there added explanatory remarks about organizations, activities, and events not sufficiently explained in the journal. All of these are set off from the text to identify them as of current origin. At the end of the book is an alphabetic glossary of the almost innumerable acronyms, code names, and abbreviations abounding in the journal.

Quite a few eliminated entries deal with my travels for public speeches. I made twenty-three formal speeches and declined invitations for sixty-five more. This formal speechmaking caused me continuing anguish and the entries left in the diary reflect it. A few remaining unflattering remarks about my speechwriters, however, are unjust. In 1959 I was unprepared for and intensely disliked constructing bland texts on nontechnical subjects several weeks in advance of delivery so that they could be cleared by the White House press secretary's office. Therefore I gave totally inadequate instructions to those who tried to write my speeches—and naturally got inadequate texts in return. My belated apologies to them. The text also contains many less-than-enthusiastic assessments of various individuals with whom I came in contact; a few comments have even been toned down. Although these were personal judgments, they are too connected with substantive issues to be dropped even when harsh. Naturally they represent a view that those cited will find partial, but a diary can be valued only if it expresses the participant's perspective.

The majority of dropped entries merely note that I had a meeting with one or more of our staff people. Almost without exception the staff was an outstanding group of dedicated and highly competent individuals on whom I relied greatly and who should be listed here as major contributors to what successes this journal describes. I list on the left those who were with me throughout the period covered by the diary; on the right, those who either joined or left during that time:

David Z. Beckler
Spurgeon M. Keeny, Jr.
Robert N. Kreidler
George W. Rathjens
Eugene B. Skolnikoff
Harry G. Watters

Donald R. Chadwick
James B. Hartgering
Frederic Holtzberg
Robert Kelman
Douglas R. Lord
George D. Lukes
Emmanuel G. Mesthene
Jesse L. Mitchell
John L. Westrate

The text that follows gives a fair idea of who among these were especially helpful in discussions of matters of policy, but I saw almost all of them frequently on more detailed matters.

We had an excellent secretarial staff; I must single out Irene Benik (now Mrs. Lee J. Haworth), whose untiring work as my executive secretary and knowledge of the inner workings of the White House Office were of inestimable help to me.

The members of the President's Science Advisory Committee played a greater policy role in the activities of our office than is reflected in the journal because I was frequently lax about recording what was discussed, so to speak, in the family.

Members of PSAC (eighteen, including the chairman) were appointed on staggered three-year terms. In my time as chairman the following members served, a few to the end of 1959, some others only in 1960.

Robert F. Bacher
William O. Baker
John Bardeen
George W. Beadle
Hans A. Bethe
Detlev W. Bronk
Harvey Brooks
Britton Chance
James B. Fisk
Donald F. Hornig
James R. Killian, Jr.
Edwin H. Land
Robert F. Loeb

Wolfgang K. H. Panofsky
Emmanuel R. Piore
Edward R. Purcell
Isador I. Rabi
Howard P. Robertson
Glenn T. Seaborg
Cyril S. Smith
John W. Tukey
Alvin M. Weinberg
Paul A. Weiss
Jerome B. Wiesner
Walter H. Zinn

The monthly PSAC meetings included the chairman's report on his recent activities, which provided an opportunity for policy discussions by members of PSAC. These normally executive sessions, were usually followed by one or more presentations by PSAC panel chairmen about their findings and conclusions, and then we heard and argued briefings from various parts of the executive branch.

In the periods between PSAC meetings various PSAC panels met. Of these about a dozen were active in my time in office, some more and some less vigorously. Each was chaired by a PSAC member and included a half dozen or so other individuals, listed as consultants to our office. Most of them were nongovernment people, which distinguished PSAC panels from several interagency panels with which I was involved.

Starting in the summer of 1960 and accelerating in the fall, the activities of the White House began to slacken except for the election campaign, from participation in which I was excused personally by the President. The journal reflects this slack—and also my declining interest in making entries because of increasing difficulties in accomplishing substantive progress, above all in the larger questions of test-ban negotiations and military research and development.

The entries become less frequent and end altogether shortly after the election of President Kennedy. I was still quite busy in Washington, did some speechmaking, and was spending more weekend time with my students at Harvard trying to rebuild my research group. Gradually, as 1960 was approaching its end, our office became engaged in the transition process and demands on me personally became minimal—I became a lame duck.

To end this preface perhaps I may be permitted a bit of self-analysis. I joined PSAC and then assumed the office dealt with in the journal, seeing myself as a technician whose task it was to execute the general policies set by my superiors. I believe the journal shows the growth of a skepticism about these policies, especially those of the Pentagon, as my term of office progressed. Conversations with the President, not all of which are here recorded, were especially influential in making me more of an independently thinking citizen, interested in the meaning and objectives of policies more than in their detailed execution. I left the office liking and respecting Dwight Eisenhower greatly. These feelings, however, do not extend to all senior members of his administration with whom I had to deal.

Contents

Science, Politics, and Defense in the Eisenhower Era

THE DIARY that George Kistiakowsky has prepared for publication covers a period of government service of only a year and a half. Like all diaries, it has the great virtue of recording immediate events and reactions and the limitations imposed by a particular and embattled perspective. For a full picture of the late Eisenhower administration, the arms race, the "scientific establishment," it can serve only as a beginning. Yet at this time it is a documentary contribution of the first rank.

Part of the reason lies in an exposure of the government's decision-making process that we cannot glean from any other written sources. The literature on the Eisenhower administration is still relatively circumscribed and rests largely on secondary accounts, interviews, and explanations of policy offered to Congress. The records of the Pentagon, the State Department, the National Security Council (which Eisenhower relied upon systematically and extensively), the Atomic Energy Commission, and other offices concerned with security and foreign policy remain classified. Declassification of documents was eased in 1972 by Executive Order No. 11652 (as well as the 1974 amendments to the Freedom of Information Act of 1966) and the staff at the Eisenhower Presidential Library in Abilene, Kansas, is helpful and obliging. (I wish particularly to thank Dennis Daellenbach, the archivist in charge of material pertaining to scientific affairs.) It should be frankly stated that the Eisenhower Library records used in the preparation of this introduction are those already declassified; the publication timetable did not permit an effort to declassify the massive materials pertaining to security affairs that are still secret and will probably remain so for many years. The procedure of opening up documentation requires specification of individual papers and a long, arduous process of clearances; only with painful slowness will the veil of secrecy slip from the material to which Kistiakowsky refers. In the meantime, the diary affords unrivaled glimpses of discussions with the President, bureaucratic and political rivalries over arms control issues, space technology priorities, and the organization of science.

On the other hand, the diary is not meant to be sensational or to contribute to what might be termed the voyeuristic side of policy formation. It concerns the public consequences of public issues, even if to illuminate these it accelerates discussion beyond what the glacial pace of declassification usually permits. If we judge from the fragmentary declassification already under way on the part of researchers, none of the material in this diary would, in principle, be kept under wraps by the various Washington offices. To record sensitive data about weaponry or about particular intelligence work is not the purpose of the journal or the goal of publication. Originally the diarist wanted merely to have his own record of interaction with the powerful men among whom he operated, so that agreements secured and disputes remaining would not be distorted. Kistiakowsky has not rushed editing of the diary, and he has reflected about possible adverse effects on national interests as well as on personal feelings. But he has decided (rightly in my opinion) that the debates in which he participated were important ones, belonging to the public domain, and helpful for intelligent weighing of contemporary problems.

The importance of the diary, moreover, transcends mere revelation of specific details and personalities. The events it illuminates occurred at a critical juncture in recent American development, even if their significance was obscured by the lackluster public image of the waning Eisenhower administration. It is helpful to make a comparison with the *Forrestal Diaries* that were published after the death of the first secretary of defense in 1949.[1] Forrestal's records rested upon an immense documentation preserved by a government figure with longer service and a more powerful

---

[1] *The Forrestal Diaries*, Walter Millis, ed. (New York: Viking Press, 1951).

role than Kistiakowsky. At their most basic level, they reflected the coming of age of American power. They intimately revealed the political pressures, collective reasoning, and ideological justifications that led the United States to assume a continuing peacetime role of enforcing world security. George Kistiakowsky served as special assistant to the President for science and technology only during 1959–1960; his position never involved direct supervision of the same massive bureaucracies and weaponry as the secretary of defense; his diary represents a single man's record and does not include the minutes and memoranda that are part of the Forrestal papers. Nonetheless, his diary permits some similar insights. Through Kistiakowsky's eyes the reader can witness part of the transition to the missile age, in which the United States and the Soviet Union would revolutionize their atomic delivery systems and inaugurate an era of new weapons rivalry and satellite intelligence.

The diary is particularly revealing, moreover, since it spans a period of policy ambiguity during which the Eisenhower administration felt compelled to press forward with advanced missile development in the wake of the Russians' Sputnik I success in 1957, while at the same time the President himself sought to check the nuclear arms race. Simultaneous encouragement of arms control and of qualitative improvement of weaponry necessarily stimulated policy advocates of opposite viewpoints. It made the opposition among intragovernmental interests and spokesmen all the more unrelenting. The special assistant of 1959–1960, moreover, was in an especially "privileged" position to participate in, and to monitor, these fierce and not always muffled rivalries. The personal trust that Kistiakowsky elicited from Eisenhower is attested to by several sources.[2] Furthermore, the relatively new position of the special assistantship had not yet lost influence to the other White House advisors concerned with national security issues. The special assistant for science and technology might command no "on-line" power, but in the post-Sputnik period—at a moment too when the interests of the Pentagon and the demands for arms control seemed to collide—Kistiakowsky's office retained

[2] I am grateful to David Z. Beckler, Gen. Andrew Goodpaster, Spurgeon M. Keeny, Jr., James R. Killian, Jr., George W. Rathjens, Eugene B. Skolnikoff, and Herbert F. York for opportunities to discuss Kistiakowsky's work, the role of science-advising in the Eisenhower administration, and general political background. I have relied on their insights at many points in this essay, but have footnoted only specific quotations or opinions. Naturally, errors of fact or deficiencies of interpretation are my sole responsibility.

the potential for critical influence. Likewise, it promised to become a decisive mediator among the diverse constituencies of scientists that the government now sought to coordinate and mobilize. Consequently, the diary plunges us into the central currents of policy-making during a period that few other sources can so directly clarify.

A word about my own minor role in bringing this record into print. After reading the manuscript of the diary, I encouraged Kistiakowsky to press ahead with his hopes of publication. I have helped with minor prose editing and assisted in the division into chapters and themes for the sake of easier reading. As a non-scientist I pressed Kistiakowsky for explanations of technical passages so they might be formulated for the lay reader. As a historian I have contributed these introductory remarks, which are intended, first, to evoke briefly the events and ambiance of the late Eisenhower administration; second, to fill in background about the progress of the missile race; then, to review the status of arms control negotiations; and finally, to discuss the role of Kistiakowsky's office in the structure of the American scientific effort and its relation to the agencies concerned with the country's defense. Although this essay concentrates heavily on the national security background of Kistiakowsky's work, I do not mean to imply that the organization of science represented a lesser effort for the special assistant. Certainly it was an unwieldy and preoccupying task, and my few concluding pages are meant only to signal the important transition that the scientific enterprise was then undergoing. Specialists in each of the areas discussed below doubtless will find the treatment superficial; nonetheless, my objective is not to write a full monograph, but to suggest an intelligent historical placement of this valuable documentation.

THE EISENHOWER ADMINISTRATION that Kistiakowsky joined in midsummer 1959 had a different mood and a sense of direction than it had in 1953. The approach of the President to questions of foreign and economic policy had mellowed; his priorities had shifted. Although Eisenhower's nomination in 1952 had deprived the Republicans of a far more ideologically consistent conservative, GOP members who came to Washington in the winter of 1953 still enjoyed the heady sensation of leading a political crusade. They felt themselves to have recaptured the government after twenty years of dangerous social experimentation, domestic

suffusion by Communism and corruption, a lack of national resolve as reflected by the Korean stalemate, long-term fiscal
slovenliness, and a general erosion of American free-enterprise
virtues. After the electoral victory one enthusiastic Senator bran-
dished for news photographers a new broom, one that supposedly
would sweep Washington clean of all the accumulated New Deal
and Fair Deal social ills.[3]

Eisenhower himself embodied more than an appeal to rollback
of the Democratic Party regime. Without his personal warmth,
the assurance of stability that he generated, his tolerance and
common sense, the Republicans who so thirsted for victory might
have been frustrated once again. Certainly the GOP congressional
triumph lagged behind the President's 55 percent of the record
vote: Republicans gained a modest 22 seats in the House to reach
a majority of 8 and, after Wayne Morse had defected to the
Democrats, retained no more than a tie in the Senate. Subsequent
polls would erode the Republican margins in the legislature, to
the extent that Eisenhower would face a House and Senate controlled by the Democrats from the midterm elections of 1954
through the remaining six years of his presidency.

But if Eisenhower returned the GOP to control of the White
House in 1953 only by virtue of subduing the fervor of the Right
to capitalize on the sentiment that it was "time for a change,"
he still effectively gave the Republicans much of what they
emotionally sought. The new President shared their sympathies
even if he sought to realize them in a less systematic and doctri-
naire way. Above all, he accepted the policy preeminence of John
Foster Dulles and George Humphrey, the persuasive and systematic representatives of an internationally minded but refrac-
tory anti-Communism in the one case, and a tough fiscal conservatism in the second. The conferences of the President-elect
and his major cabinet nominees aboard the cruiser *Helena* immediately after his December 1952 visit to Korea outlined future

[3] For general narratives of the Eisenhower administration, see Herbert S.
Parmet, *Eisenhower and the American Crusades* (New York: Macmillan Co.,
1972); Charles C. Alexander, *Holding the Line: The Eisenhower Era,
1952–1961* (Bloomington and London: Indiana University Press, 1975); the
President's own memoirs, *The White House Years*, vol. 1: *Mandate for
Change, 1953–1956* (Garden City, N.Y.: Doubleday & Co., 1963) and vol. 2:
Waging Peace, 1956–1961 (Garden City, N.Y.: Doubleday & Co., 1965);
Sherman Adams, *Firsthand Report: The Story of the Eisenhower Adminis-
tration* (New York: Harper & Brothers, 1961); and Emmet John Hughes,
The Ordeal of Power: A Political Memoir of the Eisenhower Years (New
York: Atheneum Pubs., 1963).

policy preferences that fit together as a coherent whole. The talks reaffirmed a commitment to containment, an urgency about dropping price controls and cutting the budget (though early hopes here were quickly disabused), and the establishment of a defense posture according to the overriding priorities of economy and static containment. Finally the *Helena* discussions suggested the general predominance of business and military policy advisors.

Much of the new President's policy drew upon the steady influence of Humphrey and Dulles. George M. Humphrey, chairman of the board of the Mark A. Hanna coal and steel company in Cleveland, was initially unknown to Eisenhower; originally recommended by Lucius Clay, he increasingly impressed the President as competent and forthright. Humphrey's Thomasville, Georgia, plantation where Eisenhower could retire for quail-hunting no doubt served further to buttress its owner's genial but determined advocacy of American enterprise. Perhaps because Eisenhower had matured within a military service that was hermetically sealed in the American establishment, he remained immensely sensitive to the camaraderie, loyalties, and collective thinking generated in the informal, clubby milieus he sought out —whether on shipboard, in Camp David, or at Burning Tree. In this responsiveness he resembled his predecessor Harry Truman, even though both presidents sought to clarify and strengthen formal departmental lines of executive authority. Humphrey's quiet conviction also served him better in influencing the President than did Defense Secretary Charles E. Wilson's tendency to insistent argument punctuated by unfortunately caricatured aphorisms. If Wilson, as chairman of General Motors, possessed the executive authority to take over the Defense Department, Humphrey was better fitted to impress his concepts upon the White House. He developed an informal and easy access to the President, and by March of 1953 Eisenhower had invited him to participate regularly in the National Security Council forum so that the voice of fiscal prudence would temper military and foreign policy from the start.

Both men shared the same economic orthodoxy. Excessive government spending precluded a healthy economy, which in turn was central to the preservation of American individual and collective well-being. High taxes, even without deficits, channeled money from the productive private sector to the sterile public purse; and with deficits, the danger of inflation added a further economic threat. Eisenhower obviously did not need the outcry that followed Humphrey's impromptu remark in January 1957 that without budget reduction the United States in the long run would undergo "a depression that will curl your hair." On the other hand, he shared the belief that the government claim on national output was a regrettable necessity imposed by national security: somewhere in the future he envisaged a return to a simpler and more modest fiscal polity.[4]

Eisenhower's relation to Dulles was subtle and complex. Like Herbert Brownell, whom the President admired for intellect and loyalty, Dulles offered such compelling qualifications and long-term party preeminence that no other serious candidate emerged for secretary of state. Dulles was impressive in his conspicuous moralism; he seemed to provide an overarching sense of purpose to the defense against Communism and he complemented his Manichaean world-view with an extraordinary grasp of detail in specific negotiations. On the other hand, his monolithic vision of America's mission tended to become so sweeping that it encouraged a disturbing hands-off vacuousness when specific policy choices had to be faced: at home, Dulles allowed Scott McLeod and Joseph McCarthy to savage State Department morale; abroad, the Dullesian summons to "liberation" and "rollback" could not be translated into a more militant policy without risks that Eisenhower (and Dulles) were unwilling to run. Consequently, even when such close allies as the French, to say nothing of the Hungarian revolutionaries of 1956, capitalized on the anti-Communist theme, they found the Eisenhower administration evasive. Dulles's historical role was to firm up the perimeter between "the free world" and "the slave," to use the rhetoric of the cold war, to corset as many gray areas as possible in the Mideast or Southeast Asia into anti-Communist alliances, while making sure that the core Western allies, including the West Germans, maintained a firm anti-Soviet posture. From this point of view, statements of intent and actions of symbolic value—such as the refusal to evacuate the offshore Chinese islands of Quemoy and Matsu in 1954 and again in 1958—proved to have a major role in the administration's calculus. The important objective was to keep the boundary between light and darkness a sharp one, and to this end the defense of even the least significant non-Communist territories appeared critical. Where concessions had to be made, as for instance at the 1954 Geneva conference that consecrated the Viet Minh victory in North Vietnam, Dullesian concepts proved unserviceable; the United States had to fall back

[4] Cf. Eisenhower, *Waging Peace*, pp. 127 ff.

on what looked like a semihypocritical, ostentatious noninvolvement. Underlying this policy there remained the hope that in the long haul the Communist system would be forced to evolve from within: liberation was not entirely an empty concept, but it signified a prolonged historical evolution. Unfortunately, electoral politics assured that Republican speakers would exploit the theme to raise far more immediate and ungrounded hopes among Americans of Eastern European origin. For Dulles, however, liberation remained a meaningful eschatology, as assured as ultimate redemption in the Christian sense.

Recent authors have suggested that Eisenhower was skillfully Machiavellian when he allowed his secretary of state to defend harsh positions he knew could not ultimately be maintained. Somewhat unjustly, Dulles—not Eisenhower—was seen as contemplating the possibilities of Indochinese intervention; more accurately, Dulles—not the President—appeared responsible for the policy reversals on aid for the Aswan Dam that culminated in the Suez debacle. Eisenhower's basic moderation and realism did set him apart in temperament and approach from his secretary. Nonetheless, genuine affection and admiration marked the President's treatment of "Foster"; the secretary of state would join Eisenhower in the Oval Office at 6:00 PM and together the two men would ruminate on the larger goals they were pursuing, meditate on the objectives of the Soviets, weigh whether the cold war would end only in generations or by a sudden reversal. Both leaders shared a commitment to a long moral crusade: if Eisenhower was among the least bellicose of men, he still had spent his life serving the defense effort of the United States and learning as second nature (more in peacetime than in war) how much drudgery and patience this engagement entailed. Dulles, for whom all of political life embodied a grand theodicy, felt that the cold war beneficially counteracted an otherwise corrupting affluence. The President admired the architectonic structure of Dulles's policies, and there is little reason to doubt that the simplicity of the secretary's grand design did not also suit the Presidents' own inner preferences and bloc-like concepts.[5]

[5] Townsend Hoopes, *The Devil and John Foster Dulles: The Diplomacy of the Eisenhower Era* (New York: Little, Brown & Co., 1973) offers the best survey of Dulles's role; a qualified defense and emphasis on flexibility in practice emerges from Michael A. Guhin, *John Foster Dulles: A Statesman and His Times* (New York: Columbia University Press, 1972); for a more human side see Eleanor Lansing Dulles, *John Foster Dulles: The Last Year* (New York: Harcourt, Brace and World, 1963). Eisenhower's wiliness

Eisenhower was not an early "cold warrior," but by 1952 he had come to accept the conventional wisdom on the nature of America's protracted conflict with Communism. At the end of the Second World War, Eisenhower had resisted early diagnoses of inevitable enmity with the Soviet Union: as European commander he found the Russians helpful and punctual allies; he developed a confidence in the reliability of his counterpart, Marshal Zhukov; and he required Soviet cooperation to carry out his obligations in the four-power occupation of Germany. On the other hand, as the incipient antagonism with Moscow rigidified into the cold war, Eisenhower accepted its ideological postulates without great subtlety or intellectual refinement. Called upon to chair the Council on Foreign Relations' high-powered study group on Aid to Europe in 1948, Eisenhower dutifully weighed the opinions of the academic and business participants. When challenged for a definition of a "friendly" country, the general answered any nation that was against Russia; and he defined the ultimate Soviet goal as "world revolution of the proletariat," which struck some of his colleagues as simplistic.[6]

Similarly, with Communism at home: the general explained in his farewell talk to the faculty of Columbia University that the world was engaged in a "war of light against darkness, freedom against slavery, Godliness against atheism," and that he would not have taken the job if as an American university Columbia was harboring even a single known Communist. In contrast to Senator McCarthy, however, the general did not feel that America's universities offered a haven for Communists; instead they were filling their function well of educating citizens for democratic participation. And in contrast to the Republican Old Guard, he did not believe that it was sufficient to resist Communist encroachment from a fortress America. In the Council on Foreign Relations' study group he had declared that the United States could not remain an island of freedom and democracy if the rest of the world were dominated by Communism. At this First Inaugural Address he emphasized the American linkage with free peoples everywhere, and in the Second Inaugural he reaffirmed: "No

---

is stressed by Parmet and also by Garry Wills, *Nixon Agonistes: The Crisis of the Self-Made Man* (New York: Houghton Mifflin Co., 1970), chap. 6. For the Eisenhower-Dulles reflections and the secretary's longer-term views, see Eisenhower, *Waging Peace*, pp. 367–368, 373.

[6] Council on Foreign Relations archives: Records of Groups, XVIII—a digest of discussion of first meeting, January 10, 1949, and second meeting, January 18, 1949.

nation can be a fortress, lone and strong and safe.'' While Dulles might outrun the President in the almost martinet-like compulsiveness of his anti-Communist commitment, Eisenhower could not have quarreled with the underlying ideological conceptions.[7]

Containment of Communism and the limitation of public-sector claims thus represented Eisenhower's early fundamentals. For defense policy these priorities suggested emphasis upon retaliatory capacity and downgrading of the conventional role of the army: the ''New Look,'' as Admiral Arthur Radford coined the phrase in 1953. At the beginning of his administration, Eisenhower decided that any major war would be defensive, that deterrence by means of nuclear weapons would be decisive, and that economic ''strength'' had to remain a criterion of strategic options. In the wake of Korea and the rapid expansion following the NSC-68 study of 1950, the administration found a situation where, in the words of George Humphrey, six different strategies competed: a conventional and a nuclear one for each Service.[8]

The frustrations engendered by the Korean stalemate buttressed the economic arguments that called for cohesive military priorities which would avoid bogging down American troops in any future engagement. While Eisenhower disappointed Robert Taft in not further reducing the defense budget at the outset, he imposed an approximate upper limit on military spending of about $40 billion for fiscal year 1954 and about $35 billion for fiscal year 1955. The cuts would come at the expense primarily of the Army, but the President believed a nuclear deterrent strategy would maintain a level of preparedness most suitable for ''the long pull.'' Indeed, Eisenhower thought in terms of a half-century effort. The long pull, in turn, meant avoiding a level of spending that would impose fiscal stress and strain, budget

[7] Eisenhower's views at Columbia are cited in Parmet, *Eisenhower and the American Crusades*, p. 161. See also the Council on Foreign Relations archives: Records of Groups, XVIII—a digest of discussion of June 27, 1949; and for the inaugural addresses: *Public Papers of the Presidents of the United States: Dwight David Eisenhower* (8 vols., Washington, D.C.: GSA, National Archives and Records Service, 1960–1961), *1953*, pp. 4–7, and *1957*, pp. 63–64.

[8] Cited in Adams, *Firsthand Report*, p. 398. On the New Look see Glenn H. Snyder, ''The 'New Look' of 1953,'' in Warner R. Schilling, Paul Y. Hammond, and Glenn H. Snyder, *Strategy, Politics, and Defense Budgets* (New York: Columbia University Press, 1962), pp. 379–524; Eisenhower, *Mandate for Change*, pp. 446–455; Samuel P. Huntington, *The Common Defense: Strategic Programs in National Politics* (New York: Columbia University Press, 1961), pp. 64–106.

deficits, or an inflation that would sap the free-enterprise liberalism which in the last analysis was the purpose of national defense. In his first State of the Union address, Eisenhower outlined the objective that would remain constant throughout the administration: "Our problem is to achieve adequate military strength within the limits of endurable strain upon our economy. To amass military power without regard to our economic capacity would be to defend ourselves against one kind of disaster by inviting another."[9]

Eisenhower's concept was logical but vulnerable to serious challenge. Strategically, the New Look required thinking through just what capabilities the United States could bring to bear; it fitted in with Dulles's efforts to throw an umbrella of American protection over as much of the non-Communist world as would accept our leadership. A "trip-wire" army presence in Europe, which would ensure U.S. response to any aggression against its allies, and the implicit threat that Soviet or Chinese expansion would be answered by a strike at the Communist heartlands, offered—so it originally seemed—the most watertight protection.

Dulles outlined massive retaliation in May 1952, calling it "the will . . . and means . . . to strike back where it hurts." He elaborated the implications in his address to the Council on Foreign Relations in January 1954: the aggressor could not choose the battlefield that suited him, but must expect an American response "at places and with means of our own choosing." In light of the presumed disadvantage in conventional manpower, massive retaliation was a logical response if Washington was determined to maintain an alliance system. With Dulles's apocalyptic rhetoric, moreover, the United States gave the appearance of being sufficiently committed to the single-minded and automatic resolve that massive retaliation required for credible deterrence. Even Dulles did not believe the Soviets were lunatic enough to risk their industrial achievements by an attack in Europe or a nuclear exchange. Still, as Samuel P. Huntington has pointed out, the equilibrium and protection that the strategy provided was fleeting and confined at best to the middle years of the administration.[10]

[9] *Public Papers of the Presidents: Eisenhower, 1953*, 17, "Annual Message to the Congress on the State of the Union," February 2, 1953.

[10] Richard A. Aliano, *American Defense Policy from Eisenhower to Khrushchev: The Politics of Changing Military Requirements, 1957–1961* (Athens, Ohio: Ohio University Press, 1975), pp. 39–43; Eisenhower, *Waging Peace*, p. 369; and Huntington, *Common Defense*, pp. 88 ff.

When first enunciated, the concept of massive retaliation implied that the Soviets were so behind in nuclear delivery capacity that the United States retained decisive superiority; but by the mid-1950's this assumption was questionable. Would Washington really incur massive destruction on its own territory in a nuclear exchange? The all-or-nothing alternative now appeared so perilous that it was less credible than in 1953–1954. Limited warfare on a third country's soil once more became a real possibility; thus, too, Army advocates of concentration on the appropriate strategic requirements became more vocal. The Soviet achievements in rocketry would move these challengers center-stage.

The economic postulates of the New Look also came under attack. Eisenhower persisted in his anti-inflationary priorities. "Playing with inflation," he continued to believe, even after watching the Kennedy tax-cut from the postpresidential calm of Gettysburg, "is the height of reckless folly."[11] If expansion and flush times brought economic growth, the President felt, then the government's duty was to amortize the national debt. And when in 1959–1960, trusted advisors such as Gabriel Hauge warned against a quick effort to go from a budget deficit to surplus lest a new recession be precipitated, Eisenhower still felt that he must act emphatically to move into the black in order to staunch a growing balance-of-payments problem. It would be wrong to depict the President's fiscal conservatism as too absolute. Within the administration, the Council of Economic Advisors, chaired by Raymond J. Saulnier during the final period of Eisenhower's presidency, was more flexible than George Humphrey's successor at the Treasury, former secretary of the navy Robert B. Anderson. And as Kistiakowsky's own record of the National Security Council session of October 31, 1960, vividly suggests, Eisenhower could take Anderson's dour predictions on the fate of the dollar with a grain of salt. As in the case of Dulles, what separated the President from the Texan he generally admired was more a question of tone and approach than of intellectual dissent: the objective of public frugality remained unabating.

On the other hand, critics felt that in the wake of the 1957–1958 recession and the fear of a new slump in 1960, the administration's economic postulates were out of date. Eisenhower's homey explanations that a government, like a household, could

---

11 Cf. Eisenhower, *Waging Peace*, pp. 461, 463–464. For the differentiation between the Treasury and the Council of Economic Advisors, I am drawing on General Goodpaster's observations.

not live indefinitely beyond its means seemed to betray a lack of economic sophistication. Administration spokesmen insisted that normalcy could prevail, that the American population could have a strong space program without having to renounce the good life of "the outdoor barbecue and two cars in every garage."[12] At the same time, however, Eisenhower apparently envisaged economic output as finite and feared the destruction of incentives if the public sector claimed too much, especially if it did so through deficit-finance gimmickry. Democratic Party critics and Keynesian economists countered by suggesting that the productive system might work at a higher level to the benefit of military hardware, housing, hospitals—and without cost to the private economy. Although some writers were willing to prescribe private retrenchment, or at least a plateau on household consumption, on behalf of collective expenditures—the argument to be advanced most forcefully in John Kenneth Galbraith's *Affluent Society* of 1958—most critics did not pose a great debate over priorities. Instead they indicted fiscal inefficiency. Defense and space expenditures could almost pay for themselves under a growth strategy. If this was the case, then the economic postulates that underlay the Eisenhower spending ceilings, the reluctance to embrace crash programs, the insistence on the static "long pull," were discredited.

Even as the assumptions upon which the search for military stability rested were coming under attack, the President himself attempted to move decisively away from cold-war stalemate. While the death of Dulles grieved the President, it also liberated him. Efforts at arms control or even détente had not been absent from the early Eisenhower years, but they had remained sporadic, although dramatic and inspiring, gestures. The "Atoms for Peace" address of December 1953 and the "Open Skies" proposal for aerial military surveillance made at the Geneva Summit conference in July 1955 were designed to demonstrate that the United States did not wish to remain in a stalemate forever. Eisenhower sought also to rally third countries and to reassure the current of American opinion that yearned to take advantage of any flexibility Stalin's successors might demonstrate. The proposals did not directly collide with the powerful bureaucratic interests of the Pentagon, and indeed little Soviet response was

[12] T. Keith Glennan, NASA administrator, on "Face the Nation," September 20, 1959, CBS News, *Face the Nation*, vol. 5, *1959* (New York: Holt Information Systems, 1972), pp. 296–303. The phrase about barbecues and cars was the questioner's.

expected. When in March and April 1957, Valerian Zorin indicated to the UN Subcommittee on Disarmament that the Soviets might be prepared to accept the principles of aerial surveillance and ground inspection demanded by the United States, the administration seemed embarrassed and backed away. At the same time the early nonproliferation treaty proposals of Harold Stassen, Eisenhower's disarmament advisor, met disabling opposition from the British, who needed time to produce their thermonuclear arsenal. By the time Washington worked out counterproposals, Moscow grew impatient and in late August reverted to truculence.[13]

By 1958 and 1959, however, the President was willing to turn more resolutely toward questions of arms control. No doubt he perceived that the rapid Soviet advances in missile technology doomed all thoughts of a costless massive retaliation policy. The "Gaither report" of outside experts summoned by the National Security Council and the Office of Defense Mobilization in the summer of 1957 projected that the Russians' weapons buildup could deal a devastating blow to the United States by 1959–1960. The vulnerability of both the Strategic Air Command (SAC) bombers and the civilian population was undermining deterrent credibility. With relatively inexpensive measures (largely met by the administration through supplemental fiscal year 1959 budget appropriations) security of the SAC forces could be significantly enhanced. To shield the population from radioactive fallout and reduce potential deaths from perhaps 120 million to 55 million would have involved a fiscal strain that Eisenhower dreaded, an outlay of from $5 billion to $12 billion more per year. Even if American retaliation could follow such an attack, what quality of life would the survivors experience; what would be the meaning of victory? Although some of its recommendations were leaked, the Gaither report was officially interred by classification. Its message, however, was so stark that the alternative of arms limitation must have struck the President even more urgently.[14]

[13] See Alexander, *Holding the Line*, pp. 204–206; Eisenhower, *Waging Peace*, pp. 467–474; United States Disarmament Administration, Department of State, *Geneva Conference on the Discontinuance of Nuclear Weapons Tests: History and Analysis of Negotiations* (Washington, D.C.: U.S. Department of State, 1961), pp. 9–11.

[14] NSC 5724: "Deterrence and Survival in the Nuclear Age: Security Resources Panel of the Science Advisory Committee, Office of Defense Mobilization," November 7, 1957. The report, given its name from panel chairman H. Rowan Gaither, Jr., has only recently been declassified. For a

Finally, less tangible factors entered in. Dulles's passing removed the impressive argumentation behind eternal immobility. The desire to achieve a significant step toward ending the cold war became more pressing as the end of the President's eight years came into sight. The diary entries below, as well as other printed and oral testimony, indicate how compelling became the President's hope to move beyond the balance of terror.

As military spokesmen—most notably Gen. Thomas Power, who remained a gadfly in the Strategic Air Command, and generals Maxwell Taylor and James M. Gavin, who resigned their commissions to speak out—increasingly criticized what they perceived as a deterioration of preparedness, Eisenhower vented his own impatience with the advocates of heavy new defense expenditures. Economic convictions and concern about the effects of an open-ended arms race converged. As the President told a group of science advisors (including Kistiakowsky) on May 20, 1959, he always came back "to the basic truth." The United States must find ways of cooperating with other countries, "or else." He "did not think we could carry forward the heavy burdens for armaments that we are now supporting. Inevitably we would have to go to controls throughout our society, or at the least to bureaucratic domination."[15]

Despite the President's determination, the administration reached a difficult impasse during its last two years. While Eisenhower sought to move beyond the perpetual confrontation that appeared to have become Dulles's single-minded objective, public concern focused increasingly on deficient American power and technological achievements. Could the nation accelerate rocketry and defense preparations, and simultaneously unwind the cold war? The diary reveals some of the White House's efforts to respond: the contrapuntal encouragement of research and upgrading of the military alongside the vigorous round of personal diplomacy. Unfortunately, the momentum of presidential negotiation was to be shattered by the events surrounding the U-2 affair and the collapse of the Paris Summit meeting in 1960, leaving little more than custodial foreign-policy tasks.

---

study of the conflicts within the government to which it gave rise, see Morton H. Halperin, ''The Gaither Committee and the Policy Process,'' *World Politics* 13 (April 1961), 360–384.

[15] Eisenhower Presidential Library: DDE, Papers as President (Ann Whitman file), Diary Series, Box 26, F, ''Staff Notes, May 1959 (1),'' Goodpaster notes: memorandum of conference with the President, May 20, 1959.

Focused as it is on the White House, the diary is less informative about the Democratic opposition's response during the last two years of the administration, as Lyndon Johnson impressively conducted the Senate Subcommittee on Military Preparedness, listened sympathetically to Pentagon misgivings about spending strictures, and explored the reasons for lagging missile development. The Soviet successes in space, the approach of a presidential election in which Eisenhower's personal magnetism would no longer be directly at stake, impatience with the laggard economic performance since 1957, the widespread uneasiness that equilibrium might mean only stagnation—all stimulated new appeals for movement and activism. The objectives that the electorate had endorsed in 1952—a return to private accumulation and family well-being, goals that Eisenhower felt were compatible with a healthy public community because they still mobilized entrepreneurial and organizational energies—now seemed insufficient to intellectuals and to the press. The longing for a public mission revived.

What principally distressed critics was the administration's failure to respond to Soviet gains with a sense of urgency. In the wake of Sputnik and the Gaither report, Eisenhower refused to introduce crash programs. His continuing conviction that the President had best stay above some political battles, his lack of peturbation when the United States seemed to be losing "prestige," his frequent recreation on the golf course, signaled to his opponents first a failure of comprehension, then a failure of energy. Certainly the diary entries below and the memoranda available at Abilene (aside from testimonials of those who wish to preserve his reputation) suggest, however, that the President purposefully chose and understood the stakes of his cool response in the aftermath of Sputnik. But if deliberate calm represented a sane approach to the arms race, it was a method tarnished by having served the President in confronting McCarthyism, racial integration, and recession, issues in which dramatic action might well have been more appropriate. Composure appeared as complacency. While the steady equanimity of the President in security matters was probably statesmanlike, to his detractors it appeared as a continuation of long-standing reluctance to deal with harsh and unpleasant issues.

While Senator John F. Kennedy would campaign successfully on the theme of getting America moving again, his supporters often seemed to yearn more for motion itself than for movement in a given direction. Ultimately they too faced the dilemma of objectives that were not entirely consistent. Impatient earlier with Dulles's nay-saying, the Democrats now stressed the weakness of the country and the so-called missile gap. By the last years of the administration, therefore, both the White House and its critics seemed to pursue aspirations that often conflicted: invulnerability and détente. Such dual goals—paralleled, too, by the mercurial objectives of the Soviets under Khrushchev—made the administration, and afterward the Democrats, less likely to achieve success in overcoming the cold war. At the same time, such policy ambivalence emerged almost inevitably out of the balance of terror internationally and the two-party competition at home. Security and disarmament proved beyond the President, although security and an edgy peace did not.

REFLECTING UPON his service as secretary of the navy, undersecretary of defense, and finally as defense secretary from late 1959 through January 20, 1961, Thomas Gates later stressed:

> All of a sudden the scientists became very important . . . They had great veto power. They became very important people. You paid a lot of attention to the extreme people, like Edward Teller, [also to] Johnny von Neumann who died, Wernher von Braun, Jim Killian, Kistiakowsky—these people became very important . . . The world really completely changed, in terms of military affairs. And foreign policy changed with it.

Gates recalled at a subsequent interview how rapidly the change was thrust upon the administration:

> The development of nuclear weapons, the development of technology, the development of everything came so rapidly that we could see whole new requirements in light of what was happening. In 1953 no [operational] supersonic planes, no [long-range] missiles. It all happened in two or three years.[16]

The Soviet launching of Sputnik I on October 4, 1957, catalyzed uneasiness about American preparedness. It followed over half a year of renewed political quarrels and erosion of the confidence in Ike registered by the electoral triumph of November 1956. The year 1957 brought Democratic and conservative GOP efforts to pare foreign aid and other budget items after George Humphrey's

[16] Thomas Gates, Jr., Oral History interviews, August 1967 and October 1972; Columbia University Oral History collection and DDE Presidential Library. Quoted with permission.

awkward warning about a hair-curling depression. It also saw the mounting dissatisfaction of black leaders with the President's reluctance to use his position for a moral appeal against segregation and the racist incidents that accompanied its defense. In the wake of the Sputnik launching, the *New York Times* wrote that the Russians had an ICBM capability and might soon be able to threaten American cities with annihilation. The editors then pointedly asked, "Is the policy of putting domestic budgetary and political considerations ahead of security considerations in allocating funds for defense still a tenable policy in the present situation?" For labor leader Walter Reuther, Sputnik represented "a bloodless Pearl Harbor."[17]

Eisenhower sought initially to assure the American public that the United States had not fallen irremediably behind in space technology. At his news conference of October 9, 1957, he asserted that his scientific and military advisors believed defense programs need not be drastically accelerated. Unfortunately Sherman Adams exposed such reassurances to attack when he flippantly remarked that the U.S. space program aimed at advancing science, "not high score in an outer space basketball game." With more sensitivity to public uneasiness, vice-president Richard M. Nixon, on October 15, warned against brushing off Sputnik as a scientific stunt. The Russians launched a canine cosmonaut in a half-ton satellite on November 3. Eisenhower delivered a television talk four days later in which he emphasized again that the American space program was not designed as a competitive race; this was a theme he would cling to throughout the next three years and which the diary reveals him reiterating at many junctures. In his speech he insisted that the United States program was proceeding on schedule, and announced the appointment of James R. Killian, Jr., president of M.I.T., to be—if not exactly the awaited "missile czar"—special assistant to the President for science and technology. The President also emphasized that "as of today the overall military strength of the free world is distinctly greater than that of the communist countries," and he warned against trying "to ride off in all directions at once," although he did concede in Oklahoma City six days later that the $38 billion defense spending ceiling was not absolute.

Critics were not satisfied. The President had sanctioned no crash program; he seemed unable to grasp the seriousness of the psychological setback, much less the technological one. Behind

17 Citations from Aliano, *American Defense Policy*, pp 49, 58.

the scenes the business and political leaders who had cooperated on the Gaither report found the President relatively unresponsive at their major National Security Council presentation on November 7. A policy of complacency toward Soviet industrial and scientific achievements seemed unshakable. On November 26 Eisenhower suffered a minor cerebral stroke; a few days later, the first scheduled American satellite launching with a Vanguard rocket failed spectacularly directly after liftoff.

Sputnik naturally encouraged critics of restrained defense budgets and fiercely intensified interservice jockeying for the privileges of missile development and deployment. The resultant conflicts were to become daily fare for the early special assistants, Killian and Kistiakowsky, as Eisenhower relied on "his scientists" to adjudicate the Service claims. Nonetheless, it would be misleading to see the problems that this diary stresses as merely post-Sputnik issues. They were rooted in the longer-term development of new technologies and disarmament impasses that Sputnik dramatized but did not originate.

Satellite successes and missile launchings had long been interlocked fields of endeavor. The Navy had proposed development of a satellite as early as 1946, although until 1954 the Joint Research and Development Board's Committee on Guided Missiles rejected the proposal as of insufficient military value. When the American delegation to the Committee for the International Geophysical Year (July 1957 to December 1958) accepted a commitment in 1955 to strive for a satellite launching as the major U.S. contribution, the administration endorsed the research effort and turned it over to the National Science Foundation as the American liaison with the IGY. Although the Vanguard satellite project was to rely upon a Navy booster, keeping it separate from military missile development was designed to avoid subsequent sharing of data pertinent to defense matters. The decision further reflected Eisenhower's underlying commitment to keep space exploration a civilian enterprise. In the aftermath of Sputnik and the early Vanguard failure, Army spokesmen did not fail to announce that had its Redstone Arsenal team under the direction of Wernher von Braun (then moving from completion of the 200-mile-range Redstone rocket to testing of the Jupiter-C booster) been allowed to launch the satellite, America could have beaten the Soviets into orbit.[18]

[18] See Enid Curtis Bok Schoettle, "The Establishment of NASA," in Sanford A. Lakoff, ed., *Knowledge and Power: Essays on Science and Government* (New York: Free Press, 1966), pp. 162–270; also Michael H.

The divorce of satellite and missile programs was one reason for the relative American delay in matching the Soviet achievement; another lay in the way American rocketry had itself developed. Both sides, of course, had pursued missile technology since the capture of much of the German V-2 Peenemünde team at the end of World War II. But the Soviets consistently worked to produce the large rockets needed to carry payloads into earth orbit. While in 1946 the Air Force had turned to the Convair Corporation with a contract for a project MX-774—a huge rocket assembly analogous to that of the Soviets—the contract was allowed to lapse a year later. The Peenemünde crew under von Braun remained under Army control at Fort Bliss, Texas, developing guidance systems and experimenting further with captured V-2's. The Air Force leadership remained relatively uninterested, as its generals concentrated single-mindedly on development of the B-36, then the B-47 and the B-52: a passion that has continued with the unsuccessful quest for the B-70 in the 1960's and the proposed B-1 of the present day. In 1947–1948, however, few advisors had called urgently for further missile development; even Vannevar Bush, science advisor to Roosevelt and for a time to Truman, believed that ICBM's would be uneconomical and unlikely deterrents.[19]

By the end of the Truman administration, this cavalier evaluation of long-range missiles had begun to change: the Army had installed von Braun's crew at the Redstone Arsenal at Huntsville, Alabama, and set them to work developing a ''tactical'' ballistic missile. Long-range rockets had been reserved for the Air Force by agreement of the Joint Chiefs of Staff and directive of the secretary of defense in March 1950. The explosion of a Soviet atomic bomb in 1949, which starkly revealed how brief any American monopoly of nuclear power was to be, had also prodded missile development. Still, attention had not immediately turned back to large space vehicles.

Following intense debate within the science community and the Atomic Energy Commission, Truman had decided to place highest priority on development (though not necessarily full pro-

---

Armacost, *The Politics of Weapons Innovation: The Thor-Jupiter Controversy* (New York: Columbia University Press, 1969). For an impassioned attack on Vanguard and a defense of the Huntsville Army Ballistic Missile Agency by its commander, see John B. Medaris, with Arthur Gordon, *Countdown for Decision* (New York: G. P. Putnam's Sons, 1960).

[19] For this material I have relied extensively on Armacost's excellent monograph, as well as discussions with Kistiakowsky.

duction) of a hydrogen or superbomb. At Eniwetok Atoll in the fall of 1952, the Ivy series of weapons tests indicated the feasibility of a hydrogen bomb in a package that could be carried by rocket and, because of the immense blast effect, assure destruction even with accuracy limitations.[20] As late as 1951–1952, the Science Advisory Board of the Air Force had received only a lackadaisical response to the recommendation to expand ICBM activity, which its panel on missiles headed by Clark Millikan of the California Institute of Technology (with Kistiakowsky as one of its members) had submitted. In light of the Eniwetok hydrogen bomb tests, by 1953 the energetic assistant secretary of the Air Force, Trevor Gardner, determined that missile development must be expedited and appointed a Strategic Missile Evaluation Group—the so-called Teapot Committee—under John von Neumann. Von Neumann was an activist; he had allied earlier with Edward Teller and Luis W. Alvarez in advocating a crash program for the hydrogen bomb, and his computers were facilitating solution of the equations needed to design the new weapon. The von Neumann committee decided that the Air Force should press ahead with an ICBM, but no longer the monster rocket earlier assigned to Convair. In light of the Ivy tests, the von Neumann study ascertained that a smaller missile could still deliver a warhead of sufficient size (1,500 lbs rather than the earlier 9,000) to carry a hydrogen bomb; Jerome B. Wiesner's calculations indicated that the requisite guidance system could be packaged within a smaller rocket; Kistiakowsky suggested that propulsion and reentry challenges were also surmountable. On February 10, 1954, therefore, the Teapot Committee urged development of an Atlas-type missile and preparation of a second Titan-type rocket as backup and successor. Later in 1954, when defense secretary Charles E. Wilson moved the committee from its Air Force jurisdiction to his own office, it further suggested construction of a solid-fuel missile suitable for submarine launching—the future Polaris—rather than unwieldy liquid-fuel rockets. Even after Trevor Gardner's sponsorship and the acceptance of the report by Harold Talbott, secretary of the Air Force, the Air Force staff rejected the findings. Full-dress testimony by the committee

---

[20] See Herbert F. York, *The Advisors: Oppenheimer, Teller, and the Superbomb* (San Francisco: W. H. Freeman & Co., 1976), pp. 88 ff.; Warner R. Schilling, "The H-Bomb Decision: How to Decide without Actually Choosing," *Political Science Quarterly* 76 (March 1961), 24–46; Robert Gilpin, *American Scientists and Nuclear Weapons Policy* (Princeton, N.J.: Princeton University Press, 1962), pp. 73–111.

members finally led to a reversal on the part of the Air staff and the champions of the manned bomber.

Gardner established the new ICBM program under a Western Development Division of the Air Research and Development Division in the hands of Maj. Gen. Bernard A. Schriever. While the Army had its own experts in-house, Schriever called in the Ramo-Wooldridge Corporation (encountered frequently in diary entries below) as a virtual staff and contract manager. Because the corporation fell into conflicts of interest, both supervising and bidding on contracts, it in turn split, leaving the Space Technology Laboratory as its managerial consultant for the Air Force —but not without bequeathing difficulties still reflected in Kistiakowsky's record as special assistant.[21]

The von Neumann committee stressed development of the ICBM, believing that production of an intermediate-range missile would result almost as a by-product. By a year later the administration had become increasingly anxious about Soviet rocketry, especially in view of the failure at this time to secure adequate intelligence. A second committee, the Technical Capabilities Panel, under the leadership of Killian and including Edwin H. Land, inventor of the Polaroid camera, took a broad look at America's disturbing deficiencies. In Killian's judgment, this panel represented a major reinvolvement of the scientific community with the government after the scarring experiences of the Oppenheimer hearings in 1954. Specifically, the panel recommended pressing ahead with what became the U-2 reconaissance plane and—fearful of the Soviet lead in rocketry—it advocated that priority in missile development be devoted to the more quickly attainable IRBM's, rather than the technically more challenging ICBM's, which might be subject to grave delay. For Eisenhower this recommendation was persuasive. Given the availability of European bases for an IRBM, the President now authorized a major program to develop the midrange rocket.[22]

This series of decisions only aggravated the interservice rivalries that were vexing the White House. An intermediate-range missile might indeed fall into the Army's defined area of strategic competence. In the wake of the New Look, army theorists (notably General Gavin, assistant chief of staff of the Army for

[21] Kistiakowsky information; also Armacost, *Politics of Weapons Innovation*, pp. 57–58, 160–163, 244–249.

[22] Interview with James R. Killian, Jr.; cf. Eisenhower, *Mandate for Change*, p. 457; Armacost, *Politics of Weapons Innovation*, pp. 50–53.

research and development) had been elaborating concepts of a battlefield in depth in which infantry fronts and air-support areas ranging far beyond the lines were conceived as a tactical unity. Nuclear-tipped missiles with ranges of several hundred miles and under control of the ground commander were an essential component of this design. The strategic innovation was coupled with the quest for budget support. As Gen. John B. Medaris, ambitious commander of the Huntsville complex, told Ordnance officers, "If you increase your demands for guided missiles, I think there is a chance you can get a decent budget. Why don't you accentuate the positive and go with what is popular since you cannot get the other stuff anyway?" For the Army, the roles assigned under Charles E. Wilson's auspices were frustrating and demeaning. They found the secretary unsympathetic and uncomprehending: Gavin recalled a chief of staff saying, "He was the most uninformed man, and the most determined to stay so, that has ever been Secretary."[23]

Actually, Wilson's managerial style may have served the Army's missile program. At the least, it encouraged a rip-roaring rivalry among the Armed Forces for new roles and weapons. For the defense secretary, competitive rocket development appeared the quickest way to produce a workable missile—a conviction Eisenhower shared, even though he feared the duplication of effort. The Wilson Pentagon, which looked to the model of business competition, encouraged proliferation of programs as the price of attaining results. What is more, programs were initiated with little regard for ultimate cost; this meant a harsh conflict between and within the Services for limited resources and encouraged officers to lobby enthusiastically with their respective congressional friends. The chairman of the Joint Chiefs, Admiral Radford, who maintained a close relationship with Wilson, found it easiest to let each Service have "a chance to do something. And that in some cases has cost us a lot of money." Nor was the National Security Council prepared to interpose its judgment. In November 1955, Wilson stipulated that the Army (representing also naval requirements) and the Air Force would each be authorized to press ahead with IRBM development, the missile projects christened Jupiter and Thor respectively. Wernher von

[23] James M. Gavin, *War and Peace in the Space Age* (New York: Harper & Brothers, 1958), pp. 137–138, 155, 161; Medaris, *Countdown for Decision,* p. 65.

Braun's team at Huntsville had the satisfaction of seeing a new assignment come their way.[24]

In fact, the Army's mission was precarious. As the Teapot Committee had foreseen, liquid-propellant missiles such as the Army had been working with presented huge problems for shipboard fueling and maintenance. Naval participation might thus be cancelled. The Army, with its own in-house production team, did not have the vast network of industry contractors with their lobbying potential. Furthermore, it was not even to gain strategic command of the missile it was building, for in November 1956 Wilson decreed that the Air Force would retain operational control over all missiles with a range greater than 200 miles.

Nonetheless, the confusion left for the second Eisenhower administration was considerable. The Air Force was working to develop two long-range and one intermediate-range missile. The Army was building an intermediate-range missile over which it was not to have operational control. Its ambitious theorists felt that the latest decision was only part of a succession of short-sighted downgradings of role. Increasingly they questioned the whole strategic premise of the New Look: if there were to be a balance of terror, the battlefield would again become decisive, but it would be a new battlefield of tactical nuclear weapons. If the administration were to seek force levels adequate for a "long pull," then research funds had to be more ample or qualitative sufficiency would be sacrificed. On top of all these demands for missile development and defense spending came the recommendations of the Gaither panel in the fall of 1957; as we have noted, these pleaded not merely for rockets and for an SAC less vulnerable to surprise attacks, but for a costly fallout-shelter program as well. (In fact, the fallout shelters represented a compromise recommendation; for the Gaither panel had been summoned to evaluate a Civil Defense Administration proposal for a $40-billion blast-shelter program and decided that people might never reach the bunkers in time.) By describing protection of the civilian population as a step toward making the deterrent credible, the Gaither panel consultants probably oversimplified a difficult strategic issue. Later deterrence theorists would claim that shielding cities signaled a more aggressive intent than merely protecting retaliatory forces to assure second-strike capacity.

[24] Armacost, *Politics of Weapons Innovation*, p. 76. See also Paul Y. Hammond, *Organizing for Defense* (Princeton, N.J.: Princeton University Press, 1961); Huntington, *Common Defense;* and John C. Ries, *The Management of Defense* (Baltimore, Md.: Johns Hopkins University Press, 1964).

John Foster Dulles also sensed the difference, when he argued that Washington's allies would take alarm at a shelter-building program. Eisenhower accordingly did not move to establish the report as the basis for an emergency program, but his resistance made the coalition of Democratic Party critics in Congress and influential industry spokesmen restive and frustrated.[25]

The President faced long-term and intractable problems when the Soviets launched Sputnik I. Most obvious was a crisis of confidence among the public, intensified by specific misgivings of the Gaither coalition. At the same time, the Services continued their competition for prestigious missile programs. It was logical that Eisenhower should turn to the scientists as an organized profession in a way that he had not earlier. They proffered their advice; they seemed less tainted by particular interests and constituencies; they offered reassurance to the public; and they were seen as possessing a technical expertise that could recommend efficient and correct solutions. Within two months after Sputnik the President had centralized the science-advising function at the White House. At the same time, he sought to balance the Pentagon rivalries that had grown so intense and confusing.

Supported by the new special assistant for science and technology, the White House was moving to strengthen the central authority of the Defense Department vis-à-vis the individual Services. As Eisenhower was to admit to Killian and Kistiakowsky in February 1958, he had come "to regret deeply" that the missile program had not been set up in the Office of the Secretary of Defense (OSD) rather than in each Service.[26] On November 15, 1957, Defense Secretary Neil H. McElroy announced that he was establishing an Advanced Research Projects Agency (ARPA) under the OSD to develop antimissile and space technology projects. Formal chartering took place in February 1958. ARPA, however, as Kistiakowsky's diary suggests, remained relatively ineffective; the White House, on the one hand, wished to preserve civilian control of space initiatives, while the Services

25 Cf. Halperin, "The Gaither Committee"; also Eisenhower, *Waging Peace*, pp. 219–223. The implications of civilian defense remain contested: see Albert Wohlstetter, "Strategy and the Natural Scientists," in Robert Gilpin and Christopher Wright, eds., *Scientists and National Policy-Making* (New York: Columbia University Press, 1964), pp. 223–230.

26 DDE, Papers as President (Ann Whitman file), Diary Series, Box 18, F, "Staff Notes, February 1958," Goodpaster notes: memorandum of conference with the President, February 4, 1958. Killian urged halting the Jupiter; in actuality, both the Thor and the Jupiter were developed.

were unenthusiastic about the growth of OSD power at the expense of their own initiative.

More potent than ARPA was the new Office of the Director of Defense Research and Engineering, created under the secretary and deputy secretary as part of the Defense Reorganization Act of 1958. This bill sought to solve multiple problems that had arisen since the restructuring of 1953. It aimed at unified strategic commands over all the Services, since nuclear weapons and missiles had eroded functional specialization of the Army, Navy, and Air Force. The legislation also sought to strengthen the Office of the Secretary of Defense with respect to the military, as well as to cut down interservice rivalry. Along with the establishment of DDRE, the bill subordinated the three Service secretaries more clearly by removing them from command functions and cabinet participation. At the same time it emphasized the collegial role of the Joint Chiefs rather than their individual ''line'' leadership over the respective branches. The post-1958 collective authority of the Joint Chiefs presented a different challenge to the secretary of defense; but for the moment the imperative was to recentralize the initiative over strategy and weapons systems that had been devolving upon the most energetic of the junior generals in the Army and Air Force. When Gates succeeded McElroy as secretary in 1959, he was to push integration one step further by demanding coordination of nuclear targeting—a reform that would deeply involve Kistiakowsky. On the whole, the 1958 act represented a major improvement and functioned well, not least because the director of DDRE, Herbert F. York (who also served for an interim as chief scientist at ARPA), fostered harmonious relations with the special assistant for science and technology.[27]

Finally, Eisenhower moved to systematize space exploration under as much civilian control as possible. With the advice of the President's Science Advisory Committee and the Bureau of the Budget, he introduced legislation that would create a National Aeronautics and Space Administration on the basis of the half-century-old, loosely organized network of laboratories linked

[27] Eisenhower, *Waging Peace*, pp. 244–253; Schoettle, ''Establishment of NASA,'' pp. 195–199 and passim; Armacost, *Politics of Weapons Innovation*, pp. 226–237; Ries, *Management of Defense*, pp. 167–192 (with a useful reminder that no reorganization should remove political choice, and an emphasis on Congress's stake in maintaining independent Service viewpoints); also interviews with Killian, York, and Kistiakowsky.

together in the National Advisory Committee for Aeronautics (NACA). The Senate, especially, sought to preserve the role of the military if defense applications were envisaged; nonetheless, Eisenhower's basic idea of civilian leadership prevailed. Moreover, in contrast to ARPA, NASA became not merely a development group in search of a mission, but an operational agency with its own contracting and management centers. As the reader of the diary will note, NASA was soon to inherit the Redstone Arsenal team: a bequest which Eisenhower had considered from early 1958, but which administrator T. Keith Glennan apparently accepted with some reluctance. Medaris and the Army were not happy to lose von Braun's crew, but the alternatives seemed to be splitting them up or releasing them to industry. The mode of their transfer suggests that by 1959 the costs of the Army's ambitious programs had finally become a factor to be given full weight: parsimony was catching up with ebullience.[28]

By the time Kistiakowsky took office as special assistant, then, the process of clamping controls on the missile effort was under way. In this painful imposing of direction, the science advisors— Killian and then Kistiakowsky at the Executive Office, York at Defense Research and Engineering—played an important role. While the Bureau of the Budget was ubiquitous in its effort to pare expenditures, it was unprepared to set strategic or space priorities. A full integration of budgetary and strategic preferences had to await Robert McNamara and the Kennedy administration, when fiscal evaluation of programs was imposed at the outset. On the other hand, this supposed OSD rigor was purchased at the price of a significant expansion of funding. In a sense, the McNamara regime traded the funds that the impatient Services had been demanding in return for a more complete control than Eisenhower's secretaries had achieved. The new effort to secure greater versatility also dampened the interservice rivalry centered on prestigious rocket programs. Not that it necessarily

[28] DDE, Papers as President (Ann Whitman file), Diary Series, Box 19, "Staff Notes, March 1958," Goodpaster notes: memorandum of conference with the President, March 10, 1958, on the future of the Huntsville ABMA; also August 4, discussion with Killian concerning transfer to NASA or to ARPA, in Box 22, "Staff Notes, August 1958," Goodpaster notes: memorandum of conference with the President, August 4, 1958; Armacost, *Politics of Weapons Innovation*, pp. 238–244; see also T. Keith Glennan to the President, October 20, 1959, on the funds needed for transfer of ABMA to NASA, in DDE, Papers as President (Ann Whitman file), Administrative Series, Box 17.

diminished the influence of the military as a collegial body, or precluded rasher confidence in military solutions from infecting civilian policy-makers. Nor did the McNamara regime quell technological acquisitiveness; the swing-wing F-111 was to become a monument to the continuing weapons-systems fad, and once again scientists sought to act as a coherent policy group to slow down vast expenditures—this time, on the projected antiballistic missiles (ABM) system.

In retrospect, the Eisenhower administration—the President in the first instance, seconded by McElroy and then Gates at DOD, and relying increasingly on the special assistant for science and technology and DDRE—sought to carry out perhaps a more difficult assignment than their successors attempted, that of centralizing authority even as they declined to reconcile the military Services with new programs, missions, and funds. In retrospect the task appears doubly difficult; from reasons of both conscience and politics, the usual advocates of arms limitations and civilian supremacy among the Democrats and liberals were actually calling for major new expenditures and encouraging the special pleadings of the most discontented (albeit often the most talented) military spokesmen.

This is not to argue that the critics of the Eisenhower defense posture simply surrendered to an ill-conceived militarism. Their concern with being boxed into a nuclear retaliatory strategy was legitimate; their feeling that the Eisenhower administration was dangerously running down even its deterrent capacity probably had a less secure foundation. The "missile gap" feared by critics of the administration and used to good effect politically never materialized.

There remains the question of whether Eisenhower's advisors ran risks they should not have incurred and gambled with the defense forces that would be available in the early 1960's. This does not really seem the case either. Defense experts had already lived through the false alarm of the "bomber gap" in the mid-1950's, when potential Soviet aircraft production and actual output had not been distinguished. In early 1959, the President recognized that in a few years the Soviets conceivably might possess missiles enough that they would not fear retaliation, but he downplayed this fear. As he had stated a year earlier, when Kistiakowsky as head of the PSAC missile panel had reported to him on missile progress, evaluation of intelligence material required an assessment of relative probabilities. Until an enemy could destroy most bases simultaneously, the deterrent would remain effective. It would be a mistake to credit an antagonist with total capabilities.[29]

Intelligence findings in any case were suggesting that Soviet progress had slowed. Radar trackings from Turkey and Iran indicated that Soviet firings had been suspended for more than six months in late 1958 and early 1959, perhaps because the massive Soviet vehicles required further refinement before deployment as weapons. Films taken from U-2 planes allowed further lowering of Soviet deployment estimates, and in early 1960 Secretary Gates presented Congressmen with low assessments of probable Soviet capacity rather than alarming ones of possible missile output. Because Gates did not specify on what basis his "probable" estimates were formed, he encountered sharp skepticism.

Even if somewhat reassured, the administration did not react passively. In part, the lag in American deployment stemmed from the decision to devote resources to quicker development of the second-generation, hard-silo rockets rather than to the vulnerable and cumbersome Atlases. The President also called for overtime work on the Saturn "superbooster." From the diary it is clear that for 1961–1962, administration officials, including Kistiakowsky and later Gates, foresaw a transitional lag that the Soviets might exploit but ordered short-term responses to prevent it from becoming decisive.

Still, catching up was the easiest task. Other intractable dilemmas remained, rooted as much in the logic of arms races as in any failings of the administration. In an era of high technology and rapid obsolescence, was it feasible to use budget ceilings to establish force levels? Since security involves a psychological

---

[29] DDE, Papers as President (Ann Whitman file), Diary Series, Box 18, "Staff Notes, February 1958," Goodpaster notes: memorandum of conference with the President, February 4, 1958," and Box 24, "Staff Notes, January 1959," memorandum of conference, January 12, 1959. In early 1959, the conference with Eisenhower suggests, Kistiakowsky was more concerned than the President about the impending operational capacity of the Soviets. Richest source on the missile-gap issue is James C. Dick, "The Strategic Arms Race, 1957–1961: Who Opened a Missile Gap?" *Journal of Politics* 34 (November 1972), 1062–1110. Cf. Colin S. Gray, " 'Gap' Prediction and American Defense: Arms Race Behavior in the Eisenhower Years," *Orbis* 16 (spring 1972), 257–274; Roy E. Licklider, "The Missile Gap Controversy," *Political Science Quarterly* 85 (December 1970), 600–615; and Edgar M. Bottome, *The Missile Gap: A Study of the Formulation of Military and Political Policy* (Cranbury, N.J.: Fairleigh Dickinson University Press, 1971).

component, did the President ever sufficiently explain to Americans the distinction between retaliatory sufficiency and superiority? Eisenhower sought to tread a fine line between relying on expertise and shunning what he rebuked as "parochialism." The science advisors gave him expertise without parochialism; the military, alas, too often tended to purchase the former at the price of the latter.[30] Ultimately the President felt that in security matters his own expertise should be the best guarantee of proper defense policy in the eyes of the American people. If he relied on deterrence, he did not exclude discrete uses of force, as in Guatemala and Lebanon. In many pressing areas the administration sought to avoid confronting the great social changes occurring in the America of the 1950's: race, environment, urban services, for a while even civil liberties, were low on an agenda that valued the private satisfaction of private wants. But in the area of security and military technology, the President—perhaps not without prodding—did seek both to push ahead and to control, and at the end even came to sense the social and political perils that might arise from the new technocratic establishments in defense, industry, and science.

DRAWING APACE with the Soviets in rocketry represented an urgent objective for Eisenhower, but only a partial one. By the time Kistiakowsky took office as special assistant, the President hoped to devote major effort to halting the arms race and not merely winning it. Although the administration did not initially approve separating the question of nuclear weapons tests from the more general aspects of arms control, an agreement to end testing increasingly appeared the most feasible starting point for any negotiations. Nuclear tests and fallout had preoccupied scientific and lay opinion ever since the irradiation of the hapless Japanese fishermen of the *Lucky Dragon,* who had been a few hundred miles downwind from the test area of an American "dirty" (presumably fission-fusion-fission) bomb on March 1, 1954. Public concern continued to mount as first Albert Einstein and Bertrand Russell, then 52 Nobel laureates, and finally Linus Pauling backed by 9,000 scientist-petitioners, pleaded for an end to testing.

With the Open Skies proposal of July 1955, Eisenhower himself began to move from the earlier notion of eventually eliminat-

---

[30] Interview with Gen. Andrew Goodpaster.

ing nuclear weapons to the more feasible prospect of arms control with an acceptance of mutual deterrence. Throughout 1956, Washington and Moscow fenced over the appropriate next steps to be taken. In his letter of September 11, Nikolai Bulganin emphasized the urgency of halting tests, whereas Eisenhower stressed the need to control arms buildups in general. Although any freeze would perpetuate a larger American atomic arsenal, Washington regarded such a disparity as a legitimate counterweight to superior Eastern Bloc manpower. Often unremembered in the dismal record of progress, in the spring of 1957 the Soviets in reference to the cessation of tests did suggest establishment "on a basis of reciprocity of control posts in the territory of the Soviet Union, the United States of America and the United Kingdom and in the Pacific Ocean area." In response, Dulles warned that the problem was "not so simple as many have believed." Indeed it was not; nonetheless a useful opportunity to test Soviet sincerity may have been lost. Instead, Eisenhower's disarmament advisor, Harold Stassen, was authorized to propose a suspension of tests after agreement on control measures and contingent upon acceptance of wider arms-limitation steps such as a cutoff of weapons production and control over fissionable materials. In August 1957, the Russians angrily rejected the Western response and boycotted the UN Disarmament Committee, which suspended its sessions after September 6. Pressure continued, however, for an end to testing, and the good will that Washington had banked by the Open Skies proposal dwindled. In light, too, of Sputnik and the growing American anxiety about defense, Eisenhower seemed faced with major military expenses, including those for tactical nuclear weapons, if progress toward arms control remained unattainable.[31]

Drawing upon his newly activated battery of science advisors in early 1958, Eisenhower asked a panel of experts appointed by James Killian and chaired by Hans A. Bethe of Cornell University to weigh the risks of a test ban. Given the assessment that a reciprocal suspension of testing would not handicap the United States, the major problem became the possibility of evasion offered by underground testing. On the basis of the 1.7-kiloton Rainier test in the fall of 1957 and other data, the Bethe panel suggested that the risk was surmountable and that effective monitoring was feasible.

[31] U.S. Disarmament Administration, *Geneva Conference*, pp. 3–13; Gilpin, *American Scientists*, pp. 162–177.

Eisenhower moved cautiously. Each side wanted to choose the most advantageous moment to suspend testing. In response to the December 1957 Nehru plea for a halt, the President had defended the American tests then in progress. On January 12, 1958, he moved further and suggested to Marshal Nikolai Bulganin that scientific experts from both sides review the problem of enforcing a test ban, but without committing the respective governments. Only after the Russians finished their own test series did the Supreme Soviet announce at the end of March that it was suspending explosions, with the proviso that it would feel free to resume them if the other atomic powers did not follow its lead.

When Nikita Khrushchev (now Bulganin's successor) asked Eisenhower to suspend tests in a letter of April 4, the President suggested that the Soviets were cynically exploiting the fact that they had just completed a test series. Again he asked that both sides first consider control measures. Khrushchev countered that present instruments sufficed to record any underground nuclear explosion. For Americans, however, the issue was not simply registering the shocks but distinguishing them from natural earthquakes. On April 28, Eisenhower pleaded again for technical studies as "the necessary preliminaries to putting political decisions actually into effect." Finally Khrushchev broke the deadlock on May 9; although he lamented the delay a technical analysis would cause, he agreed to a conference of experts who would study means of detecting possible violations.

This step entailed more for both sides than was immediately apparent. Khrushchev feared that inspection of possible explosion sites would reveal to the West how far behind in terms of military strength the Soviet Union really lagged. But any enforcement scheme was likely to require some degree of inspection, even though the Soviets were to seek consistently to minimize its extent. On the other side, Eisenhower wanted rigidly to separate technical studies from political negotiations. As the course of the upcoming negotiations was to reveal, it was difficult to keep the two components apart. Moreover, even though Dulles was left to repeat that the United States would still seek a test ban only in connection with a cutoff of nuclear weapons production, Eisenhower was moving toward a readiness for agreement that dismayed many cold-war strategists. Even as the President was to welcome the initial achievements of the technical negotiations the following summer, Henry Kissinger, for instance, would somberly ask "whether insistent Soviet demands for a complete ban on nuclear testing are not designed to paralyze the free world rather

than to bring about peace as we understand the word.'' The technical discussions on which Eisenhower and Khrushchev finally
agreed thus held the potential for a significant unwinding of cold-
war policies.[32]

In 30 sessions at Geneva during the summer of 1958, the experts united on a potential feasible control system. Specially
equipped aircraft and about 10 shipboard stations could collect
any radioactive debris produced by explosions above ground,
while 160 to 170 seismic posts could record underground shocks.
With this number of seismographic stations, from 20 to 100 earthquakes per year (the respective Soviet and American estimates)
might still be confused with nuclear explosions of 5 kilotons or
above and would call for on-site inspection. The lower the desired
detection threshold, the more earthquakes might be confused with
explosions. Advisor Harold Brown of the Livermore Weapons
Laboratory—generally skeptical about controlling a test ban—
felt that a one-kiloton threshold would require 650 stations for
adequate discrimination, and originally the Americans suggested
including this option. Since the number was flatly unacceptable
to the Russians, the American scientists accepted the intermediate British proposal of 180 stations with a 5-kiloton threshold
that was ultimately adopted as the basic scheme. While some
cheating might escape detection, the risk of being caught as a
violator was high enough, Western experts felt, to deter evasion.
And what seemed supremely significant was the fact that scientists from both East and West had been able to unite on a
program that promised arms control and retained security.
Eisenhower greeted the experts' report of August 21 with heart-
felt optimism and called for three-power negotiations for a test-
ban treaty. As an interim measure he suspended American testing for at least a year and perhaps longer if meaningful control
measures evolved.[33]

[32] For the exchanges of Eisenhower with Bulganin and Khrushchev, see
U.S. Disarmament Administration, *Geneva Conference*, pp. 12–15; Gilpin,
American Scientists, pp. 177–185. Cf. Saville Davis, ''Recent Policy-Making
in the United States Government,'' *Daedalus* (Arms Control Issue), 89 (fall
1960), 951–966. The Russian premier's concerns about inspection are indicated in Nikita Khrushchev, *Khrushchev Remembers*, vol. 2, *The Last Testament*, p. 536. The Kissinger citation is from ''Nuclear Testing and the Problem of Peace,'' *Foreign Affairs* 37 (October 1958), 7.

[33] U.S. Disarmament Administration, *Geneva Conference*, pp. 13–21; see
also Gilpin, *American Scientists*, pp. 186–222, for an extensive and perhaps
overly critical discussion of the Geneva system. On Eisenhower see Adams,
Firsthand Report, pp. 318–326.

On November 4, the Conference on the Discontinuance of Nulear Weapons Tests convened at Geneva, but not before Washington and Moscow had traded further accusations and the Russians had tested two large thermonuclear bombs. The new parleys yielded little. The Russians insisted that the control commission authorizing on-site inspection should take decisions by unanimous vote, which meant that the Russians had veto power. Senior Soviet negotiator Semyon K. Tsarapkin also asked that the control posts be staffed by nationals of the country in which they were located, the teams to include at best one foreign observer.

The President remained intensely interested as well in an agreement to reduce the dangers of sudden or surprise attack. Under his prompting, the State Department asked Killian to make a technical study of the measures needed to detect and deter this fearful possibility. In a conversation with Killian on August 4, 1958, Eisenhower admitted that the Open Skies surveillance measures were less relevant in an age of rocketry, although for the next several years planes, not missiles, would still be dominant.[34] Hence he hoped that surveillance agreements might yet serve both sides' security. Within a week after the November discussions on the test ban opened, technical experts convened for five weeks of fruitless exchanges on surprise attacks. The Soviets again sought to tie the issue with the more general questions of abolishing atomic weapons and eliminating foreign bases.

But dilemmas lay in the very concept of defensive surveillance. Kistiakowsky, who served on the panel, concluded that the approach was counterproductive. He wrote to Killian: "As a fresh conscript into the diplomatic army, I acquired very soon after November 10, the feeling that I was an Alice in Wonderland. The same words used by the two sides had opposite meanings and neither side was willing to state its clear intentions." The nub of the difficulty was that inspection systems in the Soviet view favored the potential aggressor and undermined defensive retaliatory strength. What the United States called instruments of surprise attack were also instruments of deterrence. As Kistiakowsky concluded, "We have not solved this problem; we only kept saying for the record that we knew the answers and they were favorable to inspection systems of the kind we advocated." Inspection without arms limitation "might actually lead to an intensified arms race and provide advantage to an aggressor."

[34] DDE, Papers as President (Ann Whitman file), Diary Series, Box 22, "Staff Notes, August 1958," Goodpaster notes: memorandum of conversation with the President, August 4, 1958.

Kistiakowsky told the President personally on January 12, 1959, that he doubted the value of the system. Only coupling limitation with inspection might provide some reassurance, perhaps by setting a ceiling on the number of airborne planes at any one time or on missiles. Eisenhower accepted the caveats but still insisted that initial agreements were required to establish momentum for arms control.[35]

Even as the negotiations to guard against surprise attack were abandoned, the problems of monitoring a test ban also became more intractable. The high-altitude shot of Project ARGUS in the summer of 1958 suggested that space tests beyond a height of 100 kilometers were not easy to detect; soon opponents of a test ban were producing scenarios of Soviet tests beyond the sun, on the assumption that if the Russians could cheat they would cheat. The Hardtack test of October 1958 also indicated more disturbingly that seismographic readings might be less sensitive than foreseen, allowing more room for confusion of earthquakes and explosions. The reliance on "first motion" of the seismograph—the telltale indication that the seismic wave was moving away from the epicenter rather than across it or toward it from some perspectives—seemed invalid. On January 5, 1959, when the Geneva discussions resumed, the American team suggested that the new data might involve ten times as many unidentifiable seismic events above a 5-kiloton equivalent, which Tsarapkin claimed meant that Washington wanted only to send ten times as many inspection teams.[36]

Those at home who were hostile to cessation of tests mobilized further arguments. Encouraged by Edward Teller, Albert Latter of the Rand Corporation issued a report on decoupling on March 30, which theoretically demonstrated that a nuclear weapon set off in an underground cavern would produce a seismic wave corresponding to only a fraction of its true power because of the air separating it from the rock walls of the chamber. At first this theory aroused the scorn of pro–test-ban scientists such as Bethe. Still, Bethe had to concede the force of the theoretical argument and had the unhappy duty of presenting it to the Soviets at Geneva. As Bethe also understood, however, decoupling was not an easy practical strategy. The task of washing out the salt domes

[35] George B. Kistiakowsky to James R. Killian, Jr., December 15, 1958; Kistiakowsky's files; and DDE, Papers as President (Ann Whitman file), Diary Series, Box 24, "Staff Notes, January 1959 (2)," Goodpaster notes: memorandum of conversation with the President, January 12, 1959.

[36] U.S. Disarmament Administration, *Geneva Conference*, pp. 30–33.

ꞁ that were the likely source of the large caverns needed would involve cost and time, and the preparations or brine disposal would
produce their own identifiable signs. Although he considered cutting the admission from his article published in the *Atlantic
Monthly* in the summer of 1960, Bethe wrote that he would have
understood if the Russians had regarded the theory as an insult
and walked out. Spurgeon M. Keeny, Jr., who as a staff member
of the President's Science Advisory Committee coordinated the
team of scientists, also admits that the Russians probably became
genuinely puzzled over the conditions Americans kept raising.
On the other hand, the negotiators were convinced that airtight
agreement was needed if any treaty was to survive the opposition
at home—not merely the hostility of a Teller or the skepticism of
the new Atomic Energy Commission chairman, John A. McCone,
but the doubts too of the Joint Congressional Committee on
Atomic Energy. The reliance on on-site inspection itself came
under a chilly review in the spring of 1959, as a new panel chaired
by Robert F. Bacher of the California Institute of Technology
and influenced by Harold Brown's preoccupations suggested that
inspection was not likely to decide absolutely whether an underground explosion had or had not taken place.[37]

In light of these difficulties and the Soviet refusal to permit
unencumbered inspection, Eisenhower suggested in April 1959
that the conference at least work to halt atmospheric tests, with
postponement of the more difficult issue of underground or outer-
space testing until adequate monitoring was assured. As he had
told Killian in late February 1959, if tests below a given size were
barred (he cited 10 kilotons), but without the capability of monitoring the ban, "the tendency for suspicion to arise will be very
great and our whole nation will become more and more jittery."[38]

[37] Hans Bethe's galleys of "The Case for Ending Nuclear Tests" for the
August *Atlantic Monthly*, sent to Kistiakowsky June 29, 1960, in the Eisen-
hower Presidential Library, U.S. President's Science Advisory Committee:
Records, 1957–1961, Box 6, Folder B (3). For the theoretical controversy see
A. L. Latter, E. A. Martinelli, and E. Teller, "A Seismic Scaling Law for
Underground Explosions," January 14, 1959 (a Rand Corporation docu-
ment), and Bethe's original attempt at disproof, "Theory of Seismic
Coupling," March 24, 1959, in the Eisenhower Presidential Library, Records
of the White House Office "Project Clean-Up" papers, Box 29, F, "Science
and Technology—General." I have also drawn on interviews with Spurgeon
M. Keeny, Jr., and George W. Rathjens.

[38] U.S. Disarmament Administration, *Geneva Conference*, pp. 49–52, 354–
355. For Killian's gloomy assessment of the difficulties of monitoring tests
underground in light of decoupling possibilities, or in outer space in view of
using lead shields ("in summary, a system with a low threshold cannot be

Khrushchev responded to Eisenhower's letter with apparently equal yearning. But the Russian leader did not want to restrict the agreement to atmospheric testing, which, he argued fervently, would not limit the arms race. He suggested that the Soviet Union might in fact allow a fixed quota of inspections, when they were justified by objective instrument readings. The efficacy of the instrumentation and the unwillingness of either side to be exposed as a violator would ensure that "no large number" of inspections would be required. On the other hand, the Soviets would not veto inspection within the agreed quota.

The Khrushchev reply was encouraging, but the technical aspects of the question remained clouded. Some hope might be drawn from the March report of a new Panel on Seismic Improvement, chaired by Lloyd V. Berkner, that the 180-station system envisioned by the experts the previous summer could be increased in efficiency. Installing more seismographs at each station would allow the original level of discrimination to be partially restored. The Geneva experts had claimed that no more than 100 seismic events above 5-kiloton equivalents would remain unidentifiable; the somber reassessment after Hardtack had said that every shock below 20 kilotons would be of uncertain origin and thus would have to be exempt from monitoring if only 100 unidentified disturbances per year could be allowed; the Berkner panel now suggested that this quota would sufficiently identify seismic disturbances down to 10 kilotons. Moreover, burial of the instruments in "deep holes," would permit reception of clearer and more identifiable seismic signals. And finally, if unmanned robot stations could also be employed, 75 to 98 percent of all signals above a one-kiloton equivalent might be sorted out. With further experimentation and use of the computer to sort out extraneous signals, detection capacities should continue to improve. Nonetheless, all these modifications would involve great expense and, as it turned out later, aroused deep Soviet hostility.[39]

The Berkner report and the apparent Soviet willingness to drop its insistence on a veto permitted Eisenhower to press once

guaranteed") see DDE, Papers as President (Ann Whitman file), Diary series, Box 24, "Staff Notes, February 1959," Goodpaster notes: memorandum of conference with the President, February 25, 1959; and Box 25, "Staff Notes, March 1959," Goodpaster notes: memorandum of conference with the President, March 13, 1959.

[39] Findings of the Berkner panel are included in U.S. Disarmament Administration, *Geneva Conference*, pp. 335–353.

more for a wide-ranging test ban as a priority task for the special assistant and the ad hoc Committee of Principals assembled to review the outlines of new American proposals. The diary reader can follow the arduous course of further negotiations. The Russians consented to send delegates to a new Technical Working Group (I), which convened in June of 1959 and reviewed the possibilities of testing in space. Russians agreed with Americans that blasts of up to 400 kilotons might be tested 200,000 miles out from earth with little prospect of detection. At the second Technical Working Group in the fall of 1959, American scientists led by James B. Fisk and Wolfgang K. H. Panofsky asked the Russians to ponder the implications of the Hardtack test and the decoupling data. The Soviets, for their part, wanted the Americans to define the "objective" criteria that would justify sending out an inspection team. Yet as one American scientist explained, if the criteria could be defined objectively, inspection itself would not be necessary.[40]

Despite all difficulties, the impetus of the negotiations seemed heartening enough to let the editors of the *Bulletin of the Atomic Scientists* in January 1960 move their "clock of doom," which ominously stood near midnight on each cover, a few minutes backward.[41] Nor did the administration wish to lose momentum. In view of the difficulties exposed by the working-group discussions, Kistiakowsky was to urge setting an easier objective. Instead of trying to police an underground kiloton threshold which, because of variations in the rock or purposeful decoupling, might register variable seismic magnitudes on the Richter scale, the United States would suggest a threshold in terms of Richter scale readings alone. The number chosen (4.75) corresponded to the capacities of the 180-station Geneva system endorsed in August 1958. In addition, there would be a common program of testing to improve detection capacities and progressively lower the threshold. America's own development program toward this end, however, Project VELA, encountered Soviet objections: how could one distinguish testing for the improvement of detection from testing for the improvement of weaponry? As the reader will see, all sorts of ingenious schemes were devised to respond to this dilemma: the sharing of blueprints, weapons depositories, and so on.

[40] U.S. Disarmament Administration, *Geneva Conference*, pp. 52–81, reports of TWG (I) and TWG (II), pp. 367–413. Cf. Gilpin, *American Scientists*, pp. 228–252.

[41] "Editorial: The Dawn of a New Decade," *Bulletin of the Atomic Scientists* 16 (January 1960), 2.

By the spring of 1960, Eisenhower was ready to accept a fixed quota of inspections as originally proposed by Khrushchev and now urged by British Prime Minister Harold Macmillan. Twenty inspections would cover 30 percent of the unidentified findings above the Richter-scale level stipulated. (The issue of how many square kilometers could be inspected, however, remained difficult.) The Russian response of March 19, 1960, was somewhat encouraging: if both sides agreed to a moratorium on small underground blasts during the research period, they would negotiate for a monitored test ban on blasts above the threshold. The quota of allowable inspection visits, they insisted, however, must be regarded as a mere political negotiation; it was not to be arrived at as a percentage of the unidentifiable seismic incidents.[42]

The gathering momentum for negotiations was interrupted by the U-2 incident in May 1960. Khrushchev's subsequent behavior at Paris and the Soviet hardening at Geneva—Tsarapkin would answer America's demand for twenty inspection visits with a quota of three—suggested that for internal reasons the Russian leader was being drawn back toward a harder line. No doubt, within the Soviet government a delicate balance between those favorable to arms control and those who wished to catch up with America may well have existed. Khrushchev's exchange with Eisenhower and Macmillan in the spring of 1959 was not without its domestic risks, and the length of time he could afford to negotiate the manifold evasion devices and defenses that preoccupied Washington's team was problematic. If Khrushchev was to keep his footing, it appeared that he needed either to reach a quick agreement without conceding major inspection rights or else to regroup in a posture of defiance.

Under President Kennedy parleys resumed as did the Soviet testing of giant weapons, until in 1963 both sides settled for a ban on tests in the atmosphere. In Kistiakowsky's opinion, this agreement had the disadvantage of legitimizing continuing advances in the arms race, while providing deceptive assurances that a major breakthrough had been made. Nonetheless, Eisenhower himself had contemplated settling for an above-ground test ban in April 1959 and had seen the value of even partial agreements in the momentum they created for more inclusive controls. Kennedy was not to have the time to reap the fruits of any momentum, and not until the end of the decade were bilateral

[42] U.S. Disarmament Administration, *Geneva Conference*, pp. 81 ff.

strategic arms control negotiations to resume again. The ground that was lost is suggested by the fact that the treaties that have finally sought to deal with underground testing in 1974 and 1976 still exempt explosions below a given threshold: but now one of 150 kilotons and no longer a mere 5 or 10.[43]

In hindsight, it can be seen that the new scientific resources of the President provided a two-edged blade. Eisenhower himself clearly belonged among the supporters of a test-ban treaty. He did not personally subscribe to Teller's dire warnings, nor did he share Teller's visions for the peaceful uses of atomic explosions (namely, the Project Plowshare that would dig the canals of the future and in the interim forestall the test ban). The President also tended to believe Khrushchev's assurances that the Soviets were not interested in testing small weapons.[44] Nonetheless, his style of administration and careful delegation of authority constrained his own convictions; he would not short-circuit the National Security Council or the special Committee of Principals in which John McCone could slow the progress of American bargaining. The late secretary of state, John Foster Dulles, had also helped lock the administration into an a priori skepticism about negotiations with the Soviets that was hard to overcome. This is not to argue that caution should have been abandoned, but the search for loopholes became all-consuming. Here the scientists' ingenuity hobbled the President even while it aided him. Killian similarly believes in retrospect that the scientists accepted the constraints upon detection too absolutely, that they framed recommendations within the state of the art and hesitated to press the idea that the technology of detection might improve with the technology of evasion.[45]

[43] See the *New York Times*, March 17 and 18, 1976. The reader who wishes to follow subsequent negotiations should see, in addition to the Disarmament Administration volume cited herein, Cecil H. Uyehara, ''Scientific Advice and the Nuclear Test Ban Treaty,'' in Lakoff, *Knowledge and Power*, pp. 112–161.

[44] See Edward Teller to President Eisenhower, May 11, 1960, proposing that a trans-Isthmian canal be dug by nuclear weapons, preceded by a pilot project of hollowing out an Alaskan harbor. (''I have no doubt that Plowshare will ultimately become one of the most important applications of nuclear energy.'') Letter included in U.S. President's Science Advisory Committee: Records, 1957–1961, Box 4, Folder ''Nuclear Plowshare.'' On Eisenhower's acceptance of the Khruschev claim see DDE, Papers as President (Ann Whitman file), Diary Series, Box 25, ''Staff Notes, March 1959,'' Goodpaster notes: memorandum of conference with the President, March 13, 1959.

[45] Killian interview. On the other hand, the Berkner panel did predict in-

Scientific expertise, then, could suggest a map of plausible alternatives in terms of what behavior might be controlled under various arrangements that would have to be negotiated politically. Like the politicians, the scientists remained bitterly divided over the test ban and arms control. Some scientific estimates separated them, although those could be narrowed, as was the case with Bethe and Latter in the spring of 1959. Where they bitterly differed was in assigning relative probability to different political outcomes, especially to future Soviet behavior. Between 1958 and 1960 the scientists were able to offer the President a matrix of choices that was far more extensive than earlier contributions. They could reassure him, for example, that with 180 seismographs, 90 percent of 20-kiloton equivalent shocks in granite might be identified, or that with 650 seismic stations, 95 percent of one-kiloton shocks and above might be identified. At the same time they could caution that digging big holes would let the Russians reduce a 20-kiloton bomb signal to the magnitude normally emitted by one of only 2 kilotons. In the flush of scientific adulation that surrounded Sputnik or discussions of C. P. Snow's "two cultures," such information was often confused with the answering of policy questions. But the new scientific advisory panels could only multiply choices or frame them more intelligently; they could not resolve them.

JAMES KILLIAN takes pleasure in recalling how when he visited Eisenhower during his last illness, the former President asked after "my scientists," adding that they comprised the only group who had come to Washington to serve the country rather than their own interests.[46] The confidence in science as a mode of inquiry and a guide to policy untainted by material interests helps explain the impact of the PSAC and the special assistant in the late 1950's and early 1960's. Nevertheless, conscious and organized direction was necessary also to win the spokesmen of science a voice in policy-making. No simple technological inevitability brought the scientists into council; and even though scientific issues remained inextricably woven into the choices facing goverment, the influence of the science advisors at the presidential level was to dissipate by the mid-1960's. Killian, Kistiakowsky,

creasing improvement of seismic techniques, especially through reliance on computer analysis.

[46] Killian interview.

and afterward Kennedy's special assistant, Jerome Wiesner, were to enjoy—each in a different way—an influence that their successors were not to have.

Advising on military or technological issues constituted only part of the new task. Organizing the collective American scientific effort represented the other half. Harvey Brooks has trenchantly differentiated the role of a "policy for science" from that of "science in policy."[47] Eisenhower and other political leaders recognized that scientists simultaneously had to help shape high policy and to develop their disciplines from an invertebrate status into a cohesive national resource. Hitherto, political and military leaders had called upon scientists for specific urgent tasks, most spectacularly the Manhattan Project of the Army Corps of Engineers, which had built the atomic bomb. The advisors of the late 1950's and early 1960's, however, sought to move beyond organization of the ad hoc project; they aspired to structure scientific education, research, and relations with the government so that the generation of technical knowledge became an ongoing process, continuously to be woven into the major areas of national endeavor. Science itself—not merely the individual brains that had to be selected, encouraged, and educated, but the very process of discovery and development—was henceforth to be treated as a collective resource, comparable almost to oil reserves or protected forest land. In part this emphasis represented extension to a new domain of the preoccupation with conservation that had been so central a motif in American reformist politics since the turn of the century.[48] Like the conservation movement (similar, too, to the current stress on ecology), the demand to marshal science combined elitist and technocratic overtones with a concern for the vulnerability of American democracy.

The President's Science Advisory Committee was intended to contribute both science in policy and a policy for science. The first goal had been a long-standing one, but before Sputnik the approaches had remained fragmentary. In 1940 a National Defense Research Committee had been established at the urging of scientists, with Vannevar Bush at its head. A year later the committee was absorbed into the Office of Scientific Research and Development with Bush as director. As part of the Office of Emer-

[47] Harvey Brooks, "The Scientific Adviser," in Gilpin and Wright, *Scientists and National Policy-Making*, p. 76.

[48] See Samuel P. Hays, *Conservation and the Gospel of Efficiency: The Progressive Conservation Movement* (Cambridge, Mass.: Harvard University Press, 1959).

gency Management, the OSRD became a unit of the Executive Office of the President and enhanced the influence Bush could exert vis-à-vis the military or other agencies. Bush, however, never established a satisfactory relationship with President Truman and returned to private life. When the Republican administration took over, Bush encouraged the new National Security director, Robert Cutler, to appoint a bright young man within the White House who could draw upon the best scientific talent throughout the country, although he remained wary of an extremely close advisor: "This would be too likely to lead us into the difficulties that are exemplified by Churchill and Cherwell."[49]

The idea remained dormant, however. A Science Advisory Committee, established by Truman in 1951, reported to the Office of Defense Mobilization from 1951 to 1957 and was chaired successively by Oliver Buckley (1951–1952), Lee A. DuBridge (1952–1956), and Isador I. Rabi (1956–1957). Under DuBridge the Science Advisory Committee helped supervise organization of Killian's Technical Capabilities panel and the Gaither report. Nonetheless, direct influence on executive decision-making remained tenuous.[50] Within the Defense Department, moreover, Charles E. Wilson's attitude toward research remained downright disdainful.

With the launching of Sputnik in 1957 the President turned more urgently to politically involved scientists. He sought the assistance of Detlev W. Bronk, president of the National Academy of Sciences, in preparation of the statement on American progress that he delivered on October 9. Six days later the Science Advisory Committee convened with the President for a wide-ranging discussion of America's research capabilities and the need for scientific advising at the White House. As chairman of the Science Advisory Committee, Rabi noted the lack of a full-time science advisor and any supporting staff. Eisenhower responded to Rabi's concern and suggested a body analogous to the Council of Economic Advisors. Outside the Science Advisory Committee, Killian, Rabi, Fisk, and others discussed how to establish more

[49] Bush to Cutler, May 4, 1953, Eisenhower Presidential Library, "Project Clean-Up" papers, Box 29.

[50] David Z. Beckler, executive officer of PSAC, "Background Statement of the Activities of the President's Science Advisory Committee," September 4, 1959, in Eisenhower Presidential Library, U.S. President's Science Advisory Committee: Records, 1957–1961, Box 5, F, "PSAC (2)"; Don K. Price, *Government and Science* (New York: New York University Press, 1954), p. 45; Lee A. DuBridge, "Policy and the Scientists," "*Foreign Affairs* 41 (April 1963), 571–588.

leverage for scientists. Subsequently, Killian and Rabi break-fasted with presidential assistant Sherman Adams and staff secretary Andrew Goodpaster to press their suggestions for a White House advisor and transfer of the committee to the White House office; they then discussed the concept with Eisenhower once again. On November 3 the President announced the appointment of Killian to the position of special assistant for science and technology. The transfer of the earlier committee to the White House as the President's Science Advisory Committee followed on November 21.[51]

As the diary amply illustrates, the new PSAC was built around a strong committee system. Members were expected to be in Washington several days a month for consultation and to contribute to the individual panels. By February 1960 PSAC had incorporated a number of working groups: antiballistic missiles chaired by Wiesner, antisubmarine warfare under Harvey Brooks, arms limitation and control chaired by Killian, basic research and graduate education under Glenn T. Seaborg, chemicals in food (Bronk), continental air defense (Emmanuel R. Piore), early warning (Wiesner), high-energy accelerator physics (Piore), life sciences (George W. Beadle), limited warfare (H. P. Robertson), missiles (Hendrik W. Bode), science and foreign relations (Bronk), space sciences (Edward M. Purcell), an ad hoc study of missiles under Donald P. Ling, and a panel to review military communications under William O. Baker. While the panels tapped participants outside PSAC, membership was densely intertwined and overlapping. Hans Bethe, for example, participated on the ABM, early warning, and high-energy accelerator physics panels, and most PSAC members accepted similar multiple duties. Social scientists did not sit in PSAC itself, but did serve on various panels. Each panel was assigned a PSAC staff assistant. In view of the talent commanded, the central staffing remained remarkably economical. In fiscal 1960 the committee was allocated $365,000 for an organization of 22 people (including secretaries) and apparently had trouble spending all the money.[52]

[51] For the events, Goodpaster and Killian interviews; also Eisenhower, *Waging Peace*, pp. 210–212, 224: Killian to Eisenhower, November 15, 1957, urging transfer of the Science Advisory Committee to the Office of the President, in Eisenhower Presidential Library, U.S. President's Science Advisory Committee: Records 1957–1961, Box 3, F, ''Administrative.''

[52] List of panels, February 12, 1960, in Eisenhower Presidential Library, PSAC records, Box 5, F, ''PSAC (1)''; budget in Joan Hardy memo for

It was clear that PSAC did not try to provide "representation" for science departments and universities of any but the first magnitude. The scientists who played the formative role averaged around 50 years of age. They had moved beyond the traditionally brief decades as creative researchers to serve more as intermediaries and administrators. A certain camaraderie united many : a common participation in the hermetic ambiance of the Manhattan Project or the M.I.T. Radiation Laboratory, then service on one of the military "summer studies." As Harvey Brooks wrote, "some degree of political sophistication," a tendency to avoid rash or relatively extreme counsel, was tested in various panels before invitation into PSAC itself. Being able to emerge with a consensus was deemed more fruitful than paralyzing divisions that faithfully reflected bitter opposition.[53] If the Oppenheimer experience had deeply hurt many participants, they were still ready to presume that the alliance between their long-term research efforts and the aspirations of the United States as a world power was logical, defensible, and meritorious.

As coordinators of this advisory effort Killian, Kistiakowsky, and Wiesner (under Kennedy) brought different talents to bear and played different roles. In retrospect there is general agreement that Killian probably performed the function of organizing the PSAC effort with more tact and conciliatory capacity than Kistiakowsky would have brought to bear. Not a scientist himself, as president of M.I.T. Killian had the knack of eliciting high confidence from the specialists with whom he worked. Killian himself probably recognized that once PSAC was organized and the American space effort under way, the special assistant should himself be a scientist. As might be expected in such a situation, Killian sought primarily to get consensus and unified recommendations from his major scientific advisors, whereas Kistiakowsky strove more to master the technical side of an issue. Kistiakowsky also possessed political gifts, namely the ability to establish a relation of trust with Eisenhower and to assert an effective voice

---

David Beckler, PSAC records, Box 3, F, "Administrative." By October 1, 1963, the staff—now the Office of Science and Technology—had climbed to 38. See also Carl William Fischer, "Scientists and Statesmen: A Profile of the Organization of the President's Science Advisory Committee," in Lakoff, *Science and Power*, pp. 322–327.

[53] Brooks, "Scientific Adviser," pp. 81–83; Robert N. Kreidler, "The President's Science Advisers and National Science Policy," in Gilpin and Wright, *Scientists and National Policy-Making*, pp. 113–121; Keeny interview.

in the broad range of security issues that a less effective advisor might have seen pass by.

Kistiakowsky was probably less successful in promoting the subcritical areas of American science than in establishing his authority in the salient issues affecting diplomacy and armaments. Perhaps the former held less interest. The committee sessions and long presentations were often ordeals to be endured; as Kistiakowsky's diary indicates, he could muster little affection for those who wasted time. On the other hand, staff members testify that he allowed them meaningful participation in the preparation of policy recommendations. He employed cohesive panels effectively, although the hydra-headed Federal Council for Science and Technology—a 1959 reform suggested by PSAC to supersede the relatively feeble Interdepartmental Committee on Scientific Research and Development and to better coordinate the government's own research activities—proved too unwieldy to become a vital center of policy.[54] Kistiakowsky probably worked best by mediating the views of experts on a focused range of questions to a President who craved orderly administration and informed advice. Eisenhower liked the idea of having the scientists as a staff resource, and he evidently responded warmly to the direct style that Kistiakowsky presented. That response permitted the special assistant valuable leverage in negotiating with men who commanded immeasurably larger bureaucratic domains.

With the end of the Eisenhower administration the advisory role of the special assistant was to alter and, after Kennedy, to degenerate. It is revealing that a policy position seemingly called forth by the imperatives of technology should quickly wane in importance. As a systematic administrator, Eisenhower was bound to draw on his special assistant. This was not a question of ignorance; the memoranda of conferences with Killian now available demonstrate that the President understood the implications of taxing technological questions. Kennedy, on the other hand, possessed a more free-wheeling, almost playful, intellect. For him science became a new area to be mastered along with many others. And while Jerome Wiesner kept him abreast of the test-ban issues with a closeness and contact that probably exceeded even Kistiakowsky's with Eisenhower, still for Kennedy the science-advising function lost its special mystique.

[54] The proposal for the Federal Council for Science and Technology was a major constituent of the early PSAC report, "Strengthening American Science" (Washington, D.C.: U.S. Government Printing Office, 1958).

In a 1962 address to scientists, McGeorge Bundy stressed that for science advice to be effective, single dramatic presentations did not suffice; the scientist had to be "there" through time.[55] Still, as a National Security Advisor who wished to centralize the flow of data and judgments, Bundy did not make science-advising easier. Nor did such powerful intermediaries with the President as Robert McNamara, and later Walt Rostow, or finally Henry Kissinger. Eisenhower's last special assistant for national security affairs, Gordon Gray, had viewed his own task as essentially one of assembling and channeling information; his successors were more ambitious and jealous of alternative sources of counsel. Moreover, as Herbert York has suggested, Killian and Kistiakowsky represented the intellectuals as well as the scientists in the Executive Office, whereas Kennedy was surrounded by intellectuals. Wiesner thus enjoyed an easy, close relationship with the White House advisors, but less as the spokesman for PSAC or a scientific consensus than as one more of a number of bright and informed men. Indeed Richard Neustadt, whose suggestions influenced the organization of the White House, feared diffusion of advice and sought to subordinate the offices purveying expertise to a few trusted deliberators on general executive policy. In 1962 President Kennedy finally gave the staff of the special assistant a new status by reorganizing it as the Office of Science and Technology (OST) outside the White House Office, though in the Executive Office of the President. This restructuring facilitated OST's ability to advise Congress by removing the umbrella of executive privilege—perhaps thereby forestalling a more drastic congressional incursion—but it also made White House links more formalized.[56]

At the same time, the assignments for PSAC became more diverse. With the issue of missile disparity fading away in the early 1960's, PSAC focused increasingly on health, environmental, or other nonsecurity questions. As Eugene B. Skolnikoff recalled in an interview, innovation was encouraged and participants felt a sense of excitement in being able to pursue a broad range of concerns. This meant simultaneously that the central

[55] McGeorge Bundy, "The Scientist and National Policy," address to the American Association for the Advancement of Science, December 22, 1962, in Lakoff, *Knowledge and Power*, p. 423.

[56] York interview; on the Office of Science and Technology, see Sanford A. Lakoff, "The Scientific Establishment and American Pluralism," in *Knowledge and Power*, pp. 382–384; also Don K. Price, *The Scientific Estate* (Cambridge, Mass.: Harvard University Press, 1965), pp. 242–245.

role of the science advisor on security matters would ebb. The broadening of issues meant that the committee membership had also to grow more catholic; the esprit de corps of the atomic physicists and Radiation Laboratory engineers tended to diffuse. Furthermore, when space and weapons issues were paramount, the government was the direct "consumer" of advice and contractor, but when, say, health care, was the problem, PSAC had ultimately to deal with multiple authorities and governments over which it had less influence.

Some of the eventual changes in the role of the special assistant were likely outcomes in any case. The fact that the special assistant had to chair the Federal Council for Science and Technology as well as PSAC tended to cut into his time and diffuse his advisory function.[57] The creation of DDRE—which Killian, assisted by James Fisk and Piore, helped work out with Charles A. Coolidge's reorganization committee in the Defense Department—gave OSD its own office for project evaluation. The policing of missile development and interservice rivalry henceforth was to be shared by Herbert York and the special assistant; as York recalled in an interview, for example, his office did not need the political aid of the special assistant in transferring the Army Ballistic Missiles Agency to NASA. (Cf. Chapter IV on this issue.) Likewise NASA itself took on a major science-advisory and management function. The special assistant thus helped solve many of the problems that had led to the creation of his office.

After Kennedy, more personal factors undermined the office. For Lyndon Johnson, preoccupied with domestic welfare on the one hand and Vietnam on the other, science could not regain a special mission. As York remembers from his stint on PSAC during the mid-1960's, Johnson repeatedly addressed the question: "What can I do for grandma?" Deserving as grandma was, she probably had few occasions to turn to PSAC for coping with the difficulties of American life. In Richard Nixon's White House, scientists like all others had to overcome the suspicion of the staff advisors and the massive authority of Henry Kissinger. The President finally did away with the office under a reorgani-

---

[57] Skolnikoff interview. For some of the difficulties with the Federal Council, see Robert N. Kreidler to Killian, March 23, 1959: Aide-Mémoire for the first meeting of the FCST, in Eisenhower Presidential Library: Office of the Special Assistant for Science and Technology, Additional Materials ["New OSAST"], Box 9, F, "Aide-Mémoire for the Chairman of the FCST."

zation plan in 1973. Only now in 1976 is it being revived in similar form, its future prerogatives to be defined anew.[58]

SCIENCE IN POLICY represented only half the assigned task for the special assistant. In his conference with the Science Advisory Committee on October 15, 1957, the President had just as urgently canvassed suggestions for a policy for science. After Edwin Land argued eloquently that the Russians were teaching their youth to enjoy science and focus on basic research, while "in the United States we are not now great builders for the future but are rather stressing production in great quantities of things we have already achieved," Eisenhower said he would like "to create a spirit—an attitude toward science similar to that held toward various kinds of athletics in his youth—an attitude which now seemed to have palled somewhat."[59]

Thus while most PSAC panels focused on security questions and defense technology, the new committee still devoted major effort to mobilizing research and teaching activity. The operational question in bolstering science was who was to get money and how was it to be allocated. Here the disinterestedness of the scientific community naturally had built-in limits. As Don K. Price has noted, "By the nature of their discipline [scientists]

[58] For an important retrospective evaluation of the accomplishments and weaknesses of PSAC see David Z. Beckler, "The Precarious Life of Science in the White House," *Daedalus* (Science and Its Public: The Changing Relationship), 103 (summer 1974), 115–134. See also the report by the National Academy of Sciences, ad hoc Committee on Science and Technology, "Science and Technology in Presidential Policymaking: A Proposal," (Washington, D.C., June 1974), especially p. 16 on the abolition of OST and pp. 19 ff. on the proposals for a new council. (Congress, however, has insisted on one official.) Cf. also Kistiakowsky's article in *Science* 184 (April 5, 1974), 38–42, and Eugene P. Skolnikoff and Harvey Brooks, "Science Advice in the White House? Continuation of a Debate," *Science* 187 (January 10, 1975), 35–41. Proposals for a Department of Science have also come under consideration. At the time the FCST was created in 1959, a Humphrey-chaired Senate committee was weighing such legislation (S. 3126), but Eisenhower had misgivings, as did the Bureau of the Budget. Indeed, it is hard to see how science as such could be effectively administered apart from the activities of the respective cabinet departments that had technological components. See Bureau of the Budget, "Organization of Science in Government," September 16, 1960, paper prepared for FCST, in Eisenhower Presidential Library, "New OSAST," Box 9, F, "FCST."

[59] DDE, Papers as President (Ann Whitman file), Diary Series, Box 16, F, "Staff Notes, October 1957," Goodpaster notes: memorandum of conference with the President, October 15, 1957, 11 AM.

remain dissenters, with no interest in founding a new political dogma. But they need a great deal of money from the government.''[60] By and large they received their money. According to OECD estimates, U.S. research and development expenditure (excluding education) rose from 0.3 percent of the Gross National Product (GNP) to 1.0 percent in 1950, 1.6 percent by the mid-1950's, 2.8 percent in 1961, and 3.0 percent by the mid-1960's. Educational expenditures also climbed from about 4 percent of the GNP before the war to 5 percent by 1950, to 8 percent by the close of the Eisenhower administration. Along with the establishment of NASA and DDRE, the year after Sputnik saw passage of the National Defense Education Act, which inaugurated a major program of federally financed college and graduate scholarships.[61]

Just as revealing as the spurt in funding was the growth of the federal contribution. Washington's share of R&D expenditures rose from 0.16 percent of the GNP in 1941 to about 0.4 percent in 1950–1951, to 1.8 percent in 1961, to about 2.2 percent by the mid-sixties. By 1965, the federal government was financing over three-fifths of the nation's research and development : as high a percentage as in a traditionally statist or directed economy such as the French, and higher than under Sweden's social democratic regime. The characteristic private-sector preferences of the United States were revealed by the fact that private industry and universities remained the major beneficiaries, receiving almost 80 percent of the government's disbursements.[62]

What sanctioned, indeed impelled, the new federal role was the integral conection of projects with perceived national challenges. Between 1954 and 1967, over 80 percent of federal R&D funding went to DOD, to NACA-NASA, and to that half of AEC expenses roughly designated as defense oriented. Reflecting the preoccupa-

[60] *Scientific Estate*, pp. 118–119.

[61] Organization for Economic Co-Operation and Development, *Reviews of National Science Policy: United States* (Paris: OECD, 1968), p. 30. See also Alan T. Waterman, ''Federal Support of Science,'' a statement to Congress in 1964, reprinted in Lakoff, *Knowledge and Power*, pp. 395–405. Waterman estimated that only 10 percent of the R&D expenditure was for basic research, perhaps another 20 percent went for applied research, and 70 percent for development; the OECD estimated two-thirds as a consistent share for development (p. 38).

[62] OECD, *Reviews*, p. 35. Of the almost 80 percent, about 7 percent went for laboratories that were effectively government installations run by universities. On the importance of private institutions see Price, *Government and Science*, chap. 3, and *Scientific Estate*, pp. 71–77.

tions of the society, this federal funding of research and development designed to confront the "external challenge" rose from a low of 35 percent of all U.S. R&D outlay, to close to 40 percent in the late 1950's, to between 58 and 66 percent by the mid-1960's. Between 1958 and 1967, the federal government spent about $112 billion on research and development, about $93 billion of which (roughly 83 percent) went to "external challenge" categories. If one turns to human resources, doctorates in natural sciences, mathematics, and engineering rose from approximately 4,000 in 1954 to 7,500 in 1964. Master's degrees in the same disciplines rose from about 12,000 to 30,000.[63]

Flush times, indeed! Generally conservative about expenses, Eisenhower told his science advisors that research was an area that justified greater federal spending.[64] Like many episodes of economic growth, the spurt in the late 1950's and 1960's reinforced, or at least did not diminish, unequal distribution of resources. Of a total of 223 universities in the United States that awarded doctorates in the years 1962 to 1964, about 120 spent virtually all the money that might be identified as university expenses on research. Twenty-one alone spent two-thirds of the total. The same 120 received almost all the federal supporting funds available, with just the top 21 pulling in about 57.5 percent of the federal funds. The same elite group of 21 awarded over half the doctorates. As Kistiakowsky explained to the cabinet when presenting the Seaborg report in the fall of 1960: "In the advancement of science the best is vastly more important than the next best, and both government and university administrators should be firm in their support of institutions and departments or laboratories which are first-rate, even when this requires hard choices."[65]

Because after Sputnik so high a priority was placed on technical expertise and scientific promise, the men helping to steer research efforts enjoyed a freedom to cultivate elite institutions that was new in the history of the American republic. Nonetheless, even ample funds could not resolve all problems of inchoateness. The role of the National Science Foundation remained relatively ill defined and ineffective—in part a legacy of

[63] OECD, *Reviews*, pp. 38, 56, 109.

[64] See Eisenhower to Alan T. Waterman, then director of the NSF, January 15, 1959, in Eisenhower Presidential Library, U.S. President's Science Advisory Committee, Box 4, F, "National Science Foundation (1)."

[65] OECD, *Reviews*, pp. 222–224; Eisenhower Presidential Library, "New OSAST," Box 5, F, "Cabinet," GBK presentation, October 7, 1960.

the conflicts that had marked its founding after World War II. The research programs within the government, according to a November 1959 memo by PSAC panel members Paul A. Weiss and William O. Baker, remained uneven. In low-technology agencies, such as the Weather Bureau and the Bureau of Standards, the going research objective was "barely enough to keep abreast with the immediate needs and the current art." In DOD, NASA, and the AEC, the goals were "so vast and so far beyond present conceptions that sensible selectivity for funding [had] become impossible. (ARPA is, of course, the pathological example.)" Long-range planning capacity still proved elusive. Although the NSF provided the Federal Council with five-year estimates of agency research funding and needs, the projections were questionable. Kistiakowsky was to try tightening up the five-year estimates, but in 1959, at least, the NSF estimated that government R&D expenses would rise from $5 billion to $9.2 billion in 1964, whereas by 1964 the bill was actually to be $14.7 billion.[66]

The PSAC and the special assistant probably had a more effective voice in shaping the directions of research outside the government. The agencies within each department enjoyed their connections with the respective congressional committees that approved their funds. The scientists outside were not directly connected to Congress, save perhaps in the quasi-dependent laboratories that were run almost as a franchise by a few of the universities. Instead, the scientists outside saw the special assistant and PSAC as their advocate and court of appeal even when they could not claim colleagueship. General consensus prevailed that the peer group must judge the worth of research and the validity of claims for funds. As the NSF reiterated, "The scientists themselves know best what can be done and how to go about it."[67]

[66] On the NSF, see Lakoff, "Scientific Establishment," p. 381; Price, *Scientific Estate*, pp. 239–240. On research objectives see Eisenhower Presidential Library, U.S. President's Science Advisory Committee, Box 3, F, "Basic Research," "Review of Research Plans of Federal Agencies . . ." November 9, 1959. On R&D expenditures see Eisenhower Presidential Library, "New OSAST," Box 9, F, "FCST (Reports) Vol. I," NSF, "Projections of Funds, Facilities, and Manpower . . . Fiscal Years 1959–1964," October 27, 1959.

[67] "New OSAST," Box 9, F, "FCST," NSF paper for FCST, "Support and Utilization of Basic Research," September 1960, repeating statement from 7th annual NSF report, 1957. Cf. Kreidler, "President's Science Advisers," pp. 130–135.

There was also consensus on the value of basic research, while the crisis in missile development and related security questions allowed a shared emphasis on the physical sciences. Of course, as the diary amply confirms, this did not eliminate hard conflicts over limited funds, especially when new laboratory facilities had to be built, as in the case of "materials" research—the properties of solids used in nose cones or rocket engines, for example.[68] Occasionally dissenters appeared; Eugene P. Wigner argued against the emphasis on high-energy physics, and the director of the Oak Ridge National Laboratory, Alvin M. Weinberg, by 1964 was to question the utility of basic research and the whole system under which scientists decided their own research priorities. But in 1960, at least, Kistiakowsky expressed the shared premises of the scientific community when he wrote Weinberg: "What you recommend in effect is an intellectual economy of scarcity—make 'good' people go into applied work by cutting down on opportunities for basic research!" Rather, basic research provided better training—the best engineers started as scientists, not the reverse —and better results: "Mighty few really new technological ideas did not originate in basic research!" Perhaps the final triumph of the consensus and of PSAC was White House endorsement of these criteria, when a president celebrated neither for his abstract intellect nor his fiscal radicalism accepted the Seaborg report ("Scientific Progress, the Universities and the Federal Government") with its call for long-term government financing of research and postgraduate teaching expenses.[69]

Clearly the needs of the "free world" benefited science. Under the spur of national interest the role of the special assistant in the late 1950's and early 1960's proved especially rewarding, both because of his opportunity to serve as a policy advisor and because of the broad agreement on the need for research that united political and scientific leaders. Given the distrust of scientists that

[68] Besides the diary entries, see the report entitled "Coordinating Materials Research in the United States," by the PSAC staff, March 18, 1958, and included in "New OSAST," Box 9, F, "FCST."

[69] Kistiakowsky to Weinberg, May 3, 1960, in U.S. President's Science Advisory Committee, Box 6, F, "Correspondence—W." For Weinberg's 1964 view see his article, "Criteria for Scientific Choice," in Lakoff, *Knowledge and Power*, pp. 406–419. See also the intra-PSAC controversy over Wigner's dissent from the PSAC–General Advisory Committee (AEC) panel report endorsing active support for high-energy physics (including re-emphasis of the value of the Stanford linear accelerator) in December 1960; in U.S. President's Science Advisory Committee, Box 3, F, "High Energy Physics."

had been so ominous during the Oppenheimer hearings, this new influence represented a considerable achievement. Was it ultimately a tribute to science as such? Science, strictly speaking, was what scientists did: a mode of reasoning or of measurement in special facilities designed to make predictions about physical observations, or about the performance of gadgets that involved vast amounts of power or vast acceleration of communication and calculation. But increasingly, science was being elevated from an endeavor to an entity, hypostasized as a basic constitutent of national power and prestige.

By the early 1960's the fear was again growing that science might become too powerful, its very objectivity in respect to the traditional American elites serving as an argument to enthrone its own practitioners. Don Price, one of the most perceptive writers about the relations of science and government in the vast new sea of commentary on the issue, saw possible abuses from scientists, in alliance either with the executive or congressional committees, if they were placed beyond effective responsibility. By and large, however, he saw the scientists as being assimilated into an ongoing American system of checks and balances among professional estates as well as formal institutions.[70] Eisenhower himself showed concern about technocratic abuses, which Kistiakowsky had to explain in print was not directed at the scientific community as it then functioned.[71] In many ways the worry was an appropriate one for the United States, which in the absence of strong class divisions anchored in an ancien régime saw contenders periodically claim prestige and power on the basis of expertise rather than nonintellectual achievements. Yet surprisingly little of such a technocratic thrust on the part of natural scientists emerged. The Great Society and the Vietnam war diverted attention from the challenges of spectacular technology as Americans were humbled by rural guerrillas or by the persistence of urban poverty and pretechnology prejudice. The resources of the society did not easily absorb or effectively employ all the scientific manpower produced. Other groups, such as

[70] Price, *Scientific Estate*, especially chaps. 5 and 8.

[71] For this incident in the wake of Eisenhower's warning about the possible danger of a ''scientific-technological elite'' in the farewell address, see below, page 425, and *Science* 133 (February 10, 1961), 355, about which Eisenhower wrote to Kistiakowsky (Kistiakowsky file): ''I would just like to say that your letter expressed my views exactly and I am grateful to you for giving them this wide dissemination among your fellows in the scientific world.'' See also Price, *Scientific Estate*, pp. 11–12.

economists, were feted as the experts, only to falter in their collective turn.

In perspective, perhaps the period of the late 1950's and early 1960's will be regarded as one of those historical transitions in which the technologies and production processes of society show a qualitative change, that is, a change of inputs and outputs compressed enough to appear as a "takeoff" in an otherwise gradually ascendant curve. The input of the 1950's was different from that of earlier economic transformations, which involved the application of new minerals or power sources. The 1950's (resuming what the war had episodically initiated) involved instead a more extensive exploitation of men and women collectively thinking about the physical world (and later the world of social interactions). Increasingly, experimentation and theoretical formulation were to be borne by public cost. The process was accelerated by the bipolar international competition in which the United States found itself: the logic of international systems and the logic of postindustrial development reinforced each other. George Kistiakowsky among others enjoyed the heady privilege of helping to superintend that process and make it more susceptible to conscious direction. No more than the industrial revolution, which Saint-Simon and followers had predicted would change questions of power into problems of efficiency, could this transition transcend politics. The stuff of politics shifted; but, as the diary illustrates, the conflicting interests, reputations, and visions of powerful men could not be suspended.

Charles S. Maier
July 1976

A SCIENTIST
AT THE WHITE HOUSE

A glossary of abbreviations,
acronyms, and code names
appears on pages 426–430.

THE WHITE HOUSE
WASHINGTON

July 15, 1959

Dear Dr. Kistiakowsky:

Within the over-all area of your responsibility as my Special
Assistant for Science and Technology, I should appreciate
your giving particular attention to the activities set forth, in
general terms, below:

Keeping yourself informed on the progress of scientific
endeavor in the various agencies of Government, giving
primary attention to the use of science and technology
in relation to national security.

Finding and presenting to me facts, evaluations and
recommendations respecting matters related to science
and technology.

Advising on scientific and technological matters in top-
level policy deliberations; making yourself available as
an adviser on scientific and technological matters to
Cabinet members, and other officers of Government
holding policy responsibilities, when appropriate and
practical; and working in close association with the
Director of the Office of Civilian and Defense Mobilization and the Special Assistant to the President for
National Security Affairs.

Trying to anticipate future trends or developments in
the area of science and technology, particularly as they
affect national security, and to suggest future actions
in regard thereto.

Aiding in the collection of information about the relative
progress of Soviet and U. S. science and technology.

Working closely with the National Science Foundation and its Director.

Concerning yourself with the interchange, when feasible and proper, of scientific and technological information with scientists and officials, military and non-military, of our allies, and to encourage science in the free world.

By separate action, I have previously designated you a member of my Science Advisory Committee. Also, as my Special Assistant for Science and Technology, you are designated a member of the Federal Council for Science and Technology, and are further designated Chairman of the Council.

You understand, of course, that the foregoing does not limit or define your responsibilities, but is intended as an aid to you in organizing your initial work. You are authorized to participate in the meetings of the National Aeronautics and Space Council, and to be in attendance at meetings of the National Security Council, the Cabinet and the Operations Coordinating Board, and to attend or to be represented at meetings of the National Security Council Planning Board. I understand that the Secretary of State has invited you to attend or be represented at meetings of the State Department Policy Planning Board whenever matters of interest to you are considered, and that the Secretary of Defense has invited you to attend or be represented at meetings of the Defense Science Board. I understand also that the Secretary of Defense has asked the Joint Chiefs of Staff to invite you to consult with them when they consider matters which lie in your field of interest. I approve these arrangements.

It is my desire that you have full access to all plans, programs, and activities involving science and technology in the Government, including the Department of Defense, AEC, and CIA.

I have indicated to you my basic purpose in establishing in my staff the position you now hold. Your work can have immense value

CONFIDENTIAL

through assisting in developing information for me and in giving
a greater sense of direction to all who are concerned in our
Nation's technological and scientific efforts.

Sincerely,

Dwight Eisenhower

The Honorable George B. Kistiakowsky
Special Assistant to the President
The White House

CONFIDENTIAL

NOTE: As of June 1976, this letter has been declassified.

CHAPTER I

Introduction to the Nuclear Test-Ban Problems

MY INVOLVEMENT prior to the summer of 1959 in the PSAC activities related to the test-ban treaty negotiations had been very minor because during the fall and winter of 1957–1958 I had been involved in other PSAC activities, then spent the summer of 1958 chairing in Washington an interagency group that was instructed to prepare our positions for the UN ten-nations conference to reduce the dangers of surprise attack, which had been proposed by our State Department and agreed to by the Soviet Union. I then spent a goodly part of the fall of 1958 in Geneva at this totally futile conference as a member of the three-man American delegation, headed by William C. Foster and accompanied by a large staff with Jerome B. Wiesner as staff director. Before my swearing-in in July 1959, however, the President had told me that he wanted me to work hard to achieve a test-ban treaty ''consistent with our national security.''

I took this instruction seriously, but the task was difficult because the opponents of the test ban, led by John McCone, chairman of the AEC, the Air Force brass (supported by the rest of the Pentagon), and Edward Teller with his affiliates at the Livermore Nuclear Weapons Laboratory of the Atomic Energy Commission had not been idle in the meantime. They challenged the feasibility of monitoring a treaty, by inventing ever more fantastic methods of potential Soviet evasion, such as conducting tests in outer space, for instance behind the sun, or in the center of huge spherical caverns, up to thousands of feet in diameter excavated deep underground (the "Latter holes"). The air gap between the weapon and the walls of such cavities would theoretically "decouple" the explosion so that only a relatively feeble seismic signal, unidentifiable at longer distances, would result. Teller and others also extolled the tremendous economic gains that would result from proposed peaceful uses of nuclear explosions, such as making harbors and digging canals: the so-called Project Plowshare. Since these projects would require much testing of special nuclear explosive devices, Teller's group urged rejection of any test ban or at least exceptions for Plowshare explosions. The Soviets flatly rejected the latter option. (After many underground tests the United States has now discontinued Project Plowshare. The Soviet Union however now insists that it wants to go ahead with "peaceful" nuclear explosions.)

To make matters worse, the efficacy of the system proposed in 1958 for seismic monitoring was seriously questioned by the Berkner Panel on Seismic Detection in early 1959; and in the spring of 1959, another PSAC panel chaired by Robert F. Bacher of the California Institute of Technology yielded (as I was informed by a member) to the aggressive arguments of one of its members, Harold Brown, then deputy director of the Livermore Weapons Laboratory, and concluded that even an on-site inspection of a seismically suspicious event on the territory of the Soviet Union would have an exceedingly small probability of proving that an underground nuclear explosion had taken place, if the Soviets made an effort to hide the telltale marks. This report amounted to a gloomy denial of the conclusions of the original interagency panel chaired by another PSAC member, Hans Bethe, which in early 1958 had concluded that the monitoring of the test-ban treaty to detect evasions might be feasible.

After long debate at a two-day meeting in the spring of 1958, the PSAC had accepted the conclusions of Bethe's panel and endorsed its involvement in the proposed Geneva conference of experts from eight nations, which convened from July 1 to August 21, the American delegation being chaired by James B. Fisk, a PSAC member, and including Bacher as its deputy chairman. The Geneva conference's report described conceptually a worldwide monitoring system which, by observing acoustic waves, radioactive debris, electromagnetic signals, seismic waves, and hydroacoustic waves, was expected to detect all but the smallest nuclear explosions in the atmosphere, underground, or underwater. The conference noted that properly equipped earth satellites could detect high-altitude explosions. The need for on-site inspection to identify the origin of unidentified seismic signals was recognized. But now, nine months later, on top of the Berkner panel, the Bacher report suggested that the on-site inspection was far more likely to be fruitless. Killian himself strongly endorsed these Bacher findings when he talked to Eisenhower in my presence, thus foreclosing room for maneuver in preparing a test-ban proposal.

The interconnection and technical significance of these new findings have already been summarized by Charles Maier in his contribution to this volume. However, so many references to the underground nuclear tests and their seismic detection will be found in the following that an additional explanatory note for those not familiar with the technicalities of the subject may be in order here.

Underground nuclear explosions, so deep under the earth's surface that no radioactive particles escape into the atmosphere to be carried away by the winds, can still be detected from large distances through the seismic waves they generate in the earth. A subsequent on-site inspection can reveal additional and conclusive evidence from surface displacements, changes in vegetation, etc., but this becomes more difficult in proportion to the area of the land that must be surveyed.

In the first approximation the seismic waves from an underground nuclear explosion are not different from those produced by shallow earthquakes. With advanced seismic equipment even the signals from very small nuclear tests, equivalent in violence to only a few kilotons (kt) of exploding TNT, can be detected at long distances. An unsophisticated seismograph, however, can only provide evidence that an unidentified seismic event took place. If several seismographs are properly distributed over the surface of the earth, the location of the unidentified event can be determined, usually to an accuracy of a few miles at best. More sophisticated seismic detection systems, which came into existence largely after the events recorded in this journal, can distinguish between seismic waves from earthquakes and from nuclear explosions because of fairly subtle differences in the nature of the signals. The discrimination becomes less reliable as the magnitude of the seismic event—or the equivalent explosive force of a nuclear weapon—gets smaller and smaller.

The intensity of seismic waves is described on the so-called Richter scale, according to which the greatest natural catastrophes rate about 8 and events of magnitude 1 to 2 are totally trivial, even locally. The frequency of natural earthquakes increases very rapidly as their intensity decreases, there being about 20 per year in the Soviet Union with a magnitude more than 4.75 on the Richter scale (equivalent to about a 20-kt explosion) and, of course, many more that are weaker.

Under the limit of discrimination of a world-wide seismic monitoring system, a limit which to this day remains a subject of some technical and political argument, many signals would be received by a system but remain unidentified. The higher (that is, poorer) is the discrimination limit of a seismic detection system, the more unidentified (weaker) signals will be in the range corresponding to nuclear explosions of military significance (itself a hotly debated matter, but the equivalent of a few kilotons TNT is frequently mentioned as a lower limit). Thus such a system will present more opportunity and incentive for the evasion of a comprehensive test-ban treaty. The absence of such evasion might still be monitored through on-site inspections by teams of technically trained individuals dispatched to the sites, or rather to land areas in which the unidentified signals originated, soon after the event. The difficulties are further compounded by the dependence of the seismic signals from a nuclear explosion upon the rock formation in which it took place, the signals being greatest in hard granite-type rock and weakest in dry alluvial deposits. Consequently, seismic detection alone seemed unlikely to provide adequate test-ban monitoring, while the efficiency of on-site inspection had now also come under question.

When I took office my participation in the process of preparing recommendations to the President on objectives and tactics of the negotiations for the nuclear test ban was my most immediately preoccupying task, by virtue of my membership in the Committee of Principals. This group, established by the President in 1958, included Christian Herter, then Secretary of State, or Douglas Dillon, his deputy, as chairman, Secretary of Defense Neil McElroy, Chairman of the Atomic Energy Commission John McCone, Director of the Central Intelligence Agency Allen Dulles, Gordon Gray, Special Assistant to the President for Security Affairs, and the Special Assistant for Science and Technology. The committee was assigned the responsibility for hammering out negotiating positions to present to the Soviets in Geneva.

Furthermore, almost immediately after my taking office the

President had charged the PSAC with an evaluation of the urgency of his terminating our moratorium on nuclear test explosions and resuming the testing which was urged on him from many sides. A major argument for the resumption of tests was that our aircraft carried bombs (which included large charges of chemical explosives necessary to produce ''implosion'') possibly unsafe against nuclear explosion when accidentally dropped from aircraft. In March 1958 a bomb dropped by a B-47 aircraft in South Carolina detonated on impact, causing local damage. There was no nuclear explosion, but the weapons laboratories of the AEC refused to guarantee their designs against the occurrence of such explosions in future accidents, thus creating the pressure for resumption of tests.

As the reader will note, though, these aspects of the test-ban issue were only two of many problems that crowded in from the outset. Others stemmed from deep rivalries for research funds, or the jealously claimed rights on the part of different agencies to develop rockets, or the contested allocation of scarce radio-frequency bands. The first month of work introduced me to conflicts that would recur again and again.

⟳ *Wednesday, July 15, 1959*

7:30 AM   Pre-press breakfast. Focus on strike; nothing in my line. State Department memo on Geneva conference acceptable to Hagerty.

> At these White House Mess breakfasts of selected staff members, chaired by James Hagerty, the press secretary to the President, current events were reviewed. Hagerty then assigned to some of us, depending on events, the preparation of memos for the President, with whom we then met at 9:30 or 10:00 AM, prior to his press conference.

9:45 AM   Pre-press briefing. President very angry about State Department leak on the supposed advisor job for Bohlen. Phoned and chided.

> Charles E. ("Chip") Bohlen, former United States ambassador to the USSR.

11:30 AM   My swearing-in ceremony. Came off well. President very friendly. Lots of pictures, no press statement required. Lunch with family; back in office read papers.

> Prior to this date I stayed in Washington almost full time for
> several weeks to attend most of the meetings participated in by
> Dr. Killian, including those with the President.

2:00 PM   Dr. Struve and Mr. John Finney re: allocation of frequencies for radio astronomy. Hi Watters noted that Killian accomplished a lot by relaxing the stiff negative position of the State Department [allowing only the alpha hydrogen frequency for the purpose]. Struve unhappy but Finney concedes that astronomers are asking too much and too late. Agreed to see Wallace Brode for clarification of position. Agreed that British and others in August conference on frequency allocation should take the initiative.

> Otto Struve, well-known astronomer, came with John W. Finney,
> a *New York Times* science reporter, to press for greater allocation
> of high radio frequencies to radio astronomy uses at the forth-
> coming international conference on frequency allocation. I was
> warned in advance by Commander H. G. Watters of our staff
> that Dr. Killian had asked the astronomers to develop detailed
> proposals for the conference but had no response until the eve of
> the conference when this visit took place. Wallace Brode, chem-
> ist, was then the science advisor to the secretary of state.

4:00 PM   Killian and Keeny re: Principals' meeting on coming Thursday. I noted to Killian that he by his statement to Eisenhower tied my hands a bit: it was too definitive in view of the weak evidence of the Bacher panel. I hesitate to restate Bacher findings now, although the wrong impression might have been gained. Except for the pure-guess P_i factors and the Latter hole, the findings are not different from a year ago. The Latter hole is pure theoretical deduction; needs to be tested experimentally, but I don't want to appear to contradict Killian.

> The P_i factors, a term introduced by the Bacher panel, related to
> the likelihood of detecting a concealed, underground nuclear
> test during a subsequent on-site inspection. Spurgeon M. Keeny,
> Jr., was a member of our staff.

4:45 PM   Andy Goodpaster phoned that Glennan is in trouble with Army, which wants to send monkeys into space and is creating a fait accompli. Glennan opposed. The latter phoned for an appointment but this did not come off because of my meeting with Quesada.

> Brig. Gen. Andrew Goodpaster was the President's staff secretary and very much in his confidence. T. Keith Glennan, the administrator of NASA, was in constant jurisdictional conflict with the United States Army which, using Wernher von Braun and his rockets, was feverishly trying to carve a bigger role in space for itself. Retired Lt. Gen. E. R. ("Pete") Quesada was the head of the Federal Aviation Administration.

5:00–7:00 PM   Killian, Quesada, Perkins McGuire, and Jerry Morgan re: FAA 3-megacycle band allocation. Q. impassioned but not sticking to point. Raves about wanting same absolute authority over the three megacycles that DOD has over its large allocation. McGuire and Killian patient. Gradually get Quesada to talk on the subject and to accept a wording that gives him authority but requires him to utilize frequencies efficiently by revealing plans to DOD and letting them use free areas. McGuire agrees to persuade military to accept this wording. Nobody seems to take Hoegh seriously even though he stated that his order, which was unacceptable to DOD, was his final word.

> E. Perkins McGuire, deputy assistant secretary of defense for International Security Affairs; Gerald D. Morgan, deputy assistant to the President, i.e., deputy to Gen. (ret.) W. B. ("Jerry"), Persons. This conference, like that with Struve, was dealing with the carving up of the radio frequency spectrum among the many agencies asking for allocations. Present at the meeting was also former Governor Leo A. Hoegh, who was supposedly allocating frequencies as the director of the Office of Civilian Defense Mobilization.

⟿ Thursday, July 16, 1959

9:00 AM   NSC meeting. Nothing much of interest to us, but the President suggested that "his scientists" maybe should look into the problem of when and on what scale to resume nuclear tests if the test-ban negotiations fail. This was in response to McElroy's remark that they were spending $8 million monthly just to keep ready.

Afterwards visited various White House offices with Andy Goodpaster.

> Neil H. McElroy, secretary of defense, former president of Proctor & Gamble Corporation.

12:00 M My staff meeting with sandwich lunch. Discussed monitoring of nuclear test suspension, Keeny and Beckler making several suggestions, and Keeny warning of the State Department paper to come, which will propose a fallback to phased suspension of tests.

Telephoned Bethe who says that his calculations confirm the Latter hole theory, but that geological data may indicate impossibility of doing it in salt domes. He will bring papers on this subject to PSAC meeting.

Glennan came. Said he informed Army that NASA will not endorse or approve the next monkey flight. If Army is to do the test, it will have to do so on its own responsibility. He also admitted that the program of NASA is getting too large and has to be reduced. There are too many projects with insufficient difference of objectives. He proposes to concentrate on the moon and forget Venus and Mars for the time being. I heartily approved.

Piore dropped in. Doesn't like the idea of spending a day in Washington monthly without specific objectives.

> David Z. Beckler, executive secretary of the PSAC; Emmanuel R. Piore, a vice-president of IBM and a member of PSAC. The reference is to my plea to all members of PSAC to come one extra day a month to our office to help me with current problems. This practice was much in vogue during the first months of Dr. Killian's appointment. Hans A. Bethe was professor of theoretical physics, Cornell University, and a member of PSAC.

4:30 PM Meeting of Principals. Dillon started by questioning me about the Bacher panel report. Fortunately we had graphs prepared by Keeny which we displayed and which passed without challenge. Thereupon meeting devoted to discussion of State paper on tactics to be followed in case President decides to change policy. McCone objected to proposal of temporary unilateral suspension of tests. Dillon agreed to present this as an alternative only. Loper and McCone then presented plans for further nuclear tests, underground and high-altitude, but not in atmosphere. Some tests, however, were to be only 50 miles up and hence would unavoidably result in fallout, in my opinion (which I kept to myself this time). McCone has a technical report on military-technical reasons for the need of further testing, subscribed to by everybody in AEC except Teller who feels report not strong enough. Dillon stated that principals will meet with President next week. This to start with presentation of our graphs and then discussion of necessary policy changes.

Meeting ended at 6 PM.

I saw Dillon for a short while privately and pointed out that I wished the change from Killian to myself to be a continuous change : there would be no sharp differences in opinion. He agreed heartily. I then discussed the significance of Killian's statement in the previous meeting of Principals as reinterpreted by Killian and myself, and Dillon accepted that.

> C. Douglas Dillon, undersecretary of state, who chaired the Principals' meetings during the frequent absences of Christian Herter, the secretary. Of the two, Herter was more sympathetic to the nuclear test ban than Dillon who tended to side with Defense, and John A. McCone, chairman of AEC and one of the chief opponents of the test ban. Gen. Herbert B. Loper was head of the nuclear weapons office in the Office of the Secretary of Defense.

ᐴ *Friday, July 17, 1959*

9 :00 AM   Cabinet meeting. Started with a presentation by the attorney general on the new rules for the removal of papers by high government officials. He said that the high officials should not be allowed to remove government records. To which the President commented that occasionally yellow-pad notes saying simply "that S.O.B. was again in my office" became government records, and wasn't there a way of limiting that designation to significant papers?

The secretary of commerce described in detail the proposed New Delhi agricultural fair with color slides. It will be a really fine exhibit if the reality looks as good as the slides.

A long discussion on birth control in India and the need to encourage it, as otherwise a hopeless situation will result. Ambassador Lodge said it was explained to him that in India it takes a minimum of nine children to support the parents adequately because of their own low standard of living, and hence a normal family contains at least eleven children, on the assumption that at least two will die before becoming useful.

The meeting ended with a presentation of the 1959 fiscal-year results, the secretary of the treasury reporting with pride the accurate predictions made last January.

> Henry Cabot Lodge, the American ambassador to the United Nations.

12:00 M   Went to lunch with Secretary Brucker and Richard Morse. Very cordial reception. I did my promised stunt and discussed at length, but in an indirect manner, the need for giving up space activities and concentrating on more mundane research for the Army.

> Wilber M. Brucker, secretary of the army and ex-governor of Michigan; Richard S. Morse, special assistant to the secretary of the army for R&D. The promise was made to Morse in an earlier private meeting.

2:30 PM   Spent an hour with Dick Bissell getting background information on special projects in the latter's cognizance. Very impressive record. Ended our discussion by trying to figure out the reasons for the strange Russian development program of ICBM, which still makes no sense to me. Suggests that our knowledge is incomplete.

> Richard M. Bissell, Jr., a deputy director of the CIA.

4:30 PM   Bob Kreidler discussed next meeting of the FCST. We decided to invite Killian to attend these meetings as an observer from PSAC and to introduce a discussion of policy for federal support for research facilities in nonprofit institutions, which is now quite disorganized and chaotic. Learned from Bob that McCone talked to Stans about the Materials Sciences Program and the latter denied all knowledge, because apparently Staats failed to keep him informed. This is rather awkward and somehow has to be straightened out, because McCone appears to be resolutely opposed to the programs and as one of the tricks avoids meeting with Bill Baker.

> Robert N. Kreidler, on the PSAC staff, was executive secretary of the FCST.
>
> FCST: the Federal Council for Science and Technology was established by executive order on March 13, 1959, following sug-

gestions made in a PSAC report, "Strengthening American Science," which was made public on December 28, 1958. The council was made up of representatives on subcabinet level of all departments and agencies with substantial R&D activities and was chaired by the special assistant for science and technology. It was just getting under way when Killian resigned.

Maurice H. Stans was director of the Bureau of the Budget. Elmer B. Staats was deputy director, BOB.

A federally-financed interdisciplinary R&D program at nonprofit institutions, the Materials Sciences Program, aimed at improving the quality of metallic, plastic, ceramic, etc., materials used by industry, was conceived in PSAC under the leadership of W. O. Baker (Bell Telephone Laboratories) and recommended by Dr. Killian to Mr. Staats and to the FCST.

⟳ Saturday, July 18, 1959

Read documents all day in the office—no visitors. What a collection of unsolved problems! The Federal Council for Science and Technology looks extra bad; really no accomplishments to date but a lot started in the way of committee discussions.

⟳ Sunday, July 19, 1959

About an hour's talk with Beckler re: his visit to Rabi in New York. Rabi not feeling well. Is convinced we should not stop nuclear test talks, notwithstanding monitoring problems. Suggested as possible proposal during Summit meeting that we and Soviets have joint summer institutes in various science disciplines.

Afternoon in the office reading documents again; finished about two-thirds of the lot given me by staff on Friday. Met Killian and Bacher in Cosmos Club at 6:30 and talked for an hour about State paper on "tactics" to discontinue nuclear test negotiations for complete cessation of tests and go to phased cessation. Killian convinced that Latter hole concept must be disclosed and a proposal made for joint investigation with USSR of seismic detection and evasion.

I. I. Rabi, professor of physics at Columbia University and a member of PSAC; Robert F. Bacher, professor of physics at the California Institute of Technology and a member of PSAC.

∾ *Monday, July 20, 1959*

PSAC meeting. Started at 9:30 AM with executive session but I had to leave before discussing much. With the President, Bissell, Killian, and McElroy for an hour. Classified project approved, although the President expressed concern about high costs in the immediate future. Regular meeting of PSAC resumed upon my return. About 5 PM returned to executive session and discussed formation of new panels: arms limitations, limited wars, graduate education, were agreed upon. At 6 PM met with Phil Farley and Charles Coolidge in my office, convincing latter of urgency of State job to develop arms limitations policies.

> Philip G. Farley, special assistant to the secretary of state for disarmament and nuclear energy; Charles A. Coolidge, senior partner of Ropes & Gray, Boston law firm, with much former government experience, including a term as assistant secretary of defense, whose arm was being twisted to chair a crash study in the State Department to develop a coherent disarmament policy "consistent with national security."

∾ *Wednesday, July 22, 1959*

7:30 AM   Pre-press conference breakfast. Back in office at 8:30; changed some prepared question-reply papers, and to the President's office at 9:45 after a brief discussion with Goodpaster re: paper he asked me to look over. My contribution accepted by President; no other activity on my part; over at 10:30. Then saw Johnny Williams for about an hour. He explained policy reasons why McCone does not want to approve Materials Sciences Program: is afraid to commit money for buildings at colleges because of congressional opposition. Johnny said other commissioners also opposed on formal grounds. In his opinion, the problem has to be carried to the President. He is doubtful that the Federal Council will amount to much generally. Expressed desire we meet frequently to discuss problems.

> John H. Williams, director of research at AEC, professor of
> physics at the University of Minnesota.

11:45 AM Went to see Gordon Gray to show him Killian's memo to me regarding State paper on disengagement from nuclear talks; also to clarify NSC directive to us. Spent afternoon seeing people and working on paper stating my own analysis of situation. Gave this paper, prepared the evening before, to Keeny and Beckler for comment. Worked with Beckler and Keeny until 11 PM, except for short break for supper. Got both papers into satisfactory form.

> Gordon Gray, special assistant to the President for national se-
> curity affairs, business executive. This directive was, to me,
> an unexpected result of the President's casual remark about "his"
> scientists (see July 16 entry). I was trying to get authorization
> for adequate freedom of action if we had to do this difficult job,
> which was the result of the President's misgivings about the de-
> tailed AEC report on the urgent need to resume tests (see also
> July 16 entry).

Thursday, July 23, 1959

9:00 AM NSC meeting, further reviewing basic national policy paper. No fireworks; finished by 10 AM. Then met with President, Gates, Dillon, McCone, Dulles, Gray. Gray raised question of NSC directive to PSAC to study preparations for resumption of tests. The President stated that the study is not simply of financial costs of preparations but an evaluation of need for early resumption of tests. The President grew heated about atmospheric tests; implied he would not approve them. Then I presented Bacher panel findings; Dillon presented his (revised) paper, which proposes an offer of cooperative study of seismic detection; postponement of negotiations for full cessation. Very little discussion. President said, he guessed we had to give up hope for reliable underground monitoring and hence could not sign complete cessation agreement. No chance for me to speak. Difficult to believe that such a decision could be taken so readily.

After this meeting went to see Goodpaster; showed him my paper on consequences of State's plan; also clarified directive to study resumption of nuclear tests; it is to include relevance to complete weapons-system development schedules and needs, but is not to cover questions of military importance of weapons systems or comparative studies of Soviet and U.S.A. progress through testing.

> Thomas S. Gates, deputy secretary of defense, a banker. Allen Dulles, director of CIA.
>
> The meeting was partly the result of my July 22 meeting with Gray. Others than the President were trying to enforce very narrow terms of reference for the PSAC study, fearing conclusions displeasing to them.
>
> Those who were meeting with the President constituted the Committee of Principals, hence the discussion of test-ban negotiations. The State Department paper proposed to give up negotiations for a total ban, hence it meant a radical change of American policy. In the months immediately preceding this meeting with the President the United States proposed a phased treaty, the first phase allowing underground tests. The Soviet Union rejected this approach, insisting on a comprehensive treaty. The Geneva conference agreed by and by to the principle of establishing control posts on the territories of nuclear powers, the main unresolved issue being the Soviet Union's insistence on veto power (or the principle of unanimity) regarding the recruitment of international monitoring personnel and various budgetary and administrative matters. On the subject of on-site inspections of unidentified, i.e., suspicious seismic events, the new United States position was that the annual authorized numbers should bear a "proper relation" to "scientific facts" and to detection capabilities of the monitoring system. On June 12, 1959, American Representative Wadsworth introduced the report of the Berkner panel on seismic detection. This panel of experts, appointed by Dr. Killian, concluded in early 1959 that the underground nuclear explosions ("Hardtack II") carried out in the summer of 1958 by the United States showed that the detection capability of the seismic monitoring system, recommended by the conference of experts in the summer of 1958, would be poorer than then anticipated. The reasons for this conclusion of the experts' panel were several. Probably the most important was the finding that the long-range seismic signals generated by underground nuclear explosions (in Hardtack II) were not as distinct in character

from shallow earthquake signals of the same intensity as was inferred by the 1958 conference of experts on the basis of a single earlier underground test. Moreover, the signals were found to be somewhat weaker. Hence they would have to be distinguished from among the more numerous natural earthquakes of lower magnitude on the Richter scale. Finally, the new study of the seismic records indicated that the earthquakes in the territory of the Soviet Union were somewhat more numerous than estimated in 1958. Thus the panel concluded that the monitoring system proposed in 1958 would at best be capable of screening earthquakes from nuclear explosions if the latter were larger than about 20 kt. The numerous seismic signals of magnitudes corresponding to the (militarily important) range below 20 kt and down to a few kilotons would remain unidentified.

To accomplish such identification the Berkner panel recommended various elaborations of the 1958 monitoring system which would make it extremely costly and complex. Ambassador Wadsworth proposed to the USSR a joint study of the feasibility of this system and the required on-site inspections. The negotiating position of the Soviet Union was that the number of on-site inspections, being an essentially political issue, should be fixed by quotas in the treaty, and on July 9, 1959, Ambassador Tsarapkin introduced formal treaty language to that effect. In the light of these developments, especially the Berkner panel conclusions and the conclusions of the Bacher panel that the on-site inspections would be almost totally ineffective, the idea of a comprehensive treaty was severely challenged in the Committee of Principals from the point of view of preserving national security.

Lloyd V. Berkner, a geophysicist, was president of Associated Universities and a member of PSAC. Ambassador James J. Wadsworth was the head of our delegation to the Geneva Conference on the Discontinuance of Nuclear Weapons Tests from the start on October 31, 1958. Semyon K. Tsarapkin served as the head of the Soviet delegation and Sir David Ormsby-Gore as that of the British.

2:30 PM   Meeting with Glennan, Dryden, Horner (all NASA), Gray, Sullivan (State), X from CIA, Y from OCB. Glennan proposes to cancel in effect the plans for Venus and Mars shots, concentrate on moon program. No objections raised. I also like the plan; sensible concentration of effort.

Hugh L. Dryden, deputy administrator of NASA; Richard E. Horner, on the senior staff of NASA, and formerly with the Northrop Corporation; C. A. Sullivan, deputy special assistant to the secretary of state for disarmament and atomic energy.

4:00 PM   A navy captain from Joint Chiefs of Staff describing the space command to be set up under JCS to coordinate operations of missile ranges and all receiving stations for space projects. Nothing obviously wrong and is better than the present three-Service triangle. He mentioned that Army and Navy are in favor, Air Force strongly objects to this plan.

Learned a lesson to keep my mouth shut: on Tuesday Bethe told me about Los Alamos finding that their fears re: the single-point initiation dangers are exaggerated; then I phoned Gates and reported the story cautiously; Gates mentioned the information at the President's conference, so McCone sent General Starbird to find from me what the story is. Probably not the end of trouble.

> The "single-point initiation" refers to the possibility of nuclear explosions when a nuclear bomb is accidentally dropped by an aircraft.
>
> Maj. Gen. A. Dodd Starbird, director of military applications at AEC and a faithful associate of McCone.

⌒⌐ Friday, July 24, 1959

Morning spent in small chores and phone calls. Left for New York City 12 noon.

2:45 PM   Jim Fisk in the Ford Foundation office. Explained to him the directive for a panel on resumption of nuclear tests; asked him to be chairman. He is opposed on the grounds that if his participation becomes public knowledge, the Soviets will raise propaganda hell and therefore State will be opposed.

> James B. Fisk, vice-president of the Bell Telephone Laboratories and a member of PSAC.

3:30 PM   Took taxi to La Guardia and got into an awful thunderstorm. Traffic dead and reached airport 4:40. Left for New London in a Piper Apache with Mr. Chapell of Electric Boat Co. at 5 PM. Boarded nuclear submarine *Skipjack* ca. 6 PM. Everybody else on board: McCone, Goodpaster, Jack Floberg,

York, Theodore Rockwell III [Vice Adm. Hyman Rickover's assistant], and others. Inspection tour. Amazing quantity of equipment; very little free space. Submerged about 8 PM. Sub incredibly quiet at about 20–25 knots; no sensation of motion. Had good dinner. About midnight demonstration of maneuvers. Very sharp turns and dives; need to hang on with both hands. We then took turns operating rudder and planes. Former easy, latter difficult to keep steady. Slept in officers' cabins—three per, except McCone who had captain's cabin. After breakfast McCone, York, and I got into conversation. Former objects to materials program as just a scientist's trick, refers to linear accelerator as "your accelerator"; predicts no money from Congress. Then general opinion that scientists cause trouble in government. President needs men like J. J. McCloy instead. I let York argue and just listened. Rather evident McCone does not think much of "scientists."

> Herbert F. York, director of the Defense Research and Engineering Office (DDRE) in the Office of the Secretary of Defense, formerly director of the Livermore Weapons Laboratory and a member of PSAC.

> The Stanford linear accelerator (SLAC) project for nuclear physics research was approved by the President on the recommendation of Dr. Killian, much to the displeasure of Mr. McCone.

> John F. Floberg was a member of AEC.

ᴏᴄ﹏ᴖ Saturday, July 25, 1959

5:00 PM   Went to see Killian in Cambridge. His advice: arguments of Fisk invalid but Jim McRae (recommended by Fisk) will also do. Conceded that McCone is feeling thwarted by PSAC; I should not force materials program on McCone; let it simmer. Should see President alone; talk about my ideas re: test cessation monitoring (perfect monitoring impossible; evasion must be accepted as possibility, study of Latter holes an endless project). Coolidge will need help in State Department job. Berkner would be good head technical man.

> James McRae, vice-president of Western Electric Company, formerly president of Sandia Corporation.

∾ *Monday, July 27, 1959*

Arrived in Washington at 9 AM. Don't recall now (July 28) what meetings I had yesterday. Many phone calls though. Jim Fisk called to tell that Jim McRae will be able to do the nuclear test task as panel chairman. I cleared his name with McCone and Gates. General Loper called to hint that he must be on that panel; I reassured him. Had several calls from Phil Farley wanting Bacher to go to U.K. and Geneva to explain new policy regarding test-ban negotiations. Bacher called to express his almost certain unwillingness. I urged mildly he should go. Then Dillon called to ask me to apply force to Bacher, fortunately too late, as I explained Bacher had already called. I tried to explain to Dillon some consequences of State proposal, i.e., the impossibility of finishing Latter hole and related tests in any specified length of time; the certainty that there always will be new schemes for evasion, etc. His reaction, then what should we do; I excused myself from advising on policy as too green on the job.

Left office at 6 PM to look at apartments. First the "State House" apartment opposite Cosmos Club. Price reasonable ($85 for efficiency) but unfurnished and the room not very attractive, although much better than the Fairfax I saw a week ago. Then the "2500 Q Street." Lobby less impressive but apartment more attractive, and the one I saw was attractively furnished at $135 monthly rent, rather than yearly lease. Applied both places.

∾ *Tuesday, July 28, 1959*

8:00 AM   Breakfast in White House Mess as usual.

1:30 PM   FCST, myself in chair. I seem to have done all right. Meeting not terribly exciting; high point during discussion of federal support for research facilities in nonprofit institutions when Dr. Nolan stressed that before giving money to nongovernmental institutions, the needs of in-house establishments should be taken care of. York outlined plan of DOD to handle the materials project. Will attempt to circumvent the buildings problem by giving grants big enough to amortize the buildings. Williams spoke on the party line of AEC: rejection of program, no buildings support until general policy established, etc. Council agreed to consider general policy. I will set up a subcommittee to report in September.

Thomas B. Nolan, director of the United States Geological Survey, Department of the Interior.

3:30 PM Talked with Jim McRae. He agreed to chair the panel on the urgency of resumption of nuclear weapons tests. We discussed briefly the membership of the panel, which is to include all differences of opinion on the subject; also the plan of activities, which is to evaluate the proposed nuclear tests from the point of view of the urgency to meet the weapons systems requirements.

4:30 PM Saw Secretary Dillon with Farley present and presented to him a frank analysis of the technical basis of alternatives facing the State Department. They are:

1. Try to prevent atmospheric tests only. This could be done without formal treaty since weapons development is not impeded thereby. There is therefore no logical need for a monitoring system except insofar as we have one of our own already. This would mean the loss of all advantages of monitoring within the Soviet Union and will certainly put a cramp into our efforts toward arms-control negotiation procedures.

2. The State Department phased-cessation plan. It is weak because we propose to the Soviets a billion-dollar monitoring system which has little justification for them. We propose space monitoring, although the experts concluded that evasion in space cannot be fully prevented or monitored, and yet we reject underground monitoring on the ground that it cannot be completely effective. I also noted that an experimental study of the Latter hole would get us into an indefinite program to which I see no end, and the Soviets are sure to attack us on this point.

3. Continue on present basis, but insist on a large number of inspections, which almost certainly will be unacceptable to the Soviets. This will force them to back out and embarrass them, whereas we will be consistent with our earlier proposals. In the unlikely case that they will accept 100–200 inspections, I do not see a real risk to our national security because the notion of evasion through Latter holes is completely nonsensical. These things are too uncertain and too costly for any national program of evasion to be based on them. However, I recognize the political impossibility of ratification of a treaty in the face of the theoretical possibility of the Latter hole.

Farley supported my critique of alternative 2. To my surprise, Dillon evinced real interest in alternative 3, and by the end of the meeting seemed to favor it strongly. Thus, the State Department plans may be radically changed and I hope for the better.

> Alternative 2 was outlined in the State Department paper presented to the President and apparently approved by him on July 23 (to my grave disappointment). I continued searching for total ban as the only meaningful arms limitation measure and presented to Secretary Dillon alternative 3 in a form that, I hoped, would appeal to him. As the following events will show, Dillon bit.

5:00 PM   Went to listen to Wiesner's group [Air Defense Panel of PSAC] again, until adjournment at 6 PM. From the snatches heard, it is clear that the SAGE system is far better than nothing, but that perfectly feasible plans of air attack could be devised which would make it inoperable.

> Jerome B. Wiesner, professor of electrical engineering, Massachusetts Institute of Technology, and a member of PSAC.
>
> SAGE: Semiautomatic Ground Environment was a complex computer-controlled radar detection and command plan of air defense by planes and missiles against attacking bomber aircraft, then being installed. It eventually cost many billions of dollars and did not become fully operational until it became obsolete because of the growth of ICBM forces.

∿ *Wednesday, July 29, 1959*

7:30 AM   Pre-press conference breakfast. Very little covered that is of my concern. Attention mostly on the steel strike and the Nixon trip to Moscow. However, Hagerty said that the State Department had prepared a paper on the nuclear test cessation conference and asked me to check it over before coming back.

8:30 AM   Arrived at the office. Followed an hour of complete snafu—State paper was never delivered. I am afraid I was a little hard on Spurgeon. Gave up talking to Wiesner's Air Defense Panel and went instead to Hagerty's office where I got his copy of the State paper and then used it for oral briefing of the President, since only a half-readable carbon copy was available.

10:30 AM   Went to room 220 and formally welcomed Wiesner's group. Long conversation with Waterman and Staats. Latter agreed to chair subcommittee of FCST on policy for federal funds for research facilities for nonprofit institutions. Waterman agreed to this plan.

Alan T. Waterman, director of the National Science Foundation.

11:15 AM   Charles Coolidge on proposed arms control study for State: I suggested Lloyd Berkner as technical deputy. Also emphasized the urgent need for strong technical representation in his group as the only possible bridge between State and Defense members of the group. Talked about other personnel and again emphasized the difficulties of the application of the principle of effective monitoring under realistic conditions, as distinguished from abstract.

12:30 PM   Lunch with Stans. Main theme of latter: balanced budget. As a conservative, I told them, I am surely in agreement with the principle of a balanced budget and will help Mr. Stans attain such. I advised reassuringly that I am very conservative both financially and politically and within reason would support Mr. Stans. However, there are unavoidably points of disagreement. Thereupon gave a little lecture on a "must" in my opinion, namely an expansion of basic research because of the growing economy of the nation and the weakness of European basic research which, until the last war, was our main idea source for applied work. Also discussed materials problem and we agreed that Mr. Staats will lead discussions on general policy regarding research facilities financing. We both agreed that the military at times are unreasonable and seem to try to make things as expensive as possible.

2:30 PM   Secretary Wakelin and others. Went over Wakelin's coming Friday presentation to the cabinet on oceanography. I made a number of nit-picking suggestions and a substantial one about a change of the order of the presentation so that actions of the Federal Council appear at the end as a logical part of the activities of the executive branch of the government.

James H. Wakelin, Jr., assistant secretary of the navy for R&D.

⌒⌣ Thursday, July 30, 1959

9:00–11:00 AM NSC meeting. First hour and a half devoted to basic security policy paper. No fireworks. Reference to supremacy taken out of space activities paragraph. Good! Paragraph on nuclear vs. nonnuclear weapons well worded, but unfortunately a footnote says this paragraph means the same thing as the one in the earlier edition, which spoils the effect of a much better statement. Last half hour devoted to Allen Dulles briefing. Evidently the fellow was tired or forgot his lesson, because he kept talking about the USSR ICBMs of 3500-mile range and how some of the U.S. targets were really outside the range, although he also mentioned that at least one of the USSR ICBMs had flown 5500 miles (I told Scoville about this in the afternoon, and he was quite exercised).

> H. E. ("Pete") Scoville, Jr., assistant director of the CIA and the director of the Office of Scientific Intelligence.

1:00 PM Lunch with Dillon, Killian, and Farley. Dillon appears to have accepted my suggestions of yesterday and now that the schedule of planned activities has been relaxed because the Geneva conference ends too soon to talk to Gromyko, there is good chance that the "phased" plan will not be adopted. Killian is slated to go to England instead of Bacher, and after lunch we talked about his companions, etc. He is almost committed to going.

> Andrei Gromyko, foreign minister of the Soviet Union.

3:30–5:00 PM I sat in on the executive session of Wiesner's Air Defense Panel and learned a great deal. Evidently I need an intensive education on this subject, as a lot of things are still not clear to me.

⌒⌣ Friday, July 31, 1959

9:00 AM Cabinet presentation on oceanography by Secretary Wakelin. He did very well, although he spoke a little too slowly, and clearly created a favorable impression. The President emphasized that our over-all plan for expanding oceanography should include collaboration with the U.K. and Japan. Rest of cabinet meeting not exciting. Considerable joking by the President, and his rather frank evaluations of senators and congressmen.

10:30–12:15 PM   Meeting with Mr. Staats, Secretary Flemming, Mr. Floberg, and staff people concerning federal radiation council. Most of the time spent fussing interminably about words. One would think these important people had something more important to do. Finally the executive order agreed to, and thank God I will not be a member, but only an observer. However, I had to agree to provide quarters for the executive secretary for the council, but not the salary. Also a press release agreed to—not quite in the form recommended, but close enough.

> The Federal Radiation Council: public concern about the health dangers of radiation from the nuclear tests fallout became acute when the testimony of the Defense and AEC witnesses before the congressional Joint Committee on Atomic Energy appeared to be contradictory. Charges of concealment of facts became so serious that the President issued a public statement on March 25, 1959, assuring the public that nothing was being concealed and that the government was putting major efforts into studying the biological effects of radiation. He also decided on setting up an interagency radiation council and transferring to it from AEC the authority for setting safe radiation-exposure standards. Before I came on the job Killian proposed to me that the special assistant for science and technology become its chairman. I opposed this, believing that regulatory functions should not be a part of this White House Office and also fearing that I would be squashed between AEC and HEW. The council was established by executive order on August 14, 1959, and Secretary Flemming was appointed chairman on August 22. The Congress passed Public Law 86–373, September 23, 1959, which substantially increased the authority of the council and enlarged its membership.
>
> Arthur S. Flemming, secretary of the Department of Health, Education, and Welfare.

2:30 PM   Meeting with Secretary Flemming, Waterman, Lambie, and a Mr. Wilson from the Advertising Council, on the "Pursuit of Excellence" program, which by the way got rechristened and is now the "Achievement of Excellence" program. Meeting exceedingly dull, and I fell asleep momentarily. Most embarrassing. It has been agreed that there will be a four-man committee—three government people and possibly Killian, who will be asked to be chairman. Flemming and Waterman think they can raise enough money to start it going.

> J. M. Lambie, a special assistant to the President. I don't remember who started the Pursuit of Excellence project. It never got off the ground and I could never bring myself to give it a real push, being unclear about the true objectives.

4:30 PM   General Betts reporting on his investigation of ICBM base costs. His definite conclusion is that nothing can be done now to reduce the cost of the first 8 Atlas and 6 Titan squadrons.

His second conclusion is that under recent pressure (possibly from us as well as other sources), the Air Force has taken major steps to reduce the cost of the installations, with the result that a cost of a hardened and fully dispersed squadron installation is less than the cost of soft concentrated squadrons. Betts concludes that there is still a considerable amount of gold plating, some of which is tied with operational concepts, mostly with the requirement for a 15-minute readiness time. This, however, is a very difficult concept to change, and he hasn't attempted to do so. He has requested the Air Force to explain why certain other changes aimed to reduce the cost couldn't be made, such as, for instance, eliminating the skirt on the missile, which would result in burning up the elevator, but who cares since only one missile will be fired from the hole anyway. None of the suggestions he has made will make a really major difference, he thinks; he expects to have complete information by mid-September.

> Army Brig. Gen. A. W. ("Rosie") Betts, director of ARPA (Advanced Research Projects Agency of the Defense Department), was instructed by the secretary of defense (on my initiative as then the chairman of the PSAC missiles panel and through the intervention of Dr. Killian) to study the then-abuilding ICBM bases, which were incredibly complex and costly. PSAC pressed for hardening and dispersion of the ICBMs to make them less vulnerable to surprise attack. This was strongly opposed by the Air Force brass as too costly and inconsistent with effective "command and control."

ᗨ⌇ *Saturday, August 1, 1959*

Spent day in office reading reports, and also had about an hour's talk with Andy Goodpaster. He agreed with the terms of reference which I prepared for Jim McRae, and we covered a number of lesser topics.

ᗨ⌇ *Monday, August 3, 1959*

Lunch with Colonel Hayworth, who is temporarily with ARPA. He is impressed by the small staff (about 100) which ARPA uses to run its program; whereas it takes ARDC 4,000 people to run a program only twice as big. I asked him to write me, off the record, a memorandum of some of the weaknesses of ARDC research program, but I am not sure he will do so.

2:30 PM   Briefing by Bruce Old and Guy Suits on the research for the Navy in York's office. Very good presentation, arguing that the Navy should spend much more money than it now does on basic research because it should regard itself as an industry in the area of rapidly obsolescent products. Industrial companies of this type spend 15 to 20 percent of their total R&D budget on basic research; whereas the Navy is now spending only 7 percent.

> Col. O. Hayworth, an old friend from my days on the USAF Science Advisory Board.
>
> Bruce S. Old, a member of NRAC (Naval Research Advisory Committee) and a vice-president of Arthur D. Little, Inc.;
> C. Guy Suits, another member of NRAC and vice-president for R&D of the General Electric Corporation.

ᗨ⌇ *Tuesday, August 4, 1959*

11:00 AM   Appointment with the President. Only Major John Eisenhower present, taking notes. President was very friendly and talkative, raising a number of points only remotely connected to my briefing subjects, which were: the slow progress in global communications integration among the Services; the problem of ICBM base complexity and cost; and finally, my analysis of the recent State Department proposal to change the objectives of the nuclear test cessation negotiations to a phased agreement.

The President was very pleased to hear about the success of Baker, York, and McGuire in improving CRITICOMM and strongly urged I make sure that McElroy becomes familiar with the problem and supports York with executive decisions whenever possible. On base costs, the President had not much to say, except to agree with me that it is essential to prevent the Minuteman ICBM system from becoming as complex as the present-generation missile systems are. He listened sympathetically and with interest, as well as asking occasional questions, to my criticism of the State Department plan and a partial analysis of the consequences of continuing with the present negotiations using the Geneva system, although aware that it may not be fully effective monitoring. He expressed great regret that it wasn't possible for him to bring together a group of top advisors (he mentioned the secretary of state, secretary of defense, McCone, and myself, as well as a political leader from the Senate) and have this group give him an unbiased recommendation based only on the interests of the country. He seemed to appreciate my statement that I could advise him only on the technical aspects of the problem and that the problem actually was rapidly ceasing to be a technical one and becoming a complex of political and other factors. In parting I mentioned to the President that the McRae panel will start discussions this Thursday and said that it will be a bloody meeting. The President positively burst into laughter, but said he wanted us to come up with definite conclusions.

> At that time each of the Services had its own more or less worldwide communications network with much inter-Service rivalry. PSAC urged integration into a single management (which later came into existence as the Defense Communications Agency) and as a first step pushed CRITICOMM—an integrated emergency network.

After lunch I mostly killed time doing little jobs. McElroy phoned, rather bothered, because the President told him he did not like "services giving out contracts for R&D in space area under $0.5 million." McElroy thought I spiked the President.

Reassured him. Actually the President complained to me this morning about this activity. He wants to leave it to ARPA so the services don't get "vested interests."

> According to a directive of the secretary of defense the Services had to clear with the director of Defense Research and Engineering (DDRE) all R&D contracts over a half-million dollars. The Services therefore tended to chop up dubious larger projects into $499,000 pieces which then did not have to be cleared with DDRE.

ᕕ࿐ Wednesday, August 5, 1959

9:30 AM Limited War Panel of PSAC, initial meeting. Robertson, Hill, Lauritsen, and Thompson. Rather hesitant beginning. Nobody is quite clear what should be the main objectives, including me, but it is obvious that we have to obtain some competence in the area of Army technology. Went over the membership and also found that the first full meeting couldn't take place until mid-September. Agreed that files of earlier reports will be established in several places and that all participants should study them carefully before coming to the first meeting, since there have already been so many reviews in the past. I urged Robertson to prepare in writing a first attempt at defining objectives so as to guide the members.

> H. P. Robertson, professor of physics at the California Institute of Technology and a member of PSAC; Albert G. Hill, professor of physics at the Massachusetts Institute of Technology; C. A. Lauritsen, professor of physics at the California Institute of Technology; L. T. E. Thompson, consulting research engineer.

10:45 AM [Willis H.] Shapley and [Gerald E.] Pettibone [two staff members] from BOB explained their plan for analysis of next year's Defense budget. It looks to me as if it is designed more for the Air Force than for the Army. Another serious weakness is that no distinction is made between realistic weapons systems and far-off schemes that may cost much money and couldn't become realistic in a few years.

1:30 PM Doyle Northrup of AFTAC complained that he is unable to get explicit directives or money from DOD to initiate accelerated seismic development program. Urged me to get presidential directive. I talked with York on the phone and got assurances that the problem can be resolved satisfactorily by his office.

> Doyle Northrup, geophysicist at the Air Force Technical Applications Center, who was in charge of our national seismic detection system.

3:00–4:30 PM Went to airport to form small part of background at the reception for the vice-president. Clapped my hands at appropriate times. He delivered a really excellent speech which sounded completely spontaneous, but was obviously carefully prepared. Very outstanding delivery and very noble sentiments. I drove to and from the airport with Stans and Quesada. Stans much upset about the enormous redundancy of the SAGE system. Asked me if that was really necessary; he suggested it was overdone. This is clearly a matter we will have to look into, since it will be raised by the BOB in the fall.

> Vice-president Richard M. Nixon, who was returning from his Moscow visit.

∾ Thursday, August 6, 1959

7:45 AM Breakfast with General Persons. Told him about difficulties with McCone in the materials project and Stanford accelerator areas. He offered immediate help through intervention, but I suggested he delay, and he agreed, but suggested I come to him if it couldn't be straightened out amicably.

> Gen. (ret.) W. B. Persons as the assistant to the President was my administrative boss.

9:30 AM–12:30 PM McRae's panel. With authorization from General Persons, I gave the panel a very stern admonition that they have to treat the whole proceedings as a privileged matter. After that I listened to proceedings, which started in a rather tense atmosphere but began to relax.

12:30 PM Wallace Brode for lunch at the White House Mess. He was full of complaints that he is being bypassed in the State Department and is not sufficiently recognized in the Federal Council. Evidently especially unhappy with Phil Farley. Would like to become assistant secretary and "enlisted" my help in that effort. I served an ultimatum that I would challenge the State Department action in requiring "instruction" of scientists, who are not federal employees, going to international scientific meetings, where U.S. dues are paid by the State Department. Spent a long time trying to explain to him the meaning of the word "instruction" and the reaction which will be produced in scientific circles. I believe I had some success, but not complete.

> These instructions on how to vote, whom not to contact, etc., were
> causing considerable indignation in scientific academic circles
> as infringing on their rights as private citizens.

Friday, August 7, 1959

The entire day spent with McRae's panel on nuclear test requirements. No fireworks and meeting very factual. Bradley gave an excessively long presentation on high-altitude effects of nuclear explosions and afterwards privately apologized to me that he had to present the "party line" but that he himself didn't consider the matter as urgent as he sounded. I got into a little argument with Admiral Parker on the subject of safety tests by emphasizing that no reasonable amount of safety testing could prove a weapon to be absolutely safe and that people on the JCS level just had to accept the responsibility for operational use of devices that had a finite, even though exceedingly small, probability of nuclear explosion. The last part of my remarks seemed to annoy Admiral Parker and he stated that the JCS were never afraid to assume responsibility. I let it slide.

> W. E. Bradley was in the DDRE office in the Department of
> Defense, whose "party line" was that high-altitude nuclear tests
> were of the highest priority for a proper design of the ballistic
> missile defense systems (ABM). Adm. E. N. Parker was also in
> the Pentagon. The issue involved was whether to undertake further
> nuclear weapons tests to make "certain" that they would not
> explode if accidentally dropped by an aircraft. These are referred
> to as one-point safety tests.

∽ *Monday, August 10, 1959*

Seitz and Robertson for lunch. Seitz is exceedingly unhappy about "booby traps and land mines" which he is encountering in the American sector of NATO, apparently largely centered in budget people who are trying to cut him off from free use of our money. Wallace Brode assured him that this was being done without his own knowledge, but Beckler's subsequent check of messages to and from State Department indicated that Brode's office concurred in withdrawal of funds from Seitz. Lovely way to maneuver.

> Frederick Seitz, professor of physics, University of Illinois, was that summer appointed as the chief scientist to the NATO's supreme commander.

∽ *Wednesday, August 12, 1959*

Somewhat more active day than the beginning of the week. Had Waterman to lunch. He was half an hour late because, notwithstanding instructions of his secretary, he went to the Hay Adams Hotel. Talked about some NSF problems, namely the difficulties in convincing Stans that basic research deserves increased support. Then switched to the problem of broad support of science vs. support of "pressure" groups like oceanography, atmospheric sciences, materials sciences, etc. My own feeling was that all of these have only small components of *basic* research and that this should be clearly brought out in budgetary considerations, so that the large sums involved would not be charged against basic research, but would be compared with applied research and development in other fields. Waterman had an approach very different in detail, which I did not understand. We also talked about the Stanford accelerator project, and Waterman is unhappy that AEC has lost enthusiasm for this project after asking to have it given to them.

3:00 PM   Keith Glennan to consult about his idea that the President propose to Khrushchev a joint program of space research. I agreed, but pointed out difficulties, namely the possibility of Khrushchev's snooty reply that we were just trying to learn their secrets, and secondly, a possible embarrassment because much of our space activities is in the military intelligence field. However, if these aspects are made clear to the President, I see no harm in making the suggestion. Keith then spoke of the budget. Our space activities this year cost $0.9 billion between ARPA and NASA and will rise to $1.5 billion next year. I said it was too much and I wouldn't support it, so he first suggested that we cut out military intelligence activities, which I didn't agree with. So we finally agreed that the NASA program may have to be cut somewhat. Details to be discussed later.

> N. S. Khrushchev, chairman of the council of ministers and general secretary of the Communist party of the Soviet Union, was planning to make a visit to the United States that fall. The issue of which country was "ahead" in space was of major political importance then.

4:00 PM   Gordon Gray about evasive maneuvers of Gates to eliminate or postpone the four special briefings ordered by the President. Gordon much upset and had me go through copies of his entire correspondence with OSD showing clearly their perfidy. Showed me a draft of a letter he plans to send and I suggested it could be stronger. At any rate, the August 15 date for continental defense briefing is indefinitely postponed. Too bad I made Jerry Wiesner and the others work so hard.

> These briefings were requested by the President from the JCS and were to describe the over-all military planning in four areas: continental defense (of the United States), strategic offensive forces, naval plans, and the West European defense under NATO.

Phoned Marshak, who wrote suggesting that I appoint U.S. delegates to negotiate with USSR a joint program in high-energy physics. I explained that this would be quite improper for me to do; that appointment by the State Department will take much time and may be inadvisable at present. I then suggested that delegates be appointed by the Academy [NAS], to which he agreed. He told me that his and other physicists' contacts with the Russians during meetings in Moscow and Kiev indicated enthusiastic acceptance of a joint program, and gave in explanation of this change in attitude that the Russian physicists are having as much difficulty raising big money from the government for high-energy accelerators as we have. I must say, physicists are having a hard time the world over.

> Robert E. Marshak, professor of physics, University of Rochester.

∽ *Thursday, August 13, 1959*

9:00–10:30 AM Committee of Principals. Dillon outlined State Department proposal to recess Geneva nuclear talks until about October 12, because Ambassador Lodge will accompany Khrushchev on his visit to United States and Wadsworth will be needed at the UN to take Lodge's place, and at the same time to announce that we will postpone resumption of nuclear testing at least until January 1. He stated that this plan was approved by the President. This provoked wild reaction from Gates and McCone, who stated that as of July 23, the President approved an entirely different plan of action, and that they didn't consider it proper to have it changed unilaterally. Dillon was taken slightly aback, but defended the action, noting that because of the quick recess of the foreign ministers' meeting, it was impossible anyway to approach Gromyko. Moreover, that the British, in [Foreign Secretary] Selwyn Lloyd's letter, urged us not to make any new moves until the [U.K.] "government had time to digest scientific information resulting from the current visit of Killian." Dillon noted that actually the British were anxious not to do anything that could affect their election results.

He then, in response to further criticisms by Gates and McCone, noted that I was seriously questioning the wisdom of the plan [phased agreement] agreed to by the President on July 23, and asked me to present my arguments. This I did, and Gates was quite surprised to learn that the experts conceded the possibility of evasion in outer space and yet would propose in the plan to agree to the banning of such tests. There followed a reasonably amicable discussion on the significance of outer-space evasion. Nobody took it really seriously. Then both Loper and particularly Starbird emphasized the extreme urgency of resumption of tests at the earliest possible moment, stressing the safety tests. Gates informed Dillon that he will have to take the matter to the JCS and if they felt strongly, he would have to appeal to the President. Both DOD and AEC reported the importance of resumption of tests in a light very different from that in which they were discussed at McRae's panel meeting, but since the panel has, as yet, drawn no conclusions, my hands were tied, and I kept quiet. McCone finally suggested that instead of making an announcement of a flat postponement of nuclear tests, we merely announce that we will allow a reasonable time for the conference to progress after resuming the meetings in October. This seemed more acceptable to Gates, and Dillon liked it too. The meeting ended on a somewhat uncertain note. There may be presentations to the President, and maybe the meeting will be reconvened next week.

During the Principals' meeting, McCone made a statement (rather than asking approval) that he will have the results of the Bacher panel study presented to the Joint Atomic Energy Committee of Congress on Friday. There were no remarks on the subject. After the meeting I realized that it may have very unfavorable consequences and informed Gordon Gray that matters of executive privilege may be involved, since the study was done by a Killian panel rather than the AEC. McCone had discussed the presentation in OCB and Gray didn't oppose, but he said that my remarks made him very concerned and he would talk to McCone.

12:30–2:00 PM   Lunch on the invitation of Admiral Burke. The whole top echelon of the Navy was present. Most impressive and also very cordial. During luncheon itself, small chitchat about Navy's troubles, most of it almost in coded language so that I didn't get the import, but the general implication is a shortage of money and personnel. After lunch Burke and Hayward asked me to join them in Burke's office, and talk seriously. Admiral Burke again emphasized the catastrophic shortage of funds, his need to mothball 25 ships next year to provide personnel for the new ships and other activities. We then talked R&D and the admirals assured me that they were essentially stopping all new aircraft programs, etc. Very cordial.

> Adm. Arleigh Burke was Chief of Naval Operations. Vice Adm.
> John T. ("Chick") Hayward, an old friend, was deputy chief
> for R&D.

3:00 PM   The Swiss ambassador and his science attaché. Only partly a courtesy call. Terribly anxious to find out how soon we will resume nuclear testing. Tried a number of ways to find out, until I had to become almost distant and had to say that as presidential adviser I couldn't answer, but passed it off pretty well.

4:00 PM   Several men from the naval research labs tried to convince me that a certain low-cost Navy intelligence proposal, on which I had commented unfavorably to Andy Goodpaster, is really worth undertaking. They almost convinced me and I will get in touch with York about it and then with Goodpaster.

5:00 PM   Harold Brown, summarizing London meeting of U.S.-U.K. Next day talked on the phone on the same subject with Killian and got a somewhat different interpretation. Brown intimated that Britain accepted the Bacher panel findings; Killian flatly said they did not. Brown also discussed his briefing to the joint committee next morning. McCone suggested that the results be presented as an AEC finding. I pointed out that they could be caught in lies, but left it like that.

> Harold Brown, director of the Livermore Weapons Laboratory, accompanied Killian to London on his mission to explain to the British the Bacher panel findings, the problem of Latter holes, etc. It will be recalled that Brown had been a member of the Bacher panel and as such was apparently very insistent on getting the panel to reach conclusions unfavorable to a comprehensive test ban.

Friday, August 14, 1959

9:00 AM   Phoned General Persons about McCone's planned briefing of the joint congressional committee. Persons said he will discuss it with McCone.

2:30–4:30 PM   Jim Fisk and Julian West [Bell Telephone Labs] explaining the facts about air defense. The emphasis was substantially different from the report by Watters and Wiesner, and I learned more from them than I did from the report. The general weakness, which West considers almost catastrophic, is that which shows itself in so many other areas of military activities: nobody is willing to make the available things work, but keeps changing and adding new and fancier tricks. West is very pessimistic about the possibility of wedding Nike missiles and the SAGE system, and yet that is what is being done, thereby destroying even the point defense. In case of an attack, complete chaos will exist because SAC and NORAD are not on speaking terms, so our NORAD fighters will be shooting at our SAC bombers. Between this briefing and the report, I am beginning to understand the picture.

> "Point defenses" meant local antiaircraft defenses using guns and short-range missiles, such as Nike surface-to-air missiles, which were developed by the Bell Telephone Laboratories as air defense weapons for the Army.

4:30 PM   Berkner and Coolidge. (I rode with them to the airport to get a little more time.) Gave my support to Coolidge by urging Berkner to work on arms limitations. Berkner has a most grandiose scheme of starting a lot of unclassified research projects all over the country, as well as other activities, and makes this a condition of his acceptance. I expressed conditional agreement, so long as it does not become a tail that wags the dog. Chances are probably good that he will accept, and once he gets in harness will work well.

Saturday, August 15, 1959

Spent morning in office, reading. Had visit from Senator John Sherman Cooper and his administrative assistant. Senator Cooper would like to see a presidential committee set up with advisory powers to influence and improve education in schools by consulting with state authorities, etc. Sounded to me a little bit like the Achievement of Excellence proposition, and since I didn't know what he knew, I didn't say much except to express my general sympathetic interest. He proposes to speak to the President on the subject and have a law passed.

12:30–4:00 PM   Herb York. His main subject was a proposed scheme for reorganizing the entire space program in the Defense Department. At present the Titan and Dynasoar projects are under BMD. The next bigger booster, Saturn B, is in ABMA and tied to NASA. Next bigger one, Nova, which is making only slow progress because of fund limitations, is under NASA. ARPA supposedly manages military boosters at ABMA and through the Air Force. ARPA also has responsibility for military payloads. Finally, ARPA has a number of miscellaneous projects. York proposes to cancel Saturn B, to transfer ABMA to NASA, and give it supervision over Nova, which will be speeded up by the addition of funds. The Air Force will be given direct authority over all military booster developments, including a new Titan C with four booster engines, to be used for the Dynasoar project, the 24-hour communications satellite, and other big payloads. The Air Force will also be given complete authority over space transportation. ARPA will be given responsibility for all military payloads, such as communication, navigation, intelligence, and the space surveillance for the nuclear test moratorium. A classified project, now in ARPA, will be canceled. ARPA will have authority over other projects not otherwise assigned. I commented that it was a very attractive scheme and made good sense, but that in effect it meant moving virtually all major missile and space projects out of southern California and I was afraid of the political effects. This seemed to shake York a bit.

I talked then about the problems in air defense, specifically that the military don't make any real effort to make work what they have, but always want newer things. I then spoke of this as a general difficulty with the military programs. Stated my conviction that we are near the end of a tremendous revolution in military technology, the change-over to automation, and that now is the time to consolidate, learn how to use the new automatic systems, and slow down on second- and third-generation work, because of financial and talent limitations. Urged him to plan the '61 budget with this in mind, while conceding that no miracles can be performed this year. Believe that my philosophy has finally penetrated. This is, of course, not the first time I have spoken on the subject. York then spoke of attempts to coordinate and almost unify global communications. I gave him assurance of assistance.

Monday, August 17, 1959

Another dull day in the office! All my weeks seem to begin slow, but things pop in a couple of days.

Had [Assistant] Secretary E. L. Peterson [Agriculture] to lunch. Very pleasant and forceful personality. Made a strong plea for agriculture and farmers, especially for agricultural research. Congress exercises detailed control over research through budget line items. Minimizes long-range improvement efforts, but puts emphasis on getting the bugs off the cotton, etc. Now it has instructed them to find out how to grow apples economically in New England! Peterson's opinion is that the research budget is completely inadequate and particularly the facilities are in a desperate state of disrepair and obsolescence. He estimates that several hundred million dollars are needed to modernize them. We then talked about the Federal Council. Peterson not very optimistic that much can be accomplished by it but urged combination of patience and persistent pressure to achieve objectives. Conceded that if the Federal Council fails to accomplish anything at the beginning of next session of Congress, it will really mean trouble.

Keeny had a call from Senator H. H. Humphrey's staff, who have heard of the Bacher panel and wanted information on the content of the report to use in Senator Humphrey's speech tomorrow on nuclear test cessation negotiations. AEC is the only conceivable source of the leak. Other information, reaching us from General Starbird, is that briefing of the joint committee by AEC on Friday regarding the Bacher panel results went smoothly and without any reference to Bacher or this office being made.

∽ Tuesday, August 18, 1959

10:00 AM–12:15 PM   National Security Council. The President was in excellent form and did as much talking as anybody else. When the item on highest priorities for missiles and similar things was discussed, he launched into a long discourse essentially on the same topic that I heard him speak about before—that we need a unified Service approach; that this must stretch all the way from research to operational use; that he does not like the idea of General Power in Omaha flying some missiles and the Navy boys flying others, etc., etc. McElroy naturally assured him that they were doing everything possible, but the President said that wasn't good enough, and he thought he would like on January 22, 1961, after leaving office, to get up before the country and tell it what was really wrong and that in the meantime they should start pushing really hard to get true unification through Congress. Then a long discussion of our policy towards France, where again the President took a vigorous part without mincing words. At one point he suggested that he will be delighted to let anybody here in the room speak to de Gaulle in his stead—that he wasn't really relishing the idea. Later on he said that the French Prime Minister Michel Debré was just a menace. He nipped in the bud attempts of Debré to persuade him that we should change our policy and support France in its Algerian policy. Said this was totally inconceivable to him—that our basic principles of anticolonialism must be supported at all costs,

French friendship or not, regardless. Then he enunciated the principle that the time was ripe to make France and other countries that were getting rich while we are losing our gold reserves start paying for their weapons. He thought we might offer some at reduced prices. He mentioned the scandalous Mark 47 tank that the Army rejected after less than a year's use and said he would be willing to sell it for $50,000 each, even though they cost us half a million apiece. He was really in rare form.

> Gen. Thomas S. Power, successor of Gen. Curtis LeMay, was the commander of the Strategic Air Command, headquartered at Offutt Air Force Base near Omaha, Nebraska.

12:30–2:00 PM Lunch with McElroy and Gates. Began with assurances of mutual love and cooperation. McElroy welcomed unconditionally our meddling in military affairs so long as we keep him and York informed. Complained that Killian did not always inform him. Said that occasionally this made him very mad. I promised to do better. Then both urged me to make a very thorough study of Nike Zeus. If deployment plans are carried out, they will cost us over $2 billion in the next fiscal year. I discussed its technical weaknesses and they grew more eager by the minute to have us really study the problem and assured me that they would be willing to take the politically difficult step and send the project back to the laboratory if we so recommended. Then talked about nuclear test cessation. I explained my objections to the State Department "phased" proposal and didn't run into too much difficulty; also not too much on my statement that I didn't believe that monitoring systems in general could be completely reliable. We then talked about the Federal Council, in which neither of them has any interest, so my plea to impress its importance on York probably fell on barren soil.

> Nike Zeus, the first generation of the antiballistic missiles (ABM) system was then fiercely pushed for production and deployment by the Army and the Bell Telephone Laboratories, its development contractor.

Spent all afternoon with McRae's panel on the urgency of test resumption, which started work at 9 AM. Great deal of argument, in which I deliberately took very little part, except for making occasional jokes when Teller got passionate. This seems to be the best technique for deflating his pathos, certainly works better than trying to heat up and argue with him. The panel made numerous changes in the excellent draft report that McRae and Keeny worked very hard on yesterday and last night, but the changes are not drastic. More emphasis now on need for testing, but no dramatic statements about desperate and urgent need, and actually an admission that tests can be delayed, except possibly for single-point safety tests on which issue there was strong disagreement between Livermore and Los Alamos. Hence in briefing the principals and then the President I could well bring that out and then, being an expert in my own right on the subject, side with Los Alamos.

All in all, I think, we will have a very useful report which will give the President adequate freedom of action.

> While the conflicts and reconsiderations described on the preceding pages were occupying us in Washington, the Geneva conference on the nuclear test-ban treaty was in session. This session began on June 8 and recessed on August 26. It received the report of the joint Technical Working Group I on monitoring high-altitude (outer-space) nuclear tests, a subject not dealt with by the experts in the summer of 1958. Also the findings of the Berkner panel on the need for a more elaborate seismic detection system were presented and were later challenged by the Soviet delegate. The Soviet Union formally proposed a quota system of on-site inspections of unidentified seismic events, including some inspections that would be veto-free. Much of the time was spent on problems of staffing and administration of the monitoring organization, including details of a comprehensive test-ban treaty, as if the discussions of disengagement were not taking place in Washington.

CHAPTER II

The Competition for the Budget Dollar

THE BUDGETARY CYCLE in the executive branch starts early in the year on low levels of federal bureaucracy and percolates gradually upwards, to end with the dispatch of the President's budget to the Government Printing Office about the end of the year. By August, senior officials have already become seriously involved in endless conferences. Following in the footsteps of Dr. Killian, I also found myself taking part in increasingly frequent budget meetings dealing with a wide range of R&D projects and research programs.

By far the largest demand on my time and that of our staff and several PSAC panels was generated by the budgetary problems of the Defense Department and of the Space Agency. Together, of course, they accounted for considerably more than half of the federal R&D moneys. But to a lesser extent I was involved also with other large R&D programs and budgets. The exceptions were those of the Department of Health, Education, and Welfare, that is mainly the National Institutes of Health, and that of the Agriculture Department, which comprised the major portion of the life sciences programs. With these our office apparently had little to do prior to my term of office.

At the same time I had to shepherd the McRae panel report to a satisfactory completion and then to its acceptance as the evidence that the resumption of nuclear tests was not urgently needed for our national security.

These are the main topics of the second chapter.

⤳ *Wednesday, August 19, 1959*

Bruce H. Billings, who is in charge of Special Projects, York's office, complained about the unbelievable chaos among the highly classified projects—the piling up of one project on top of another without any effective mechanism for evaluating even the potential usefulness of each. Apparently the situation in that field is much the same as what I am encountering in continental defense and communications. As an example, he noted that in the satellite project SAMOS, there would be ten television channels to the ground and a library of information so complete that a general (LeMay?), sitting in his easy chair in the Pentagon, by just pressing a button will be able to see on a screen the complete display of current military activities in televised form from anywhere in the world.

10:30–11:30 AM   Teller here. The meeting started with mild mutual strain, but Teller rapidly took charge of the situation and expressed his regrets that the McRae panel report was not more forceful and that I prevented him from adding the really important part, namely a statement about the urgent necessity to continue nuclear tests ad infinitum (I assured him that my strict instructions were to limit myself to the immediate test series and not to enter into philosophical discourses). He then proceeded to develop many ideas. Firstly, he is convinced that the Soviets will cheat and will push hard the development of nuclear weapons. So far as I can tell, he seems to be the only man who really believes it. Even Dick Latter, later the same day, said to me he was quite convinced that the Soviets would rather denounce the treaty than take a chance on cheating and exposure. Teller then expressed his conviction that the President in his meetings with Khrushchev will commit himself to stop testing forever regardless of the effectiveness of monitoring systems. I assured him that I had no evidence of the possibility of that happening. Teller then went into a long discourse on the tremendous military importance of "pure radiation" small tactical nuclear weapons. I wasn't fully convinced, although they might be useful under some circumstances, but they certainly wouldn't be available militarily for quite a few years. Then Teller went on to tell me about his shocking experiences in Europe, where he found out that our troops armed with nuclear weapons had to store these weapons, because of safety reasons, in a highly disassembled state, and because of blind adherence to outmoded directions would require several hours for assembling them. He complained that Norstad, who was at first very much interested in Teller's ideas for the tactical use of nuclear weapons and their provision to the Allies, later, apparently on instructions from Washington, effectively refused to discuss this topic and insists on massive retaliation. Teller now believes that the massive nuclear deterrent has outlived its usefulness, although retaliatory forces are, of course, essential to prevent the Soviets from starting a massive nuclear attack against the U.S.A.

> Edward Teller, theoretical physicist, creator and deputy director of the Livermore Nuclear Weapons Laboratory; Dr. Richard Latter of the Rand Corporation, with his brother Albert the inventor of the Latter holes, and a loyal disciple of Dr. Teller; Gen. Lauris Norstad, supreme commander of NATO forces. The mininukes, lately in the news as the coming tactical nuclear weapons, are related to the "pure radiation" weapons here discussed.

11:30 AM–12:15 PM   Franklin W. Phillips, the secretary of the Space Council, gave me various pointers on the past history of the council and its organization, the refusal of the President to build up a large organization like NSC; the discovery that informal meetings without the President must precede formal ones to accomplish anything, and the unwillingness of Glennan to chair the former meetings because he is an interested party. More or less by default, Dr. Killian became the chairman. It was Phillips's feeling that I would have to do the same. He assured me that he would prepare for me memoranda and notes making the chairmanship bearable. Later that day I phoned Glennan, who felt I should become the chairman. I suggested he clear it with Gates, and later was told by Glennan that Gates is also in favor, so that's that. Another useless job and spoils my plans for a week's vacation.

Afternoon spent restfully in the office doing very little. Read over final draft of the McRae panel report, which is really quite satisfactory.

Thursday, August 20, 1959

8:00–9:00 AM   Breakfast at McCone's home, just the two of us. He was very cordial and we talked about a variety of problems related to AEC and Defense. McCone is enthusiastic about Admiral Rickover and insists that I must meet him as I would benefit from the experience. I naturally agreed. In the Soviet Union Rickover visited several nuclear power reactors, including those on the icebreaker, and because of their "crude" appearance concluded that the Soviets were not capable of building effective atomic submarines. Therefore McCone wants to oppose the transfer of submarine power reactor plans to our Allies in Europe (really some logic!) for fear that the plans will find their way to the Soviet Union and so accelerate their progress. I suggested that their apparent crude appearance to Admiral Rickover may have very little bearing on their performance and cited as examples Russian tanks in World War II, MIG 15s, etc., all of which were extremely battle worthy, although crude by our standards. I urged McCone to reconsider his attitude toward NATO. Toward the end of the breakfast, McCone read the conclusions of the McRae panel report and asked to keep the report, so that he could read it during his drive to Germantown. I agreed, although I suspected that it would be duplicated. McCone promised to return the report immediately, with his driver, but actually the report was not returned until 4 PM, at which time I phoned McCone, thanking him for the return of the report and was assured over the phone that it was not duplicated (which assurance was not a result of my direct inquiry).

> Vice Adm. Hyman G. Rickover was in charge of Navy's nuclear propulsion program.

At 9:00 AM, a short meeting with General Persons to discuss McCone's congressional testimony about the Stanford accelerator (SLAC). He readily agreed to put pressure to ensure that congressional action would be taken early next session. He also conditionally approved my taking a four-day vacation, September 1 to 5, with the understanding that I would be available on a moment's notice to return to Washington.

9:30 AM–1:00 PM   Meeting in Persons's office, including all top White House staff members to hear Stans report on 1961 budget. Stans is desperate because the President wants the budget not to exceed $80 billion, but unless very drastic measures are taken, the budget will be more like $81.5 billion, which is likely to convert a small surplus into a deficit. The discussion was partly political and, therefore, way above my head, e.g., how to present the budget in such a way that Congress couldn't claim it has reduced it while in reality increasing it, as it has supposedly done with the 1960 budget. Stans said that in DOD the Service secretaries are making no effort to keep the budget down and had joined the Service chiefs in a maneuver to force the secretary of defense himself to make the bold cuts. He was also very unhappy about the NASA expansion. HEW is causing him much concern because of its expansionist plans and the conviction of Secretary Flemming that if cuts are to be made they should not be in his area because human benefaction is more important than defense, space, national park preservation, etc., etc. There was some talk, largely initiated by Andy Goodpaster and joined in by Persons, of a committee made up of Glennan, McElroy, and myself, to be set up instead of the Space Council, to trim and control the overall space program. The preliminary figures which Stans passed out indicated some increase in NSF funds, but certainly not very adequate.

Possible Extreme Measures to Keep 1961 Budget Down. Possibilities discussed in the meeting were: Set a dollar limit on agricultural support. Terminate the REA program. Stop merchant ship replacement. Reduce military forces (backup forces only). Slow down federal construction work (dams, buildings, etc.). HEW not to spend budgetary increases given it by Congress. Liquidate some of strategic stockpiles. Curtail AEC production of U 235 for weapons, thus reducing TVA power requirements and new construction. Slow down the airways modernization program. Reduce the NASA program. Limit the veterans hospital program. Reduce the federal mortgages program.

2:00–2:45 PM   Phil Farley. Showed me his proposed briefing to Herter later this afternoon. It is a summary of the anticipated U.K. position, which is essentially that a treaty stipulating even only few on-site inspections in USSR is the best possible course. It is also a proposal for a U.S. position, which would be a complete reversal of the phased treaty approved by the President at the meeting with the Principals' Committee on July 23. His proposal is similar to the one I outlined to Dillon, that we insist on a large number of inspections (substantially over 100), which will almost certainly be rejected by the Soviets, and thereupon propose to them a limited treaty, forbidding only atmospheric tests or at most atmospheric and nearby outer-space tests, which

can be monitored from the ground. This would be accompanied
by the proposal that we extend the treaty when monitoring of the
underground tests becomes possible and undertake a research
program leading to that end. I warned Farley that this policy
must make allowances for the remote possibility that the Soviets
will accept our proposed large number of inspections. This proposal, I believe, would provide an adequate deterrent to keep the
Soviets from cheating, but it may put our government into a
hopeless political position, because of attacks by people like
Teller, and the secretary should be aware of it. Farley conceded
the point.

Friday, August 21, 1959

9:30–10:30 AM   Waterman's office. Spent some time selecting
two replacements for the National Science Board—one black, one
social scientist. It seems all eligible people live in New York City,
but that's just where they shouldn't come from. Then talked
about NSF budget, I telling Waterman that I will support an
increase if it is reflected in bigger support of basic research and
not only in training programs and he assuring me that he will do
so. We then came to the proposed PSAC panel on education of
scientists (Seaborg would be chairman, if he accepts). Waterman
said this was really a province of NSF and we were poaching. I
assured him that the panel will be instructed not only to start its
work with the reports prepared by NSF, but also to work very
closely with the NSF board so that it would be a joint NSF-
PSAC project. Waterman assured me that, of course, he had no
objection to the PSAC having a close relationship with the President and conceded that his board did not have it.

> Glenn T. Seaborg, professor of chemistry at the University of
> California, Berkeley, and a member of PSAC.

10:30–11:45 AM   General Goodpaster. We first discussed a
classified project and agreed that the President should approve it,
but with strong reservations and insistence that McElroy provide him with a sensible plan for over-all management organization instead of the double talk that the last McElroy letter on the
subject contained. We then took up the problem of resumption of

nuclear testing, on which subject the President received a strong memo from the JCS. Showed him the conclusions of McRae's panel and suggested that the President authorize preparations for the zero-nuclear-yield safety tests with the stipulation that he will decide as to whether to approve them or not when he returns from Europe. Goodpaster liked this plan. At parting I made a timid suggestion that I might be useful to the President in some of his conversations with Mr. Khrushchev because of my sensitive appreciation of the Russian language. We also discussed the JCS memo on underwater nuclear weapons test evasion, and I explained that as things stood, I would not be willing to call the panel together, but would ask JCS for clarification. This was accepted.

> The "zero-nuclear-yield safety tests" I proposed to the McRae panel as a way to avoid infringing on the weapons test moratorium and yet to investigate the so-called one-point safety of nuclear weapons when accidentally dropped by aircraft. These tests would involve such modification of the weapons that plutonium, while still present, could not undergo a full nuclear explosion; but by proper instrumentation it could be learned whether the normal weapon would undergo a significant nuclear explosion in case of accident. McCone, possibly to apply pressure for resumption of tests, cut down the production of nuclear weapons until such time as their safety could be established. This made the military even more anxious to start the weapons tests and so meet McCone's demands and realize the resumption of full production.

3:00–4:00 PM   Gen. Philip Strong, in with a sob story about the panel, which had been set up as a result of Wiesner's recommendation to Killian, running out of hand, and proposing a classified mammoth project, whereas he (Strong) feels that a much more modest approach will produce better information. I agreed with him wholeheartedly and said I would support his project and not the other.

> Brigadier General Strong was attached to the CIA.

∽ *Monday, August 24, 1959*

9:30–11:30 AM   Reviewed with Dave Beckler our various current activities. He suggested that I go to the President and ask his blessing for our major effort in evaluating Defense Department R&D program and that we later present to the President our findings, asking his guidance in various critical areas. This looks like a very good proposal, and I shall do so.

11:00 AM–12:45 PM   Alan Waterman. Discussed the joint research in Antarctica by NSF and Navy, where Navy is balking at assuming the responsibility for and, therefore, the expenses of the logistics required. Long discussion of FCST, from which both of us concluded that short of heroic efforts, the council is headed for oblivion. Unfortunately, neither of us can clearly conceive what actions will reverse the process. Discussed at length the decision of Flemming not to participate in the 5-year projection of R&D requirements which was agreed to by the Federal Council. Clearly he is afraid to commit himself, because of possible animosity of Congress, but it is not clear how to bring him into line, which I am supposed to do tomorrow. Then Waterman brought up again the relationship between his board and the PSAC, stating that he really didn't mind our usurping their prerogatives, but that it could be nice if we would work together. I agreed. We then talked about the Space Council and mechanisms by which that group could acquire importance. This looks even more difficult to accomplish than the Federal Council problem.

1:00–2:00 PM   Lunch with Gerard Smith, Robert Murphy, and David Beckler. Very pleasant and informal occasion, at which Murphy skillfully pumped me about my adventures as a member of the White Army, etc. Since I had a feeling he suspected me of being a hopelessly aloof and impractical scientist, I gave him lots of gory details, feeling this couldn't do any harm.

> Gerard C. Smith, chairman of the Policy Planning Council of the State Department; Ambassador Robert D. Murphy, at the State Department.

2:00 PM   Gerard Smith, talking about problems of disarmament mostly, with assurances of mutual collaboration. Smith promised to intervene on my behalf to get me to accompany the President to Soviet Russia. Smith is much more optimistic than I am about the Soviet willingness to accept arms limitations. He thinks that because they are convinced of their success through propaganda, subversion, and economic offensives, they wouldn't be interested in a gamble on the future such as a large war would be. Hope he is right. Suggested to Smith that one of the initiatives of the President could be to propose Wiesner's idea of a joint communications satellite with USSR. He seemed to like it very much, but I warned him that this shouldn't be done in September because it would require first a careful technical study to make sure there were no pitfalls.

4:00–4:30 PM Paul Pearson from the Ford Foundation. Came to ask for my opinion regarding the most effective ways of the Ford Foundation's becoming involved in scientific areas. Most charming gentleman, but so talkative that at the time the meeting was over, he hadn't even finished his own side of the story. I didn't have a chance to speak, which was just as well, as I didn't have many ideas. I did, however, suggest that the problem of providing enough senior positions in colleges was growing very serious, because of the large number of young people doing postdoctoral work on fellowships and then having "no place to go."

4:30–5:30 PM Pre-press meeting in General Persons's office. I got off scot-free. Nothing to prepare since the expectation is that the whole press conference will be devoted to the President's trip and also to Khrushchev's visit. After the meeting, I showed General Persons a letter from Senator Morton, asking me to deliver a speech at a Republican fund-raising dinner and suggested it would be very unfortunate if I did, as it would destroy the nonpartisan character of the PSAC and would cost me the good will of many people. He agreed and said he will square it away with Morton. Goodpaster showed me a statement prepared for the President in which he will postpone resumption of nuclear tests until after December 31, to be released as soon as the Geneva conference recesses formally. Evidently State stuck with its position notwithstanding British urging that we don't specify an exact date when our unilateral moratorium comes to an end. The text is not very fortunate because, unless it is read to the end, the impression may be gained that we will resume testing on January 1, regardless of the conference progress.

> The President's trip, August 26 to September 7, was to England, France, and West Germany, partly to reassure Allied leaders about Chairman Khrushchev's visit to the United States.
>
> Senator Thruston B. Morton, chairman of the Republican National Committee.

⌁⤳ Tuesday, August 25, 1959

10:30–11:00 AM   With Secretary Herter and Farley. Herter read carefully not only the conclusions, but the whole discussion of the McRae report and asked me a number of questions about the safety tests, what they were for, etc. Instead of complimenting me on a job well done, he was disappointed that the report still recommended resumption of tests. I noted that the sense of urgency was gone, that considering the people who signed the report, it was much better to get them to make this statement than to go afterwards and scream from the housetops about the U.S. being sold down the river by some presidential advisors, and also that the report explicitly doesn't consider the relation of our technology to Soviet technology, nor the political aspects of nuclear tests. It is somewhat questionable whether my arguments had much effect on him. He remarked that he just couldn't see how the Senate would ratify a treaty unless we had a foolproof inspection system. I remarked that there I couldn't help. I could only hold the scientists reasonably in line with this report.

> The point I was trying to make to the secretary was the importance of the conclusion that technically the resumption of tests was conceded to be not urgent. Hence political considerations, i.e., world-wide opposition to the resumption, could be used to prolong the moratorium. This argument would be reinforced by the consideration that the Soviets were somewhat behind us technologically and hence would gain more from the resumption of tests.

11:30 AM–12:30 PM   With Secretary Flemming, whom I got to see only after waiting for twenty minutes. Brief conversation on the Radiation Council. He wants the executive secretary of the council to be very definitely a member of my staff and said the fellow should be associated with the White House. Asked me if I had somebody in mind. I did not. We then came to the reason for my visit, the Federal Council, and I emphasized my concern about the failure of the council to act and the possibility that this may lead to objectionable legislation in the next session of Congress. I also reminded him that he was once very enthusiastic about the council, according to Dr. Killian. Flemming assured me he was enthusiastic about the council, agreed with the consequences of inactivity, and generally offered his help to make the council more effective, including attendance. We then came to the projection of R&D expenditures, and after some hemming and hawing, Flemming said that his difficulty was that the President, after signing the last (1960) appropriation bill which gave HEW much more money than requested by the President from Congress, attached several conditions under which he approved additional expenditures. Flemming said that these conditions were difficult to put into effect, that the department was engaged in effectuating them, and that this made projections very difficult. We discussed the matter some more and he agreed finally to prepare projections, but not quite according to the plan outlined by NSF, and with reservations. He also stipulated that when these projections are presented to the cabinet, the money should not be shown by departments, but by fields, such as medical research, etc., regardless of whose department is spending it. To this I agreed insofar as possible.

12:45–2:00 PM Lunch with General White and Joe Charyk. White rather stiff, although assuring me that he considers my office very important and beneficial. Conversation during lunch largely between Charyk and me, on space exploration and other such things, but White did get into it to some extent and then discussed his reason for insisting that we develop one more manned aircraft. He favors the B-70, which he believes can do long-range bombing missions and also act as an interceptor instead of the F-108. I asked him whether he thought he could have an effective force of B-70s costing at least $15 million apiece and still have money left for other important weapons systems. To this I got no reply.

> Gen. Thomas D. White, Chief of the Air Staff; Joseph Charyk, deputy secretary of the air force.

2:00–3:15 PM Army Signal Corps briefing concerning the unified communications plan. Strong plea that there should be a single agency under a single commander, presumably run by a single Service, e.g., Army, to operate the global communications network, whereas terminal facilities would be operated by the users. They strongly emphasized that the present independent Service networks were thoroughly inadequate. The plan looked very good on presentation, but Bill Baker then emphasized the real complexity of the problem, i.e., that the same facilities at most terminals are used for global communications and also for strictly Service purposes, such as ground-to-air or ground-to-ship and back. The Signal Corps people assured him that the problems were soluble, but didn't give the details. My general opinion is that this plan would run into tremendous opposition from Navy and Air Force, because it would clearly mean many of their facilities being taken over by the Signal Corps, even to the extent of replacement of their personnel by Army men.

∞⌣ Wednesday, August 26, 1959

9:45–10:30 AM   Attending USIB meeting at which an advisory panel presented its conclusions regarding the USSR missile program. A really excellent report which should do a lot to silence those who maintain the threat is imaginary.

> USIB: United States Intelligence Board charged with the preparation of national intelligence estimates. At that time the public was much concerned about the Russian ICBM program, about which the opinions in the media ranged from complete dismissal as not a threat to assertions of a "missile gap," disastrous to us. The panel's conclusions were roughly in the middle of this broad spectrum.

11:30 AM–1:00 PM   Meeting of the Principals. Most of the time was taken by my presentation of the McRae report. I did it by reading each conclusion and then ad-libbing the discussion, rather than reading the report itself, which I justified by noting that all of the Principals had already had a chance to see the report. I also emphasized that this was only one input, since there was no attempt at comparative U.S.–USSR evaluation. Both McCone and McElroy tried to emphasize greater urgency of tests than that defined by the report. Rather interesting thing —McElroy said that as far as DOD was concerned, they didn't care about safety tests. If they could only induce McCone to resume full production of weapons, they would be willing to take the chance, because, as Neil stated, York was also (as I am) convinced that the problem was largely imaginary. If we could only get him [i.e., McElroy] to make that statement to the President, McCone's position would certainly be weak, and it is obvious that the token reduction in production which he [McCone] ordered is a purely political trick to put pressure on the President. Dillon clearly avoided bringing to a head the discussion on key issues and emphasized that the conclusions of this report must be presented to the President as soon as he returns. I kept repeating that it was only a partial input, that it didn't involve comparative evaluation which may cast an even more negative light on the urgency of tests, etc. However, it is clear that the fear of the political effect of the release of information on the Latter hole would make a comprehensive treaty impossible in the minds of the other Principals. Dillon has now completely reoriented his thinking, has rejected the State Department phased treaty plan and argued in favor of my suggestion to come to the conference with a proposal for a large number of on-site inspections. He asked Defense why they objected to this, and Loper said that was because such a proposal would be only an initial one and in the course of discussion the U.S.A. would scale down the demands to 20–30 inspections in order to reach an agreement which would be unacceptable to Defense. He pointed out how State has compromised our initial position on staffing of the monitoring stations to meet the Soviets half way. Dillon assured him that in this case nothing like that would happen, and Loper said then the objections would not exist. The meeting then proceeded to the discussion of tests of the safety of nuclear weapons against accidental explosion. McCone and Loper were clearly unenthusiastic about my recommendation that only zero-nuclear-yield tests be undertaken, because they thought that awkward leaks [of information] may develop. Dillon, on the whole, liked the idea, but also was worried about the leaks. Whereupon, I pointed out that if the tests are done in the Los Alamos canyons, they would be clearly rated as research, and could not embarrass us vis-à-vis the Soviets. McCone accepted this suggestion and agreed to make inquiries. McElroy repeated that DOD did not care about safety tests, because York, like G.B.K., did not think the hazard significant, but the reduction in production by Mc-Cone was troublesome and hence tests were needed. This is quite a change from the stand taken by Gates. McCone just grinned.

2:30–3:00 PM Sir Harry Melville and Sidney Hiscock. A courtesy visit which I filled by an innocuous discourse on the importance of this office.

> Harry Melville, a chemist, was a senior science official in Great Britain. Sidney Alfred Hiscock was an official of the British Lead Development Association.

4:30–6:00 PM CW [chemical warfare] briefing, and a very interesting one, because it was quite informal and answered many of my questions. It is quite clear that the general research being done by pharmaceutical firms in the country is uncovering many compounds, the analogues of which could have major military significance and could transform chemical warfare from something unspeakably horrible to really very humane (???) warfare. A number of compounds have been found which are effective in very small doses (one mg and less) and which are nonlethal in any dose. They create such effects as complete anesthesia or paralysis, or mental aberrations, which convert well-disciplined troops into a chaotic mob, and yet complete normalcy (!) is restored after hours or days, depending on the dose. This account of the great accomplishments of Edgewood, of course, must be taken with a grain of salt, but the thing looks like a really interesting beginning.

> Edgewood Arsenal: Army's central R&D establishment for chemical warfare.

6:00 7:30 PM   Herb York. Evidently my telling him about the problems of air defense had a very good effect. He had a chance to look at NORAD and came back completely appalled about the dreamland in which these people live: in all their planning it is assumed that no missiles will be used, and York concluded that the whole system could be destroyed by a rather limited number of Soviet missiles. He then talked about his overall plans for reorienting the DOD program. An impressive plan, which shows an outstanding mind at work, and I feel that we should support him by all means at our disposal, although we may not agree with some details. It is rather clear that York intends to reduce the role of ARPA and restrict it to the field which is defined by its name [Advanced Research Projects Agency in OSD]. He wants to put all space activities directly into the Air Force except for specific missions to be assigned to the Army and Navy, but even those are to use booster vehicles of the Air Force. He feels that making that program part of the Air Force budget will automatically restrain the wildest boys, whereas at present they simply write fantastic requirements and expect ARPA to take care of them. York wants to continue the B-70 on a reduced basis, drop the F-108, and he asked me if I would support this. I said I would certainly support the latter, and was dubious about the former, but he felt that politically it was impossible to stop the aircraft program completely.

He then made a very important statement, with which I

completely agreed in principle, namely that we simply do not have the means to support all-out development efforts in all "important" areas, i.e., retaliatory forces of invincible character, perfect air and missile defense, effective submarine defense, and limited war potential. Something has to give, and his conviction is that rather than weakening everything, we have to cut some chosen areas to the bone, while leaving other R&D strong. My own feeling is that we have to cut air and missile defense R&D to a point where it is only a shadow that may make the Soviets nervous, and that the important thing is to prevent an attack by making retaliation certain. No matter how much money we spend on defense, it will not be good enough to prevent destruction of the country, so why do it and invite attack by weakening the deterrent. I asked York then to prepare a tabulation showing how much money was going to be spent next year on various projects, arranged by the year in which the products of development would become militarily important. I was worried lest we be spending too much money on things that will be useful only in the far distant future, and neglect more immediate needs. From this tabulation I exempted basic research because this can be utilized at any time. Herb initially objected, maintaining this was impossible, but finally changed his mind and agreed that this may be important and that they will try to do so. We agreed that the Nike-Zeus investigation requested of me by McElroy will be a joint one, with H. R. Skifter [assistant director, DDRE Office] being actually a member of our panel, so that we wouldn't in the end find ourselves the sole political scapegoat if the Nike-Zeus project has to be terminated.

In between these various activities, I was asked by Gray to read a letter from McElroy to him, rejecting the plan to make a presentation to the President on continental defense, and later suggested to Gray the form of the reply. Clearly, Defense is unwilling to present an integrated picture, and we should insist that the alternative proposed by them, a series of fragmentary presentations, is not what the President needs or wants.

8:00 PM Lloyd Berkner was to come to my hotel room but actually showed up only at 9:30 and had to leave half an hour later to catch his plane. I tried to explain to him that his long memorandum to Coolidge, the acceptance of which by State he made a condition of his joining the Coolidge study, was totally unacceptable. It is rather childish to imagine that direct negotiations of the National Academy of Sciences with the Soviet Academy on arms limitation measures would be either acceptable to the State Department or profitable. Similarly, it is quite absurd to get all the scientists of the country working on the problem in unclassified and uncontrolled fashion as he wants. Unfortunately, Berkner was not in a reasonable mood and grew quite emotional, so I managed to get my opinion across only partly. He is evidently very bitter about something that is not clear to me, but which can be generally described as the Republican party, and I now doubt very much that he could be useful to Coolidge.

∝⌣ Thursday, August 27, 1959

12:15–2:30 PM   National Press Club luncheon and speech by Herb York. He gave a brief 10-minute summary and then answered questions for an hour, some of which were very awkward to answer. He did exceedingly well, only once taking refuge behind security and only a few times avoiding an honest answer. Makes an exceedingly attractive appearance on the podium, with childlike innocence which hides a mind like a steel trap.

3:00–4:30 PM   George P. Sutton [chief scientist, ARPA], came to tell me of the ARPA program. Clearly they have heard that I am suspicious of what they are doing, because Sutton was very much on the defensive and succeeded in reassuring me that boondoggling doesn't cost anything like $60 million, just a few million dollars, a year. I didn't get into any details with him about the big space projects like Midas, Discoverer, etc., but did discuss to some extent the Dynasoar. Clearly York hasn't told ARPA about his plans to take ARPA pretty much out of this whole field, and I acted as if ARPA would be in permanent control. He made a pleasant impression, but obviously knows little about science, and therefore is enraptured by fancy proposals. He was taking notes during the conversation, and out of a variety of fragmentary remarks by me could probably reconstruct a rather frightening picture of my attitude towards most of the projects they are doing, even though I tried to be careful.

Friday, August 28, 1959

11:00 AM–12:30 PM   Berkner. This meeting was much more rational than his visit to the hotel. I tried to persuade him that his memorandum should be considered by the PSAC panel on arms limitation and developed to a point where concrete plans could be made, presented to the President, and upon approval put into effect; the Coolidge study was simply not relevant to his memorandum because it had to concern itself with quick changes in an obviously ineffective national policy toward arms limitations. After some argument, Berkner agreed to this, and immediately thereafter said that in that case he wasn't interested in working for Coolidge since he was obviously not the kind of man who would do the job that Coolidge was assigned to do. He was only interested in these broad, long-range studies. The same thing became obvious to me too, so I didn't put much pressure on. We then went over available people, realizing again that frightfully few would qualify for the Coolidge job. Charlie Townes, if he can be pried from IDA, looks like the best bet. With this Berkner departed.

> Charles H. Townes, vice-president, Institute of Defense Analyses, and professor of physics, Columbia University.

12:30–2:00 PM   Lunch at the White House Mess, to which I invited Elmer Bennett, undersecretary of the Department of the Interior. Very cordial, particularly after I remarked that it was my deep conviction that the Federal Council would be a flop unless every agency involved in it felt it could get something beneficial for itself out of the council. To this Bennett quickly agreed, and when I asked what the old-line agencies hoped for from the council, he said that they would hope for quite a bit. They hoped that their technological activities would receive reasonably high marks in the priority allocation planned by the council. Bennett feels that the glamor girls in the government R&D, i.e., Defense, AEC, NASA, get all the money at the expense of the old-line agencies. (In later discussion with Beckler, we reached the conclusion that actually the situation is a little different: the over-all budgets of the old-line agencies are held in tight check because of all the new demands, and internally the old-line agencies then squeeze research to the limit because that is the easiest way to get money for other things.) I agreed with

Bennett that this was a very worthy objective and assured him that I would have a very impartial attitude. We then went over the problem of national policy for the research facilities, and when I explained to him that in my opinion the text of the policy should include not only the discussion of facilities outside the federal government, i.e., in nonprofit institutions, but also refer to in-house facilities, Bennett cheered perceptibly, so we agreed that this future policy will at least contain a strong statement on the subject. The rest of the time was spent in my pumping Bennett about activities of the Department of the Interior. He discoursed on them willingly and interestingly, but nothing worth mentioning came out.

⌒ Monday, August 31, 1959

9:00 AM–1:00 PM   Informal meeting of the Space Council, at which I was presiding on request of Glennan. Most of the meeting was devoted to presentations. First by the Weather Bureau on what can be accomplished with satellites. It was an excellent job and rather convincing. Then NASA and DOD presented reports on their vehicle programs. Subsequent to that there was a discussion of the management of space facilities, specifically as it affects Project Mercury—the man-in-space program. Finally, we held a short executive session, at which it was decided to recommend that there be no formal meeting on Tuesday, September 8, presided over by the President.

⌒ Tuesday, September 8, 1959

10:00 AM   Half an hour with Dr. Donald Chadwick [an officer in the Public Health Service] regarding his candidacy for secretary for the Federal Radiation Council. He made a very excellent impression on me—intelligent and well informed.

3:00 PM   Half an hour with Mr. Akalovsky, devoted largely to his giving me his estimate of Khrushchev and his peculiarities of speech and conversation.

Alex Akalovsky, State Department, the senior Russian-English interpreter who normally worked with the President.

⌒ Wednesday, September 9, 1959

9:30–10:30 AM Meeting in Staats's office of the FCST panel to develop national policy regarding research facilities at non-profit institutions. I attended only the first part of the meeting, from which it was already clear that although the opinions are not quite unanimous, this panel can probably come up with a sensible unanimous policy recommendation. Difficulty will arise when the old-line agencies start arguing about it in the Federal Council.

12:30–2:00 PM Radiation Council meeting in Secretary Flemming's dining room. It was agreed that in addition to the Radiation Council, there will be a working group made up of top technically qualified people from the agencies involved, which will do the work, using the council only for over-all policy decisions. After some challenge from AEC, it was agreed that Chadwick will be secretary (executive secretary was changed to secretary to please AEC). As I learned afterwards, the objections of AEC, although they were addressed to Dr. Chadwick's age (34), were actually the result of his having once worked with Dr. Morgan, the executive secretary of the Advisory Council on Radiation, which has criticized AEC policies. It was agreed that Dr. [Lauriston S.] Taylor's National Committee on Radiation Standards will be continued and be used as a chief source of technical advice. Dr. Chadwick will be detailed to my office on full time by the Public Health Service at no cost to us. Flemming conducted the meeting well, but perhaps deliberately was exceedingly vague in his proposals and allowed them to be changed substantially without any protests. Most of the changing was done by Jack Floberg, member of AEC, and sensibly. I may have to do more work than I really would like to on this whole business, because Flemming asked me in a way where a flat no was impossible, whether I wouldn't attend some of the meetings of the working group and specifically to start them on their labors. He also urged that through the mechanism of the working group, the Radiation Council release to the public a number of documents explaining the nature of ionizing radiation, protection against it, health standards, etc. All this looks as if Chadwick will have a full-time job.

2:30–3:00 PM   Glennan, who had just been to see the President with a proposal that U.S. invite USSR for cooperation in peaceful space projects. The President was very sympathetic and asked that a high-level committee, including DOD, State, NASA, CIA, be set up with me as chairman to consider the proposal in detail and pass on it from a broad point of view involving such factors as propaganda effect, etc.

4:00–4:30 PM   Coolidge and Wiesner, with the conclusion that Coolidge desperately needs a senior technical man and that Brock MacMillan [of Bell Laboratories] is the only possibility. Later that same day I called Jim Fisk and asked him to release MacMillan, but Fisk was not enthusiastic, to say the least. During our discussion with Coolidge, both Jerry and I used the opportunity to impress a lot of our ideas regarding arms limitations on Coolidge. The latter took extensive notes. After Coolidge left, Jerry and I both agreed that in view of the new deadline for his report, January 1, Coolidge's effort could not get into any detailed work, and the best one could hope for was to get a set of good, but fairly general recommendations.

⌒⌒ Thursday, September 10, 1959

9:00–10:15 AM   Very uneventful NSC meeting. Matters considered were no concern of mine and weren't even of a very educational nature.

10:15–11:30 AM   McElroy and Twining stayed behind with Gordon Gray, and the latter asked me to stay also to discuss the presentation to the President on continental air defense, about which, as I noted earlier, there is a conflict between Gray and McElroy, the latter insisting that the presentation in June met the requirements and the former arguing that it did not. McElroy stated his case by insisting that as DOD conceives it, continental air defense is only bomber defense, and therefore involves only SAGE, ground-to-air missiles, and fighters like the future F-108. All of these, except the F-108 problem, were presented to the President in June. F-108 will be discussed with him in the course of budgetary presentations, thus the Defense Department will have met its commitment. Gray insisted that continental air defense includes also the problem of the survival of the retaliatory forces, but McElroy said that this was part of the presentation on strategic strike forces. I suggested that the distinction between continental air defense and strategic strike force would be washed out even more as time progresses; that I didn't care what they would be called, but that the President should be given in one or two presentations a comprehensive picture of the intercontinental warfare plans, including survival of retaliatory forces, our capability for inflicting damage before and after a surprise attack, etc. I also said that continental air defense is usually presented as if the missile threat did not exist and that therefore it should be presented again, making allowances for the missile threat. McElroy's rejoinder was that we can't do anything about the missile threat, and the ballistic missile early warning system, BMEWS, will be included in the SAC presentation. He then switched to the threat from the sea, and it was agreed that this need not be brought into the picture. The whole discussion was friendly, although heated at times, and ended on a note of compromise: Gray is to prepare a memorandum of agreement, indicating the scope of a presentation on SAC which in his view will meet the general directive of the President regarding these presentations when it is thought of as combined with the June presentation. After this discussion Gray and I, in my office, tried to define what the scope should be, and he said he will show me a draft of his letter to McElroy before sending it off.

> Gen. Nathan F. Twining, chairman of the Joint Chiefs of Staff.
> The F-108 project was later canceled.

12:30–2:00 PM   Lunch with Dr. Nolan. He is very unhappy about the standing committee of the Federal Council (FCST). The committee is too large (17 members), contains much deadwood, people who are on too low a level to speak for their agencies, and also contains people like Dr. Shannon of HEW who are openly antagonistic to FCST. The standing committee has degenerated into a discussion group, nobody being willing to do any work. We tossed the subject back and forth and agreed that tomorrow (Friday) I shall make some introductory remarks at their meeting in the nature of a pep talk. The whole situation is not very cheerful, but it is really not surprising, because, as is clear from Nolan's own reactions, the full-time government scientists are not happy about the Federal Council and the PSAC, which they consider as tied together, because they feel that part-time scientists have acquired too much influence. I kept assuring Nolan that being realistic, I know that no such body could be effective unless everybody was expecting some advantages from it, and that I was prepared to guide the activities so that this feeling will develop. Nolan stated to me strictly in confidence that they were having great troubles with Secretary Seaton because he is a great devotee of a balanced budget and feels that the best way to balance the budget is to minimize expenditures of the Department of the Interior, and furthermore that the easiest way to cut expenditures is to cut research. Nonetheless, Nolan is against a department of science and technology.

> The Federal Council for Science and Technology was composed of people on the "policy-making" level, more or less that of a deputy or assistant secretary, from all federal departments and agencies with substantial R&D budgets. Its standing committee, supposed to do much work for the council, was composed of senior Civil Service employes with technical backgrounds from the same departments and agencies as the members of the FCST.

> Dr. James A. Shannon, director of the National Institutes of Health in HEW, had powerful allies in Congress and could get all the funds he felt were needed by NIH regardless of presidential budget decisions. To him, therefore, FCST was not only unnecessary but even a threat to his freedom of action.

> F. A. Seaton was secretary of the Department of the Interior. Dr. Nolan was the chairman of the standing committee of the federal council.

In the afternoon had an informal meeting with Mr. Schaub of the Bureau of the Budget, Beckler, and Keeny, at which Mr. Schaub gave some idea of the budget requirements of the Services and spoke at length of his difficulties in pinning the Services to a point where they will explain where the money actually goes. He thinks that the Air Force is best in this respect and the Navy is worst.

> William F. Schaub, Chief of Military Division, Bureau of the Budget.

7:30–11:00 PM   White House stag dinner with the President. Present were sixteen guests—seven White House staff members, eight from executive departments and a visitor, C. D. Jackson. After watching from another room on closed-circuit television the President's TV address on his European trip, we had drinks and went to dinner—very elegantly served, needless to say. At dinner largely talked to my neighbors, George Allen, who pumped me about my life, and Livingston Merchant of the State Department, about a variety of State problems. We then proceeded to the library where a long discussion developed about the purpose of Khrushchev's visit and the reasons why he is bringing his family. I kept quiet during the evening but in my own estimation the rest were making him out to be too clever. At least one reason that he is bringing his family along is that Mama K. put her foot down. Then followed a discussion as to whether the President should return the visit or not. Opinions ranged from rather strong advice against going, except after substantial general political concessions by Khrushchev (C. D. Jackson) and guarantee of equal treatment, to a view that a visit by the President to the USSR under almost any circumstances would be advantageous to the U.S. (Nixon, whose view I share). Then followed a discussion on our policy toward Poland. Should we make strong use of Radio Free Europe, Radio Liberation, and other political warfare weapons or restrict them and rely on factual information of the Voice of America and be real friends to the Polish Government, trying to win it away from the Soviets? C. D. Jackson, during the entire evening, very much dominated the conversation, not always soundly. On the subject of Poland he spoke far more than anybody else, insisting on extreme forms of political warfare, extending to all European USSR satellites.

Charles Douglas Jackson, publishing executive and a former assistant to Eisenhower; Ambassador George V. Allen, director of the United States Information Agency.

Friday, September 11, 1959

9:00–10:30 AM   Cabinet meeting. Nothing exciting happened, but it was really amusing to see how easy it is to juggle budgetary figures. Stans gave a summary of this year's actions of Congress which, according to Senator Lyndon B. Johnson, have saved the taxpayers $900 million. According to Stans the actions of Congress have increased expenditures by approximately $1.8 billion and will inevitably result in an additional expenditure of

$10 billion later. As far as an innocent like myself is concerned, both sets of figures were pretty convincing, but the Stans figures will be reproduced and sent to all cabinet members and some members of Congress, to be used in refuting Lyndon Johnson's criticisms. Governor Hoegh outlined a proposal for handling the strategic stockpile and to reduce it by about $3 billion. Perfectly sensible, but obviously nothing much will happen in the near future. The chairman of the Federal Reserve Bank presented a report on banking and money problems after a nuclear attack. A certain sense of unreality permeated it; for instance, they are having great difficulty trying to decide about the moratorium on debts and on how to compensate people who lose property. I was impressed with how the President immediately caught on to that and said he really didn't care one iota whether U.S. Steel went broke, but only whether we would have some steel production left.

12:30–2:00 PM Mannie Piore for lunch. I talked to him about various activities that have gone on since the last meeting of the PSAC, emphasizing the failure of the Federal Council to accomplish anything. I, thereupon, rather sharply accused him and the rest of the PSAC of dropping me cold and not doing what the PSAC did for Killian, i.e., being available in Washington for advice and discussion. Mannie was visibly embarrassed, but maintained that the original members of the PSAC have done their job and that it was up to me now to find a new crew. We parted "good friends," but each firmly convinced of his own righteousness; it is of course a very serious problem for me.

2:00–3:00 PM Standing committee of the FCST. Gave them a pep talk, suggesting that they must see themselves as one of the three arms of the Federal Council, the second being the PSAC—that they were the only people who could effectively discuss problems of scientists in federal service, etc. Urged them to initiate proposals for Federal Council action and at the end more or less threatened them that, unless they got busy, some other agency would have to be set up to do the work. The result of my pep talk was rather interesting in that Shannon said he didn't believe in the standing committee, it couldn't do any work because the members were too much committed to their agency policies, etc. On the other hand, Carmichael and T. Killian endorsed my remarks and thought I injected new life into the committee. Nolan was carefully choosing the middle course. I left at three and Kreidler thought afterwards that a very useful change of tempo had been produced. For the first time those people were willing to talk about factors which inhibit their effort rather than beating around the bush as they did the last six months.

> Leonard Carmichael, secretary of the Smithsonian Institution, member of NAS; Thomas J. Killian, chief scientist of the Office of Naval Research in the Defense Department.

3:30–4:00 PM   Bascomb to enlist my support for the "Mohole" project to drill a hole from the bottom of the ocean to the earth's plasma. Wants me to come to a meeting of big oil executives to give a pep talk to get funds. I said I was very much interested in the project, but wanted to know in detail what to say. Would certainly endorse scientific merit, but may not be in a position to urge fund raising; asked him to prepare an outline that I may think it over.

> W. L. Bascomb, oceanographer, on the staff of the National Research Council of NAS.

4:45–5:15 PM   General Starbird with an account of the present status of the one-point safety problem of nuclear weapons. It is in rather bad shape because Los Alamos calculations are clearly inadequate to resolve the problem. He made a good case for the need of experimental work, but was quite obviously unwilling to accept my basic thesis that one could never make weapons 100 percent safe.

ᴕᴗ Monday, September 14, 1959

Morning in office, engaged in minutiae. Also phoned Goodpaster re: planned Vanguard launching Monday night, which concerns me because of possible failure and embarrassment to the U.S.A. Goodpaster far more exercised than I on hearing of the plan. Said he will discuss the matter with others.

> Vanguard was the first American satellite project, financed (rather underfinanced) as a part of the so-called IGY, the International Geophysical Year. On specific instructions of the President, the American satellite was to be a civilian project for scientific purposes and separate from the military. Following this decision the Vanguard project, originally a Navy operation, was transferred

to the IGY management. In 1957 the Vanguard rockets repeatedly failed on the launching pad, to the great public embarrassment of United States because of the successful launching of the Soviet satellite. One of my initial assignments in PSAC in the late fall of 1957 was participation (with H. York as chairman of the panel) in assessing for the President the prospects for the success of Vanguard. We rated it low and gave better marks to Wernher von Braun's competitive project which was using the Army's Redstone rocket. The President had disapproved of the launching of this satellite as "military" but on hearing our technical assessments he authorized its (successful) launching in January 1958. The first successful Vanguard satellite was launched only later in 1958. By late 1959 many American satellites has been launched using other (military) rockets but the Vanguard project, now in the hands of NASA, was still active. This particular launching was scheduled to occur just before the arrival in Washington of Chairman Khrushchev and a failure could be a great political (but otherwise trivial) embarrassment for the President.

2:00 PM Meeting with Goodpaster, Dryden, and Hagerty. The latter beside himself on learning about Vanguard—feels a catastrophe is in the making. This morning I talked to Dryden (by phone) regarding the Vanguard, and also Glennan (by phone), and both were insistent that it should go on, as the cancellation would be more awkward than failure. At the meeting with Hagerty, Dryden took a firm position that Glennan's instruction was to go ahead with Vanguard unless the President personally instructed otherwise.

3:00 PM Meeting with the President, Dryden, Goodpaster, and Hagerty. Dryden made brief statement about Vanguard plans. Thereupon Goodpaster stated that he, Hagerty, and I had grave misgivings. Before he was finished, the President looked at him and said: "You all are wrong. We will go ahead with Vanguard and that fellow [Khrushchev] is not going to scare me out of anything I plan to do." Hagerty raised objections, but in five minutes we were out of the office. [Technical difficulties prevented launching of that Vanguard on Monday night, but on Thursday night it was launched successfully.]

4:30–6:30 PM Farley, Beckler, and Keeny in my office, talking about nuclear test cessation prospects. Farley intensely unhappy that there were no discussions or agreement on the subject during President's trip to England. The British refuse to talk substantive matters until after elections. State Department position in total fog according to him, but can be roughly analyzed as follows: Herter firmly, and Dillon less so, believe that complete cessation is politically unacceptable in view of Latter hole theory, etc. Dillon rather strongly, and Herter so-so, think that the July 23 tactical plan should be abandoned and a new approach developed. Farley is assigned the task of preparing staff paper. We agreed that Keeny and I will both put our thoughts on paper for Farley to use. Situation generally dim.

6:30–11:00 PM Session with Fisk and others, discussing nominations for PSAC. This year the nominating committee did an excellent job. We started with about 50 names, narrowed the list down to 22, and agreed that further narrowing-down will be done by the whole PSAC membership.

> In those years PSAC prepared a fairly large, ordered, slate of nominees for the four to six annual vacancies on the eighteen-man committee. During this process the political affiliation of the nominees was never considered. The list was then submitted to the White House personnel office for political clearance, which in my day was apparently rather casual because the list was returned without deletions. Indeed, much later the President said to me that he knew most PSAC members were Democrats but that this did not bother him. After receiving the approved list I approached the candidates in the agreed preference order, explaining the duties of PSAC members and then extending the invitation to join on the President's behalf.

Tuesday, September 15, 1959

PSAC meeting, until noon in executive session devoted to review by myself of various problems facing this office and PSAC. I deliberately made the story a little gloomy and emphasized the lack of support from PSAC members, except for a faithful few.

12:00 M–1:30 PM Briefing by Wiesner on air defense, which stirred very vigorous discussion by PSAC. The main topic of it was: Does it make any sense to build an air defense system on the assumption that no missiles will be directed against it in an age when missiles may be abundant and may actually be the primary threat? The general conclusion was that the centralization, such as super-SAGE, etc., is very questionable.

At lunch in the White House Mess I was phoned by McCone, who told me he understood McRae will report to PSAC on the findings of his panel this afternoon. He stated that this was quite unacceptable to him because of the commitment to very limited distribution of this report, and that if I allowed this to occur, he would distribute the report to AEC commissioners and file it with the joint committee of Congress. I asked for a few minutes discussion before replying, but he refused to wait and demanded an immediate answer on the grounds that he was leaving his office and wouldn't be available. Since listing of this presentation in an open meeting was clearly a mistake on our part (because such consultants as Shannon, Brode, etc., would hear it), I realized that I would be in a weak position if I insisted on giving the briefing, and agreed to cancel it. This was tactically easy, because when McRae's presentation was scheduled, Khrushchev was arriving at the White House, and I suggested that we interrupt the meeting and all go to see Khrushchev, which was enthusiastically welcomed. By the time we returned, it was almost time for McRae to leave town, so everything went smoothly.

8:00 PM   White House dinner for the Khrushchev party. A glittering assembly of nearly ninety people, except that all the Russians were in plain blue business suits and their two younger ladies were in short dresses. I had a very pleasant time as a partner of Mrs. Stans, who is a charming lady, but had virtually no contacts with the Russians. Exchanged a few words with the Russian novelist, [M. A.] Sholochov, and a few with a "comrade" who identified himself as in charge of the American section of the Foreign Ministry [A. A. Soldatov].

As I mentioned my name, he said he knew a lot about me, which, I said, flattered me no end. It was an extremely interesting occasion though. As we were leaving the White House, McCone came to me and asked if I had difficulties canceling the briefing. I said none, and added that I acceded to his demand only because I felt that a mistake was made by my staff in scheduling this briefing in an open session; that I considered he had nothing to say about what I discussed with PSAC in executive sessions; that the members were as much advisors of the President as I was and, therefore, the matter was none of his business. McCone demurred, muttering something about security, etc.

ᴄᴎ⁓ *Wednesday, September 16, 1959*

9:00 AM–1:30 PM PSAC meeting, which started with an open session in which the computer panel report was presented. A very interesting and sound evaluation of our relative position vis-à-vis the Soviets. The latter appear to be a couple of years behind us, but in order to maintain our position it is necessary for us to put more emphasis on ideas and less on hardware. After that came brief status reports of all the panels, nothing notable developing. At 12:15 I asked for another executive session and presented to the PSAC my difficulties with McCone about the McRae report. Additionally I asked Keeny to report to PSAC about what Phil Farley told him. It was that McCone telephoned Farley, and told him that he stopped me from having a presentation of the McRae report and then he enlarged generally on the security weakness of the PSAC, suggesting that several security leaks were due to it. After Spurgeon reported this to the PSAC I asked for guidance. There was a unanimous reaction that I should take a strong stand, i.e., accept the fight.

[Note added later: Subsequently I learned the following about the two security leaks attributed by McCone to PSAC:

1. ARGUS—Roy Johnson of ARPA has investigated this leak and has definitely traced it to the no. 2 man in the van Allen Laboratories at Iowa State University.

2. The report on the High-Energy Particle Accelerators Program. This report was sent by McCone's office to the secretariat of the Joint Congressional Committee on Atomic Energy without any security markings and only the following day did McCone write that the report should be treated as classified. Inquiries were made of the secretariat, who admitted that in the interim they put no restrictions on the handling of this material. The obvious inference is that John Finney, who published the article, was shown the report.]

Roy Johnson, first director of ARPA.

ARGUS was the code name for high-altitude nuclear warhead tests in the fall of 1958. The report on high-energy accelerators to be built for nuclear physics research was prepared, as I recall, by an ad hoc panel of PSAC and AEC.

James van Allen, professor of physics and the discoverer of the "van Allen belt" of electrons in outer space, trapped by the earth's magnetic field.

✐ Thursday, September 17, 1959

8:30–10:00 AM   NSC meeting, which I left for a while to check with Persons and Hagerty the statement which, as a result of the pre-press conference, I asked NASA to prepare overnight for the President. The statement was OK.

10:00–11:30 AM   Meeting of Principals, at which nothing was decided, but rather frank statements were made about possible policy choices. Naturally, McCone delivered a strong, but not well-reasoned, argument for resumption of tests; so did Gates. Later Jack Irwin added his two-bits worth with a fantastic remark that if we don't resume testing, in a limited war we will have nothing but the 5- to 50-kt weapons; whereas the Soviets will develop Dove and Starling, and hence will demolish us. I corrected some of McCone's statements and also that statement of Irwin. I then proceeded to argue that the decision to abandon our efforts to have complete test cessation shouldn't be made on the simple basis of loopholes in the control mechanism; that other considerations may still make such a complete agreement more advantageous than a limited one, which in all likelihood will be nothing but just a gentleman's agreement not to test in the atmosphere. I found myself very much alone: Dillon, Dulles, and Gray not commenting or committing themselves, and the other three attacking my position. After the meeting, I spoke to Gerard Smith, who sat silently throughout the meeting, and said I was alone and clearly couldn't hold that position much longer, to which he replied that it was Dillon's method not to reveal his position until the end and that I should feel assured that the State Department would not wish at this early stage to adopt a policy which implied abandonment of our efforts to reach complete agreement. I also spoke privately to Jack Irwin, trying to dispel the effects of Teller and Co.'s sales pitch about Dove and Starling, noting that after all they will do no more than other nuclear warheads, except that they will produce little blast effects compared to radiation damage. It is quite obvious that Teller has done a good job in the Pentagon.

> John N. Irwin II, assistant secretary of defense for international security affairs.

> Dove and Starling (only paper designs then) were in a sense precursors of the mininukes which former Secretary of Defense Schlesinger was boosting in 1974–1975.

2:00 PM   Meeting with Glennan, Dillon, Dulles, Gray, Scoville, Farley, General T. D. White, Washburn (USIA), and others to discuss Glennan's proposal to the President that the latter offer to Khrushchev cooperative projects in space. It developed rather soon that the members of the meeting were not enthusiastically for this proposal because of the successful Soviet lunar shot and because space activities are an area in which we cannot lead from strength. A somewhat vague agreement was reached an hour later that the President be presented a paper prior to his meeting with Khrushchev in which a number of possible cooperative space projects in the general scientific, technological, and humanitarian area are listed. The listing to be detailed enough to identify the projects, but no more, and no recommendation be made that he should make this proposal. I agreed to prepare a preliminary list, and Farley was charged with the job of writing the final paper.

After this meeting saw several people in my office, including Dr. Chadwick, about his budget; and later attended a meeting in Staats's office with Waterman and Bronk, at which the latter explained the need for a substantial increase in the NSF budget. Staats took the position that it should be essentially the same as in 1960. I conceded that the total budget has to be kept within bounds, but maintained that in the military-space activities there is enough money in what is called RDTE to move as much as 100 million to NSF without hurting anything but boondoggling. It was agreed that Waterman will present more detailed facts to BOB later.

> Detlev W. Bronk, president of the National Academy of Sciences, chairman of the National Science Board, ex-officio member of PSAC, president of Rockefeller University, etc.

9:00–11:00 PM   (After a few drinks and a dinner at the Cosmos Club.) Executive session of the PSAC's Limited War Panel, which wanted me to tell them what was their task. They started by telling me that they were in a fog as to what to do; that the area was immense and that it was impossible to decide on the value of R&D projects without knowing what we meant by limited wars, etc. I tried to define a sensible concept of limited wars on which planning could be based; namely, an engagement of indigenous forces with forces from the Soviet bloc; the assumption that indigenous forces will be only lightly armed and that we must transport rapidly a numerically limited but powerfully armed expeditionary force to swing the engagement in our favor. I emphasized that nuclear weapons would not be used because once even tactical nuclear weapons are used, the assumption that the conflict won't develop into an all-out war becomes invalid. To my surprise, John Foster agreed with me, whereas Charlie Lauritsen did not. I expected exactly the reverse. I also emphasized that the function of the panel was not so much the writing of a formal report in which certain proposals of the DOD were condemned and others praised, as to influence York and his people into taking the right decisions and then supporting these decisions in recommendations to me, which I could carry to DOD. Afterwards I was told by Robertson that the session was very helpful to the panel.

John S. Foster, Jr., deputy director of the Livermore laboratory and a dedicated disciple of Teller.

∽ *Friday, September 18, 1959*

9:00–11:30 AM   Session in Pentagon with York, Dryden, Horner, Morse (Army), Charyk (Air Force), and Abe Silverstein (NASA). First a short presentation of the technical features of various large booster rockets (super-Titan, Saturn B, Saturn C, and potentialities of Nova). Thereupon we went into executive session with the above only and it quickly developed that a technical decision as to which of these large space booster vehicles to support was tied in with administrative decisions as to who would manage the programs and where the work would be done. Of course, the toughest issue was the fate of ABMA. Morse's position was that while the Army is willing to have ABMA transferred, it insists that ABMA be preserved for the time being at its present strength (about 5,000 people), and that it is in the national interest to preserve the large "model" shop facilities of ABMA able to manufacture space vehicles in-house. The first preference of Morse is to have ABMA in OSD; then in Air Force; and finally in NASA. York's position was that the large booster development program must be managed by a single agency; that the present split is absurd. Finally, that very large boosters like Saturn and Nova have no military use envisaged at present and, therefore, logically should belong to NASA.

NASA's position was that they would not be meeting their statutory obligations if they didn't have large booster rocket development, but that they could not accept ABMA with its present practices and costs (a couple hundred million dollars per year at least) because it would wreck the NASA structure. They prefer to see ABMA reduced to at most half its present size. Also if Saturn B is not to be continued, it should be canceled prior to transfer. York tried hard to make me take a firm position. I said that of one thing I was certain : I would not support a simultaneous development of more than one large booster vehicle; that I felt that these developments should be in sequence rather than simultaneous. Hence, of super-Titan, Saturn, and Nova, only one should be continued. It developed that a vehicle, using only one Nova engine (a million and a half pounds thrust) could be ready only a year or two later than a Saturn B of the same payload capability and, of course, the Nova Engine could later be used in multiples to provide such large payloads that manned exploration of the solar system would be feasible; whereas Saturn B couldn't go beyond two million pounds thrust. There was considerable discussion of the reliability of Saturn with its eight engines and hence incredible complexity, but a well-tested engine, versus a single untested Nova engine. As regards the executive organization, I refused to commit myself. It should be added that York suggested transferring to NASA $70 million from the fiscal 1960 budget for the support of Saturn B if NASA takes ABMA, and also offered to make maximum efforts to provide $140 million for the same purpose in fiscal 1961. These offers did not enthuse NASA people, since they felt that Congress, upon learning of these funds, would simply cut the NASA budget. The whole thing is a horrible mess, and I am going to have Ed Purcell and Don Hornig talk to me next Tuesday to clarify my ideas. I intend then on Thursday to seek guidance from the President and then go back to York and company with a firm position.

> Edward M. Purcell, professor of physics, Harvard University, a member of PSAC, and chairman of the PSAC space panel;
> Donald F. Hornig, professor of chemistry, Princeton University, and a member of the same panel.

2:30–3:15 PM   Private meeting with Persons, at which he first asked me many questions as to my evaluation of the purpose of Khrushchev's visit; and then I told him about all the difficulties with McCone, starting with the tirade delivered while we were on the *Skipjack;* then his blocking of the materials research and Stanford accelerator programs, and finally the events of this week. I emphasized that I didn't ask for any action, but needed advice. Persons took a serious view of the situation; expressed his sorrow that this sort of attitude could exist among close advisers of the President and said that he will think it over during the weekend and give me his conclusions next week.

3:15–4:30 PM   Messrs. Patterson and Vila from Convair Corporation. This was a sales pitch visit to explain why Atlas is superior to Titan. Naturally, their data proved it conclusively. I didn't challenge anything.

⟶ *Monday, September 21, 1959*

8:15 AM   Left for the Whippany Laboratory of Bell Telephone Labs to attend the beginning of a session on Nike Zeus, spent there three hours and returned to Washington at 4 PM. While at Whippany, I heard two presentations. One, by Brock MacMillan, was an attempt at a general analysis of the strategic role of active missile defense, and the other by [Donald] Ling, on a new concept of Nike-Zeus employment. Brock's analysis started from the identification of four strategic situations: 1, a deliberate, rationally planned, attack by the USSR; 2, a preemptive attack by the U.S.; 3, an irrational, but well-planned, "madman" attack by the USSR; and 4, an accidental attack without careful planning by the USSR. MacMillan's discussion considered the role of active missile defense as a preventative for each type of event, and also as a protective device once the attack took place. He concluded that in the first case, Zeus was useful only as a preventative, but that in other cases, its role would be mainly one of protection. When allowance is made for the relative cost and effectiveness of Zeus as contrasted with additional offensive missiles, for instance on Polaris submarines, it became clear from MacMillan's presentation, and I had to agree with him, that Zeus would be a poor investment as a protection for our retaliatory force, but he contended that it may be useful in the defense of our population. The latter point I have serious doubts about. Ling thereupon proceeded to develop a new concept of Nike Zeus. My general impression from this brief contact with the problem was that the whole thing isn't ready to be put into production, but perhaps should be still continued at an intense pitch as an R&D project.

7:00–12:00 PM   Dinner at Admiral Burke's house and then the Navy Relief Ball. There were ten couples at the dinner. During cocktails, Mrs. Lewis Strauss made strenuous inquiries as to the real reasons which induced Dr. Killian to resign. It was clear from her remarks that my statement, that it was his wife's health and pressure from M.I.T., was not adequate for her. After dinner, during which I sat next to General Erskine, who solemnly promised me an official invitation from the Thailand government to visit them (he is visiting them himself shortly), the high point was a lengthy discourse by Eric Johnston, who flew with Khrushchev to the west coast and was in Los Angeles during the Khrushchev visit. Johnston contended that all the eruptions of Khrushchev were carefully planned showmanship; that at no time was he actually angry, and at most was slightly irritated. He bolstered his arguments by the observation that Khrushchev could switch to complete coolness and amiability in a matter of seconds in the midst of his most violent outbursts, and that in private he was quite amused and vain about the same incidents which produced his public anger. Johnston contended that these outbursts were planned to prepare the situation for the President's visit to the USSR and also to give food for Soviet propaganda in the meantime. He sounded very convincing. The ball was pleasant, but not of an exhilarating nature; in fact it was exceedingly boring, since the amateur performance took most of the time, and the way I was sitting I had my neck twisted practically out of joint trying to see it.

> Lewis Strauss, retired rear admiral, banker, and former chairman of AEC; Eric A. Johnston, motion-picture executive and an intimate of the President.

ᕫᕽ Tuesday, September 22, 1959

9:45–10:00 AM   Principals' meeting without President. Subject was mainly the statement of McCone that AEC is unwilling to certify to the President that it is in our national interest to provide Holland with technical information on nuclear submarines. This got Dillon as near to anger as I have ever seen him, since, as he pointed out, the President in the presence of the then chairman of AEC, Lewis Strauss, and Secretary of State Dulles committed himself two years ago to provide these data.

10:00–11:30 AM   Principals met with the President. First I read the summary of the McRae panel report, following each conclusion with my comments. There were no exceptions taken to my statements and as I heard afterwards from Goodpaster, the presentation was "just right." Thereupon ensued discussion of safety tests. Notwithstanding McCone's half-hearted objections to my proposal of zero-nuclear-yield tests (he stated that plutonium from underground explosions in Los Alamos would poison the water supply, but was contradicted by Gates, who pointed out that according to Carson Mark's report to him a day earlier, there was no such danger and in fact Los Alamos has had a large plutonium dump for fifteen years), the President decided that the tests will be made at Los Alamos; that they will be treated strictly as experiments without nuclear explosions; that they will be begun before January 1, since they are not nuclear weapons tests in the sense in which the latter are defined; and that there will be no advance public announcement of them. This decision seemed to please Dillon and Gates, and is the best possible one as far as I am concerned. McCone obviously would have preferred to have the tests carried out in Nevada. Thereupon the conference came to the problem of the nuclear submarine for Holland. On listening to Dillon and McCone, the President delivered what amounted to a lecture, addressed to McCone, pointing out that the commission should take a bit more enlightened attitude on our responsibilities in the world, and that if it could do so, it would immediately see how wrong it was in its conclusion. That he expects AEC to change its mind, and that it should certify to him the need for the disclosure of the information. After the meeting, in the company of Gray, Gates, and McCone, I recommended that a public statement be prepared and held in abeyance for release when and if leaks took place about the experiments at Los Alamos. Gray enthusiastically concurred, emphasizing that the text must be approved by all the Principals. McCone agreed without protest. Thereupon, McCone asked to speak privately to me and in effect apologized for his actions of last week, which are recorded in this journal, and suggested that we be friends again. Since he made no reference to his telephone conversation with Farley, he is obviously not very sincere and thus I merely reiterated my position that I acceded to his demand because of a staff error, but that I reserved the complete right, subject only to presidential instructions, to decide on my disclosures to the PSAC. We parted coolly, but no harsh words passed.

> Carson Mark, head of the theoretical physics division at the Los Alamos laboratory.

1:30–4:00 PM   Federal Council meeting, which went quite well. The notable events were a presentation of the panel on oceanography by Assistant Secretary Wakelin; whereupon it was agreed that all agencies involved will submit to Dr. Wakelin their fiscal '61 plans on oceanography; that Dr. Wakelin will compare their plans with the recommendations of his report and then present this comparison to the Federal Council for eventual presentation to the cabinet. The question as to whether oceanography should be given special national priority was not decided, an inaction with which I cannot disagree, since so many other activities are competing and the council hasn't had a chance to compare their relative urgencies. Other interesting actions were the decision of the Federal Council to instruct the members of the standing committee that while on the committee they should act not only as representatives of their agencies, but also as individual science administrators. This is a notable defeat of Dr. Shannon, regarding whom several oblique remarks were made in the council, although his name was not mentioned. Another decision was to set up a panel on science and foreign affairs, under the chairmanship of Wallace Brode. Finally, it is interesting to note that the so-called 5-year projection in the science and technology fields, which was presented by Waterman on the basis of agency data, was received with comparatively little criticism (York the only one seriously objecting) and agreement was reached that several years of experimentation must pass before the usefulness or wastefulness of this work will become apparent.

4:00–9:00 PM   Purcell, Hornig, and later Killian and Doty. Discussion of vehicles for space program with particular emphasis on large vehicles. It was our conclusion that aside from political considerations, the most sensible thing to do is to abandon the Saturn B and to concentrate on the Nova, starting with a single-engine Nova vehicle and gradually progressing to multi-engine vehicles. This admittedly leaves the Soviets superior to us until the late 1960's, but ensures a reasonable over-all level of effort and focuses the space program as a truly civilian effort.

> Paul M. Doty, professor of chemistry, Harvard University, and a PSAC panel member.

✒ Wednesday, September 23, 1959

8:30–9:30 AM   Bacher, acquainting me with the problems of STL. It looks as if STL is really in trouble with the Air Force because of its dependence on Ramo-Wooldridge Corporation.

> STL, the Space Technology Laboratories, was a division of the Ramo-Wooldridge Corporation and was the technical manager of the ICBM projects for the Air Force. Public and Congressional criticism caused Ramo-Wooldridge in 1960 to separate the Aerospace Corporation as a nonprofit institution engaged in long-range planning for the Air Force, while the technical management of on-going projects remained with STL. Ramo-Wooldridge later coalesced with the large Thomson Products Corporation to become the Thomson-Ramo-Wooldridge (TRW) Corporation.

10:00 AM–12:00 PM   Meeting with Glennan, York, Harr, Baird (Treasury), and George Allen. Glennan started by reading to us his briefing to the President, in which he contended that either we had to be wholeheartedly in competition with the Soviets for space supremacy or should be honest about it and treat space exploration as a purely noncompetitive technological venture. He contended that his three-quarter billion budget was the minimum acceptable for a competitive proposition, and that the noncompetitive program shouldn't be even as large as half a billion. This developed into a general discussion of whether we should be competitive or not, in which Harr took the extreme position that we must be all-out competitive, and I was on the other extreme, suggesting that the Soviets smartly maneuvered us into accepting their challenge in just one narrow field in which they are superior to us, and that considering the money involved the all-out competition with the Soviets may cost us leadership in other fields of science and technology, and this would be much more serious. Glennan challenged me to have the guts to tell that to the President. I said I might do so in a week or two and asked him to accompany me. The latter part of the meeting was on the more specific problem of what to do with ABMA and the Saturn vehicle. Nothing definite was decided. York kept insisting that DOD will support Saturn unless NASA guarantees that it will have a vigorous large-booster program.

Lunch at the Mess with General Goodpaster, discussing preparations for the meeting in Camp David on Saturday. He also told me that my briefing on Tuesday went off very well.

> Karl G. Harr, vice-chairman of the Operations Coordinating Board in the NSC office; Julian B. Baird, undersecretary of the treasury.

2:30–3:00 PM   Gerard Smith of State, asking for a restatement of my position on the nuclear test cessation problem and related areas, which I gave quite frankly, emphasizing that this is a problem which requires political decisions and cannot be settled on a purely technical basis.

⌒ Thursday, September 24, 1959

8:30–11:00 AM   Frantic work on the briefing for the meeting with the President.

11:00–11:30 AM   Waiting in the waiting room, while my time with the President was consumed by the State Department. Agreed with Kendall to let him go ahead of me, and postponed my briefing until Tuesday.

> David W. Kendall, counsel to the President.

12:30–1:45 PM   Had Coolidge for lunch. Talked about his problems, returning again and again to the fundamental problem of what constitutes adequate safeguards. Suggested that in the past our monitoring objectives emphasized military operational units; the monitoring, therefore, provided a tactical warning, but a much more successful approach could be to think of monitoring in a strategic sense, i.e., placing inspectors at transportation centers, at factories manufacturing key weapons systems, and relying on unilateral measures for tactical warning. Discussed also the problem of his staff acquisition, which is not progressing.

1:45–2:30 PM Meeting with Glennan, York, and Horner. Continuation of yesterday. York somewhat softer in his insistence on a large booster. I stated that I would wholeheartedly support a vigorous program for one booster vehicle, but not for two simultaneously. It is quite clear that both York and Glennan would like me to assume responsibility for canceling Saturn B and I might yet have to do it, in the sense of going to the President and convincing him of the wisdom of that act.

2:45–3:30 PM Meeting with Governor Hoegh and others on radio frequency allocation. Extremely futile meeting, nobody having clearly thought through the proposed congressional bill and other alternatives. Hoegh speaking in a firm, resounding tone, with a warning finger; looks very impressive, but actually doesn't say much. Meeting ended with agreement to reconvene two weeks later after everybody has had an opportunity to think about the problem. I predict identical outcome.

4:00–4:30 PM Cyril Smith, who in effect told me that he was unwilling to work on the military budget and similar things, although he offered to put in a lot of time on nonmilitary projects and suggested forming a panel to consider the technological consequences of a possible general disarmament, noting that this might result in a frightful economic slump unless proper measures are taken. I gave him my philosophy on what the function of the PSAC is, emphasizing that it is primarily here to assist the President and only in "spare time" to occupy itself with very general problems. Agreed to form the panel if he can spell out its purpose more explicitly. Cyril stated that he thought he probably should resign in a few months' time because he was really not willing to pitch in on the activities of the PSAC as defined by me.

> Cyril S. Smith, professor of metallurgy, Massachusetts Institute of Technology, and a member of PSAC.

4:30–5:15 PM Mr. Walker Cisler, who has just returned from Soviet Russia after a very successful visit to see their nuclear power reactors. Tremendously impressed by their progress and convinced that benefits would be mutual from a close collaboration in the nuclear power area. He also was impressed by the friendliness of Russian technologists and by their interest in the U.S.A. He saw Koslov and Mikoyan, who had great hopes for Khrushchev's visit to this country in the sense of preserving peace.

Walker L. Cisler, public utilities executive; F. R. Koslov and A. Mikoyan were members of the Soviet Politbureau.

CHAPTER III

Chairman Khrushchev in Camp David

EISENHOWER HAD SENT a personal letter early in July inviting Khrushchev to visit the United States. The President hoped that the invitation would start relaxing the tensions of the cold war and especially that it would induce a change in the adamant Soviet position on the West Berlin issue at the foreign ministers' conference then in session in Geneva.

Although unsure of the reasons for the invitation and expecting snubs and worse (see *Khrushchev Remembers: The Last Testament*), Khrushchev accepted. However, the Soviet position in the Geneva conference remained rigid and the conference adjourned in deadlock. Moreover, Khrushchev publicly set his own deadline for signing a separate peace treaty with East Germany, which would abrogate some Allied rights in West Berlin and transfer full control over the access routes to East Germany, unless an agreement with the Allies could be reached in the meantime. Eisenhower viewed the Khrushchev procedure as an ultimatum.

Khrushchev landed at the Andrews AFB airfield on September 15, to be met with full honors by Eisenhower and Nixon. Later that day they all met at the White House, with PSAC watching the arrival of Khrushchev. Khrushchev presented Eisenhower with a model of a 300-pound projectile, Lunik II, that the Soviet Union had just hard-landed on the moon (by a lucky timing coincidence, as Khrushchev with a grin told newsmen in Washington). This gift privately annoyed the President quite a bit. In this brief initial meeting Khrushchev expressed his indignation about a recent speech of Nixon's, designed, he felt, to arouse animosity in the United States against himself and his party. The meeting was followed that evening by the state dinner, which I witnessed, and the following morning Khrushchev and his party left in the President's plane for New York where he gave a speech to the United Nations General Assembly. In this speech Khrushchev made a dramatic appeal for ''general and complete disarmament'' which provided for international controls only in the third and final phase of disarmament. From New York he proceeded to Los Angeles where two incidents aroused Khrushchev enough to threaten to cut his visit short. He was denied permission to visit Disneyland because of fears for his personal safety and later the (conservative Republican) mayor of Los Angeles delivered a dinner speech that was considered by Khrushchev highly offensive to the Soviet Union.

The rest of the trip was more congenial and included stops in San Francisco, at an Iowa farm, and visits to factories in Pittsburgh. Khrushchev delivered enough speeches, toasts, and responses at press conferences to fill a whole book, *Khrushchev in America* (translated from the Russian edition, *Live in Peace and Friendship*). He returned to Washington and on September 25 flew with Eisenhower to Camp David.

To appreciate the remark of Khrushchev to Nixon at the lunch described below, note the order in which Nixon refers to guns and then to game.

Friday, September 25, 1959

10:00–10:30 AM   Coolidge called to tell me that Herter informed him that the President is very anxious to be able to respond to Khrushchev's United Nations proposal on complete disarmament with a more realistic, but equally bold, proposal of his own, and that he would want an intensive study made of conditions under which we would offer to eliminate all nuclear weapons. We discussed various people for a study. Later the same day I asked Goodpaster about this and learned from him that Coolidge got things a little garbled in transmission, and that the President wasn't singling out nuclear weapons but wanted to have a measure of disarmament which would be bold and yet realistic, and only mentioned nuclear weapons as a possible example.

11:30 AM–12:00 M Mr. Frank Pace called, clearly to look me over, under a rather improbable offer, vaguely worded, of help. We had a cordial conversation and both agreed that a lot of things are wrong with our military weapons program, Pace emphasizing that it should be so modified that it could be of benefit to civilian economy as well; i.e., by developing new things of use to civilians even though that may degrade the weapons somewhat. He cited as an example his experience in World War II, when the Army was buying hundreds of millions of dollars worth of special trucks which were of no possible use to civilians, and yet he felt that by changing their specs, civilian trucks could later have been improved from appropriate military development.

> Frank Pace, Jr., president of General Dynamics Corporation, former secretary of the army.

12:15–1:15 PM For lunch Dr. George Weil, former student of Enrico Fermi and now a consultant on atomic energy. I arranged the lunch on the suggestion of I. I. Rabi, to whom I talked in the morning. Weil largely disagreed with Cisler, because in his opinion plutonium breeder reactors are very unpromising and couldn't be made economical for a long time to come and yet that is the direction in which the Russians are going. This is also Cisler's hobbyhorse on which he has bet a lot of other people's money. Weil's conclusion is that we may not be able to learn much from the Russians if we have an agreement on exchange of information on power reactors.

> Some time during the sixties Cisler's advanced nuclear power reactor (the Fermi reactor near Detroit) failed and had to be laboriously dismantled.

1:30–3:00 PM Studied briefing papers for Camp David conference in Goodpaster's office. One recommendation for a statement by the President rather horrified me because it is so close to being dishonest and, worse than that, could be proved to be dishonest by Khrushchev. I hope the President will not make it, because it may have rather catastrophic effects on our relations with Khrushchev. The papers generally leave me in a state of complete fog as to what my function will be in the conference and Goodpaster wasn't helpful by saying that people sit around casually in the room while conversation is carried on by the President and Khrushchev, and sometimes the secretary of state, and occasionally the others pitch in. All the briefing papers I saw were on such a general level that there seemed to be no opportunity for technical inputs, so I shall probably remain silent and in the meantime still have no idea how to prepare myself.

3:00–5:30 PM My staff meeting to discuss my participation in Camp David, but really wasn't very helpful because I didn't know what questions to ask. We discussed the three topics which will come up tomorrow. One other difficulty is developing insofar as briefing papers go. The President is to suggest that we resume *technical* discussions on nuclear test cessation, but as we analyzed our position in the staff meeting, the proposition makes no sense unless we are willing to sign a complete ban agreement and that is certainly not the agreed position of the Principals. In general the statement of the President on nuclear test cessation represents a major reversal from the July 23 position paper which he supposedly approved, because it includes the sentence that we are still ready to sign a total ban agreement. I hope that this is really so, and is not another instance of being almost dishonest.

5:30–6:15 PM Pete Scoville to give me more papers to read tonight on atomic energy and on cultural exchanges. When we were alone I told him about the item that troubled me, to which I referred above, and he was completely horrified and wanted to discuss it with Dulles, but that would obviously get me in trouble, so I asked him not to, and I would try to take it up with Goodpaster tomorrow morning. As this is being recorded, I am awaiting Leghorn, who is joining me at 7:30 for dinner. I talked to him on the phone today and will try to persuade him to work on a part-time but regular basis for Coolidge. Coolidge obviously desperately needs a man like Leghorn. After that I am going home and will work on a stack of papers for tomorrow. This was a horrible day, even though the diary is short.

Richard S. Leghorn, president of the Itek Corporation.

⟶ *Saturday, September 26, 1959*

8:30 AM   When I went out to the car, the driver said that the helicopters weren't flying because of fog and he would drive me to Camp David. He was an excellent driver and thus even though we were traveling at speeds up to 90 miles an hour, I wasn't worried. We made Camp David at just 10 AM, the stipulated arrival time. The outside door of Aspen Lodge leads directly to a paneled dining room, where I found McCone sitting at a little table. I could hear voices through a sliding partition which separated the glassed-in terrace. I joined McCone and while we were drinking coffee, he showed me his agreement with Emelyanov, as well as a memorandum from the latter, in which Emelyanov expressed concern about the cost of peaceful uses of the atom and hoped for a joint program, as well as mentioning that strong forces in the USSR objected to arms limitations and wanted to push military atomic programs over peaceful ones.

V. S. Emelyanov, head of the Soviet (nonmilitary) Atomic Energy Authority.

10:45 AM   The partition slid open and the President and Chairman Khrushchev, followed by others, came out. The President introduced McCone and me, as well as Gates and Irwin who had arrived a few minutes earlier, to Khrushchev. The President then took several State Department people into another part of the lodge while we made small talk with Messrs. Khrushchev and Gromyko, not an easy task. The President returned alone and asked Khrushchev to join him for a private conversation, which they held at a bridge table in the corner of the terrace, with just one interpreter with them. From that time on, and until about 2:30, they continued this private conversation, not raising their voices, but obviously not getting anywhere, judging by their expressions. In the meantime more Soviet people arrived, and were introduced all around. After some inconsequential conversation, Emelyanov and I sat on a settee on the terrace and engaged in a long conversation, including such topics as the need for better scientific cooperation. I broached the subject of cooperation in the field of dissemination of scientific information (abstracting, etc.) and he eagerly agreed, saying that their central institute doing this job has 12,000 specialists and still is unequal to the problem. We then spoke about disarmament; I introduced the problem by stating how important I considered it, and that I am wholly devoted to the objective of avoiding war, and that this may be difficult in a continuing arms race. He chimed in eagerly, but then started advocating the total disarmament plan, along Khrushchev's proposal, which I questioned and suggested a more cautious approach. Most of the time I spoke in English and most of the time Emelyanov spoke in Russian, although at times we both tried to use English or Russian. For a while we even tried both speaking German, but discovered we weren't fluent in the language. Emelyanov made a good impression on me. He is obviously a very intelligent and cultured person and unless he was being dishonest, he is seeking collaboration and peace.

2:30 PM   About two hours behind schedule, we sat down to lunch, which included the President, vice-president, secretary of state, two or three State Department people, McCone, and myself, as well as Khrushchev, Gromyko, and several people in their Ministry of Foreign Affairs, and Emelyanov. Initially there was small talk around the table, but eventually everyone's attention became focused on Khrushchev, who suddenly launched into a violent attack on the U.S. exposition in Moscow, accusing us of having done a very poor job. The American kitchen with its many gadgets particularly annoyed him. He thought it was very silly, that it left a very bad impression on the Russians, and that many Russians who bought tickets for the exposition later traded them for tickets to a Czech exposition. This was accompanied by rather personal remarks to the vice-president, who later stated flatly that he didn't get a bad impression of the exposition, and Ambassador Thompson stated that the exposition was a very great success to the very end.

Later the vice-president asked Khrushchev whether he preferred hunting with a rifle or a shotgun for birds or big game, to which Khrushchev replied that the vice-president obviously didn't know anything about hunting since rifles were used for big game and shotguns for birds. This reply was not due to wrong translation (as I observed) but was just one more demonstration of a chip on the shoulder. Khrushchev went on then to talk about his hunting; he said that he hunts for wild boar and also for ducks; this led him to the remark that although he hunts with Gromyko and Gromyko bags some wild ducks, he still isn't sure whether Gromyko is a good shot since he may be buying his ducks. Gromyko commented that his wife saw him shoot the ducks, so Khrushchev then said he didn't trust Gromyko's wife either. Trying to deflect him, the President remarked about the difficulty of his office—that he is never away from the telephones and is interrupted all the time and is working even while on vacation. He then asked Khrushchev (who had mentioned his vacations while discussing hunting) whether he was free of work during his vacations or was followed by the telephones. Khrushchev again became almost violent, stating that telephones were even installed on the beach when he went swimming, and that he could assure us that soon they in the USSR would have more and better telephones than we have and that then we will cut off our telephones since we are always afraid of comparisons. The last remark didn't make any sense, and his whole performance was almost personally insulting to the President and the vice-president, who tried to carry on the conversation. The President remained completely silent, but Nixon behaved as if he didn't notice the offensiveness. Akalovsky was acting as interpreter during the lunch and I could see by the expression on his face that he was rather shocked at the way in which Khrushchev spoke during his outbursts. Actually, I would say that Akalovsky distinctly toned down the offensiveness of Khrushchev's remarks without changing the substance. Even at that I could see by the President's expression that he was intensely angry and just managed to control himself.

My impression of K. was of an extreme inferiority complex exhibiting itself in arrogance, and this was also the opinion of Nixon in a post mortem which we had on the terrace after lunch, when the President took a nap and all the Russians went for a walk. During that post mortem we learned from Herter that the President kept insisting that so long as there was an implied threat over Berlin, we were not willing to undertake any substantive negotiations and Khrushchev insisted that he couldn't change the USSR position on Berlin and that we should agree on complete disarmament, which would then solve the Berlin issue automatically. Herter said that this looked like a complete deadlock and, therefore, no progress could be made on other items of the agenda.

I had also a rather useful conversation with Tom Gates, who complained about the political jam in which he is finding himself because York gave a press interview in which he described the new setup under which ARPA will be taken out of space projects and these assigned to the Services. Gates said that Roy Johnson is mad as hell and will now resign as director of ARPA and blame the resignation on this order, whereas he was resigning anyway; that he, Gates, will get it in the neck. Then he stated his concern about the development of the large space vehicle, Saturn; he said that it probably should be carried on in the Defense Department even though there are no clear needs for it at present. I talked to him about the technical shortcomings of Saturn B, pointed out that if vigorously pursued Nova would be ready only two years later, etc. It seemed to have a strong effect on him, and he is now considerably more likely to accept an arrangement under which Saturn will be discontinued and Nova pushed instead. During the conversation he expressed considerable irritation at von Braun and his political maneuvering, but conceded that politically it will be murder to try to reduce ABMA to a reasonable size, which *must* be done because 5,000 people there is something that we just couldn't support indefinitely. However, von Braun insists that ABMA cannot be reduced in size.

About 4 o'clock the President came out and went immediately out of doors to find Khrushchev, with whom he continued a walk. Toward 5 PM we got a message that the two, accompanied by just one interpreter, had taken off in a helicopter for the President's farm in Gettysburg and would not return until dinner. At that time Herter suggested to a number of people present, including myself, that there was no sense in our staying any longer. He also told me that I wouldn't be needed on Sunday. During this long afternoon everybody was very much depressed and discussed some devices to write the communiqué as if there was no actual break. There was a general feeling that the meeting will end in a nearly complete failure and hence may actually worsen rather than improve relations. There was a lively discussion as to whether the President would make a trip to the USSR or not, in which Nixon urged that the President go regardless of the outcome of this meeting, because of the impact his presence will have on the Russian people. The State Department argued that he shouldn't go unless the meeting was a significant success. (From the communiqué of the meeting, which I read on Sunday, which postponed the President's visit until spring, I concluded that the State Department point of view prevailed and that regardless of the communiqué, very little was achieved at the meeting.) I left Camp David at 5 PM by helicopter. The flight was uneventful.

A few days later my journal contained the following entry:

Today, Friday, October 2, at the White House staff meeting, Andy Goodpaster summarized the events at Camp David.

At 10:45 AM on Saturday, the meeting of the President, Khrushchev, Herter, and Gromyko broke up without reaching any understanding on Berlin. The President said the State Department would prepare a memorandum summarizing our position in the matter. Thereupon everybody came out into the dining room, as I described earlier. The long private talk between the President and Khrushchev dealt largely with a draft of the State Department on Berlin which Khrushchev found completely unacceptable. According to Goodpaster, toward lunch time Khrushchev became very tense and angry and this reflected itself in the attitude of their Ministry of Foreign Affairs people, who became totally frozen. Judging from my conversation with Emelyanov, the information simply didn't filter to him and that is why he was friendly. Things began to get better when the President and Khrushchev went to the farm after lunch, at which time Khrushchev conceded that he wouldn't insist on a time limit for negotiations on the Berlin problem and the President said this would be acceptable to him as removing the threat.

The freeze of the Soviet Foreign Affairs people Saturday afternoon was so extreme that even though they themselves had proposed the original agenda, they refused to discuss anything with Herter while Khrushchev and the President were at the farm.

According to Andy, at dinner Khrushchev was considerably more relaxed. There were no demonstrations of ill manners and neither were there on Sunday. Goodpaster's general evaluation is that during the entire meeting, Khrushchev showed extreme respect for the President and so chose at Saturday lunch to show his displeasure by attacking the vice-president.

Saturday night Khrushchev raised the Chinese issue because, as he explained, he couldn't go to China without having discussed it. He thereupon stated his position that Chiang Kai-shek was in rebellion against the government of China, that Formosa was a part of China and that, therefore, military force had to be used against Chiang Kai-shek to liquidate the situation. The President stated our position as being precisely the reverse, and that actually the Soviets at one time held a meeting with Chiang Kai-shek to discuss his gaining control of all China. Khrushchev ended the conversation by saying that he merely wanted it to be on the record that they discussed the matter. It is our interpretation that he didn't dare go to China without being able to tell them that the matter had been discussed, but that he knew nothing would come of it and needed it only for the record.

Most of Sunday morning was spent arguing whether the statement on Berlin would be included in the communiqué. Khrushchev refused to do so and finally proposed that the statement be made orally by the President the day after Khrushchev got back to Moscow, and Khrushchev would then confirm it. Our interpretation is that Khrushchev went beyond the authorization of his Politburo in making this concession and did not dare to put it in the communiqué because he had to get approval in Moscow. Another interesting sidelight brought out by Goodpaster was that Gromyko during the entire Khrushchev trip was clearly anxious to make it a success and was very positive in his approach, whereas the Soviet Ambassador Mikhail Menshikov took a very negative attitude, was domineering toward the U.S. staff, and actually managed even to irritate Khrushchev to a point where the latter said something very sharp to him in Russian, whereupon Menshikov subsided for the rest of the Camp David conferences. Lodge reports that on the trip Menshikov was persistently feeding untruths and poison about the U.S. to Khrushchev.

> Eisenhower in his memoirs, *Waging Peace,* refers to the acrimonious Khrushchev-Nixon exchange during lunch in Camp David but places the event on Sunday, September 27, just before departure and after the concession by Khrushchev on the Berlin deadline. Khrushchev in his memoirs (*Khrushchev Remembers: The Last Testament*) does not recall the lunch on Saturday but refers to lunch on Sunday, just before the return to Washington, as an event "more like a funeral than a wedding feast" because of the failure to solve the impasse.

ᖰ⁓ *Sunday, September 27, 1959*

I watched Khrushchev give his speech on television. The speech was cordial enough, but of course a strong defense of communism and a promise that it will surpass us as a superior social and economic system. It was very poorly delivered. Khrushchev was racing through his text and in many places stumbled over more difficult words and expressions, evidently not having rehearsed it.

CHAPTER IV

Controversies over Space Vehicles

BY THIS TIME most of the readers, certainly the younger ones, must be quite bewildered by the references to a multiplicity of missiles and space vehicles, most of them now defunct, over which the fierce although quasi-gentlemanly Washington struggles were fought in the late nineteen-fifties. The rest of this journal contains many more references to them and so at least a sketchy technical description and historical review may be appropriate.

In the early nineteen-fifties two Air Force long-range missile systems were under intensive development: The Snark (of Northrop Corporation), really an unmanned, small subsonic jet aircraft, one squadron of which was deployed in 1958 and was phased out a few years later as obsolete; and the Navaho (of North American Aviation Company), a supersonic (2000 mph) winged craft weighing 300,000 lbs., to be lifted and started on its stratospheric flight by three booster rockets, each generating 135,-000 lbs. thrust. The Navaho (minus the rockets) was flight-tested at low altitude in 1956 but the project was terminated soon thereafter.

Still earlier the Convair Corporation was designing for the Air Force a very large, true intercontinental ballistic missile (ICBM), the MX774, but the funding of it was terminated by the Air Force before 1953 and only token work continued.

The von Neumann committee, to which reference has already

been made, urged the Air Force in 1954 to resume the development of ICBMs but of a reduced size compared with the MX774 because by then it had become clear that a megaton nuclear warhead could be accommodated in a re-entry nose cone weighing less than 2 tons. (John von Neumann, a mathematician, was involved in many military problems, and was a member of AEC in the mid-fifties.) This project, named Atlas, was started late in 1954 by the Air Force's Ballistic Missiles Division at Convair, whose MX774 was in several respects its precursor.

At that time great technical doubts existed whether a rocket engine could be reliably started at high altitudes, that is, in vacuum. This militated against the development of two- (or multi-) stage missiles. In these the first stage is dropped off when its propellants are exhausted and the (much smaller) upper stage is ignited to complete the lift and acceleration of the re-entry nose cone to its intercontinental trajectory, or of the space payload to its orbit.

Therefore Atlas was a "stage-and-a-half" missile, all three engines being ignited on the ground. When the propellants were mostly used up in flight, two of these engines were dropped off, but the remaining one continued to burn with sufficient thrust to continue the acceleration of the residual weight of the missile.

To achieve the transcontinental range, various design compromises had to be made in the Atlas airframe for this stage-and-a-half concept. Therefore the von Neumann committee urged on the Air Force the development of a backup missile. A two-stage ICBM project, the Titan, was started at the Martin Company in 1955. Both missiles weighed about 200,000 lbs. and were powered by engines (built respectively by the North American and the Aerojet companies) using kerosene and liquid oxygen (LOX) as propellants. The use of LOX meant that on operational sites the missiles had to be stored empty and be fueled just before launch. The use of these "nonstorable" propellants in Titan was criticized by the von Neumann committee and later by PSAC, which urged that a modified Titan with storable propellant engines be built instead.

Also in 1955 a 1500-mile-range (IRBM) missile, the Thor, with only one stage weighing about 100,000 lbs., was started by the Air Force, not on the recommendation of the von Neumann committee. About the same time, the Navy began to be interested in the submarine-launched missiles, the first plan involving the use of an IRBM, the Jupiter, almost a twin brother of the Thor, which was proposed by the Army and started by ABMA in 1955.

The use in submarines of such a missile with its liquid propellants, especially LOX, created such problems that the Army-Navy partnership was rapidly dissolved. The Army however continued the development of the Jupiter as an ostensibly semi-mobile land-based missile system. The Navy, with the encouragement of the von Neumann committee, turned to solid-propellant-type rockets and the Polaris missile was developed by the Aerojet and Lockheed companies. A few squadrons of Thors were deployed in England and a few of Jupiters in Italy and Turkey by 1960 and were decommissioned as obsolete a few years later. In 1957, the Air Force began the development of the "second-generation" ICBMs, designed to be launched from hardened underground silos, the solid-fueled three-stage Minuteman.

By 1958 Sputnik I was in orbit and Thor, Jupiter, and Atlas had been flight-tested but none of the six long-range missiles was operational. The only rocket vehicle that was specifically designed to launch satellites was the Vanguard which, being a civilian project, then suffered greatly from underfinancing and the attendant technical troubles. However the Army, using its already operational short-range Redstone missile, equipped it with three upper rocket stages, ostensibly to test the re-entry nose cones of the Jupiter. It was called therefore Jupiter C. However, as designed under the leadership of Wernher von Braun at the Army Ballistic Missile Agency (ABMA) and by the Jet Propulsion Laboratory (JPL) of the California Institute of Technology, the rather Rube-Goldbergish (but successful!) contraption was fit to launch a small (31-lb.) satellite. This, called Explorer I, was indeed put in orbit on January 31, 1958.

The fascination with space for military and civilian purposes grew rapidly in 1958–1959 and led to a large number of space vehicle projects. Except for the smallest vehicles lacking sophisticated guidance—the liquid-fueled Vanguard and Jupiter C, as well as the very successful solid-fueled Scout succeeding them—all the other space vehicles made use of the military missiles, the Thor, Jupiter, Atlas, or Titan, as the first-stage or booster rockets. On top of these were mounted various upper-stage engines and appropriate guidance systems as well as the planned space mission payloads. While the Vanguard was designed to put a 20-lb. payload into a "low orbit," that is, about 100–150 miles altitude, and the Scout 150 lbs., the two-stage Thor Able could put into orbit about 300 lbs., Thor Delta and Jupiter Juno still more. As the more sophisticated upper-stage engines with greater thrust-to-weight ratio entered the development stage, the orbital payloads increased in about the order Thor Delta → Thor Agena B → Atlas Agena B → Atlas Centaur (and later the Titan Centaur, Titan C, etc.). The Atlas Centaur could place up to 8,500 lbs. of payload in low orbit (as was done in 1962) and hence could also be used for more sophisticated missions, such as placing much smaller satellites in a synchronous orbit at 21,000 miles altitude.

In addition to the vehicles listed above, other projects were started from time to time but did not progress much beyond the initial hardware stage or the test firings of the upper-stage engines. The origin of most of these projects is hard to trace, because it involved the interplay between the contract proposal-writing teams of the aerospace industry and the Air Force or NASA staffs. Each project meant not only the building of the hardware but also the construction of sometimes highly elaborate test facilities. As the projects advanced, the launch facilities usually had to be added at Cape Canaveral in Florida or at the Vandenberg AFB in California until veritable forests of launch towers began to adorn both establishments. The prime and the subcontractors involved in each vehicle and/or payload project had to have their highly trained teams of engineers in check-out buildings nearby. It was the beginning of feverish growth when the grant of authority to manage some space project was the difference between growth and decay of an in-house federal establishment, and the contract for the vehicle and/or payload meant industrial success in the aerospace and electronics industries.

In 1958 the Air Force, having a vested interest in manned flight, started the project Dynasoar, a hypersonic glider of strategic range, to be launched by a rocket booster. This was terminated in the nineteen-sixties (and has now re-emerged on a much grander, indeed monstrous scale as NASA's Space Shuttle). NASA started the one-man Mercury project using the Atlas as the booster, the first manned orbital flight taking place in February 1962 only ten months after the first Soviet cosmonaut's flight. Meanwhile the Air Force had also initiated the design and testing of a very large (1,000,000 to 1,500,000 lbs. thrust) engine (Nova) using kerosene and LOX as propellants, and preliminary activity was underway to start building a high-energy engine, also using LOX and liquid hydrogen, but much larger than the Centaur engine, for upper-stage use. Nova was to power the huge Saturn rocket for manned space missions. It was the project that appeared most promising to my office in 1959, for it would accommodate the large payloads that future missions demanded. Army's Ballistic Missile Agency, fighting to keep its fair share of space, in 1959 started a project they christened Saturn B, so as to make it sound like the initial stage of the Saturn-Nova system under development by the Air Force. To put it crudely, however, Saturn B was merely eight Jupiters bolted together, so that about the same booster thrust (over 1,000,000 lbs.) was available as from a single but then untested Nova engine. In the sixties, Saturn B was used by NASA for Project Gemini, the early part of the Apollo project.

Monday, September 28, 1959

12:30–3:00 PM Lunch with Admiral Burke and others and then the rehearsal of the Navy presentation, "Control of the Seas," prepared for the NSC. The presentation was exceedingly poor. Three-fourths of it could perfectly well have been written up as an article for the *Saturday Evening Post,* and the few more technical classified problems were unrelated to each other and not well presented. Certainly it left a very inadequate picture of the basic objective: the major tasks facing the Navy in the future and what is being done to make the Navy capable of carrying them out. After the presentation, which lasted about an hour, we all, that is Gordon Gray and I, as well as Admiral Burke, criticized the presentation freely, although politely. It was clear that Admiral Burke was embarrassed about its quality and it was agreed that the presentation to NSC will be postponed beyond the October 8 date.

Tuesday, September 29, 1959

7:30–9:00 AM Breakfast with Persons, Glennan, Goodpaster, Horner, and Staats, to discuss the fate of ABMA. Various possibilities were considered, the general conclusions being that ABMA should be moved into NASA, and that Saturn B, if it is not continued, be canceled before the transfer is completed. It was recognized that because of von Braun's attitude, the process may be a very painful one (he banged his fist on the table in Glennan's office and said he wouldn't let Glennan discharge even the last floor sweeper of his 5,000 men). It was generally conceded that the ABMA cannot be preserved in its present size, and Persons suggested that an early meeting with the President would be desirable. I kept hammering on my pet topic, namely that only one big booster should be under development. Clearly I can get myself into trouble on this. Only the future will show.

10:00–11:00 AM   A meeting with the President, who was in very vigorous form. Evidently he had talked to Persons about space, because before I even began, he launched on a long discourse which lasted half an hour, stating that he didn't want the NASA budget to go much over half a billion dollars a year; that we weren't in a race with the Soviets, but were engaged in a scholarly exploration of space, etc. He flatly said that ABMA should be put under NASA and on my warning conceded that he will have to defend Glennan publicly. He expressed highest confidence in Glennan and said he understood the pressures under which Glennan was operating. Then the President wouldn't let me complete any of my carefully prepared briefings, but as soon as I began would take the ball in his hands and carry it enthusiastically. He firmly endorsed our plan for a steering committee of panel chairmen on the military budget and expressed great interest in talking to us in November. When I suggested a study by PSAC of the area of life sciences, he expressed enthusiasm that we were getting interested in civilian technological activities of the government; he expressed concern about too much money in NIH, and upon my warning that such a study may arouse some irritations, instructed Andy Goodpaster to prepare a letter for his signature to all cabinet officers to tell them that such studies were being done on his personal instruction and requesting full cooperation. Also, the idea of a panel to look at conservation of natural resources to see if there are any weak spots in the federal program was fully endorsed. Finally, the President launched into strong appreciation of the role of the committee and gave it his full endorsement. He expressed grave concern that his successor may not appreciate its full importance and may abandon or degrade it. He suggested, therefore, that we study the problem of whether it would be better to have Congress pass a bill to establish the committee and the office of the special assistant for science and technology. It was a very pleasant meeting, to say the least, and the degree of confidence expressed in us was most heartening. After the meeting, I arranged with Glennan and Purcell and Hornig to have a session on big space vehicles this coming Thursday, because the President said he wants a high-level meeting to settle everything about ten days hence.

3:15–4:00 PM Meeting with two staffers from the Bureau of the Budget on NSF budget. I could see that they might shave a few million from the $240 million presented by Waterman, but said that I wouldn't agree to major cuts. Agreed to have a meeting with Staats tomorrow morning, at which I shall take a firm stand in defense of dear Waterman.

⌁ Wednesday, September 30, 1959

9:15–10:00 AM Meeting with Staats and others from BOB concerning NSF budget. With rather good help from Hugh Loveth I hammered at Staats that the $240 million budget for NSF was not excessive, conceding that a few million could be shaved off some items, but that the increase in basic research grants was not excessive, and that the allocation of $15 million to the renovation of graduate laboratories was particularly inadequate. I had to say at one time that if the budget were reduced much below that figure, Waterman couldn't accept it and felt he had to discuss it with the President, and that I would support him. Finally, Staats decided to instruct NSF to submit the detailed budget on the basis of $240 million.

Hugh Loveth, assistant director, Bureau of the Budget.

11:00–11:30 AM Glennan in my office. He was much upset because he thinks I am going to cut his budget to shreds. Argued with me that there wasn't any item there that could be cut out without serious damage to the over-all program. I assured him that I wanted to understand more about his budget before reaching any conclusions and committing myself before the President to a firm figure like $500 million.

11:30 AM Brief visit from a navy captain in the military division of AEC with a draft of a press release to be used in case there are leaks to the press about safety experiments at Los Alamos. Draft is not too bad, but in my opinion makes too much reference to possible nuclear reactions, which is not necessary since this release will be used before any experiments are actually carried out. I suggested to him that it was essentially a political matter whether such reference should remain or not and that I wanted to discuss it with Goodpaster, so he left me a copy. Shortly afterwards, I had a phone call from General Starbird, rather firmly insisting that it was his privilege to discuss this with Goodpaster. I said he was welcome to do so, but I certainly had the privilege of expressing my opinion to General Goodpaster, which he said he couldn't deny. I then phoned Goodpaster and explained the situation to him. He expressed the opinion that I was too gentle with Starbird, and should have insisted on my rights.

4:30–6:00 PM Sutton of ARPA and Jesse Mitchell. Sutton explained the situation with the Saturn B project, emphasizing that the eight-engine Saturn would be much more reliable and also would be ready sooner than any vehicle based on the Nova engine. He then proceeded to discuss the disastrous effect on ARPA morale of the order of the secretary to transfer space projects out of ARPA, and particularly of York's press interview, in which he supposedly stated that even long-range space research would all be given to the Air Force.

According to Sutton, this has angered Secretary Gates, who has denied that he approved such an arrangement. Thereupon, Sutton made several remarks suggesting that I was responsible for all of this, also for the planned transfer of ABMA and for the presumed cancellation of Saturn. Jesse made a very bad move by asking me to explain my position in the matter. I tried to get off as easily as I could by stating that I wasn't making any decisions. I was only gathering facts, trying to digest them, and form judgments of the consequences of various alternative decisions, to present them to the President, who makes decisions. I told Sutton, on his direct query, that I wasn't willing to tell him what went on between York and me nor what went on in our private meetings, but that I was not consulted about York's press conference or the statements he made in it. I also said that I was not the one who made the decision to transfer space projects out of ARPA. Sutton said he really didn't mind seeing the big space projects taken out, but he was very unhappy about space research being lost by ARPA. I can see that it was a sensible point of view. All of this put me in a very awkward position and since Sutton was making notes while I was speaking, it will undoubtedly be spread in the Pentagon. I am rather uncertain as to how much Jesse has been saying in the Pentagon since he mentioned only this morning that he asked ARPA to investigate the possible use of a single Nova engine in a booster vehicle and during Sutton's visit he (Sutton) told me that they have initiated such a study at ABMA. The way it was put, the indication was that it was being done at my request, and obviously if this is what is eventually settled on, we will be held responsible for the decision.

After the meeting Spurgeon Keeny dropped in to tell me that Farley of the State Department has finally come up with a recommendation regarding nuclear test cessation, which is essentially the statement of the paper Keeny and I prepared, rather than the various other schemes that were being written up in Farley's office. This proposal recommends going to Geneva with an offer to agree on a complete ban in exchange for several hundred inspections and then, depending on our internal decision and the reaction of the Soviets, either stand pat on this proposal (unquestionably unacceptable to the Soviets) and so lead to a suspension of negotiations, or go into tough bargaining, naming perhaps a hundred inspections for a real treaty. He said the Principals will have to meet shortly to discuss it.

> Jesse L. Mitchell, on the PSAC staff, loaned by NASA, responsible for space problems.

⌒ Thursday, October 1, 1959

9:00–10:30 AM   NSC meeting, minus the President and the vice-president. Herter presided until 10 and then even he left. Can't blame any of them. It was exceptionally dull and nothing of importance was discussed. Most of the time was taken up with fine and subtle shadings of words about mobilization base.

12 M–1:30 PM   Lunch at Shoreham with General Schriever, Douglas, and others. Inconsequential conversation, except for an exchange of apprehensions with Schriever regarding the managerial capacity of Martin Company to get the Titan ICBM missile operational on time. Secretary Douglas told me that the Air Force decided to reduce the total number of Atlas squadrons and not to go with hardening the deployment beyond 25 PSI, throwing the resources instead into more Titan squadrons, converting

Titan to storable propellants (at last! It has taken them two years to do what should have been done then) and to "in-silo" launching.

> James H. Douglas, Jr., secretary of the air force; Gen. Bernard A. Schriever, head of the Air Research and Development Command (ARDC).
>
> The reference to 25 PSI (pounds per square inch) meant that even hardened Atlas operational sites could not resist shock waves from nuclear explosions greater than 25 PSI in intensity. On the other hand, if the Titan missiles were placed in silos they could be made immune to far stronger shocks.

1:30–2:00 PM My speech before the Sixth Annual ARDC Symposium. I raced a little and finished it in 25 minutes. The only good laugh I got out of the audience was the result of spontaneous ad-libbing when Jerry Wiesner introduced me and said that as a scientist I had many friends, but undoubtedly had none now. I said that on the contrary, I had more than before, only I would probably lose all of them by the time the '61 budget was settled.

3:00–5:30 PM Meeting in Glennan's office with Horner, Silverstein, Purcell, and Hornig to discuss the large booster program. It seems that we have now reached a sensible agreement, which is as follows: 1. We all agree that a booster must be developed. 2. We all agree that a very large engine, e.g., one and a half million pounds, must be developed. 3. We all agree that *if we are to maintain vigorous competition* with the Soviets, we have no choice but to push the present Saturn vehicle and at the same time develop the Nova engine and vehicles based on it, thereby probably spending one-half to three-quarters billion dollars more per annum than in a more conservative program. 4. If we just choose to go our own way in a not especially competitive but a vigorous and far-sighted program, the Saturn should be canceled or rather, for political considerations, the name retained, but the project completely modified so that the vehicle would be suitable eventually for use with the Nova engine, and so that its first flight tests would be with four to six Jupiter engines. It was also tentatively agreed that if alternative 3 is selected by the President, there really is no strong argument for transferring ABMA to NASA. I actually think it ought to go to the Air Force. On the other hand, alternative 4 clearly requires transfer of ABMA to NASA. Thus, I now have two clear-cut alternatives involving both budgetary and political considerations, and the President should find it easy to make a decision.

Friday, October 2, 1959

11:00–11:45 AM   Mr. Harold L. Goodwin of the USIA staff. General exchange of views, from which I was pleased to learn that he is also opposed to picking a fight with the Soviets over their chosen ground of large space payloads. Then discussion of my October 21 speech, for which he agreed to write a draft.

2:00–2:30 PM   Undersecretary of Commerce [Philip A.] Ray, largely a courtesy call to get acquainted, since he will be the Commerce representative on the Federal Council. He made a good impression as a thoughtful person who was generally sympathetic to technical work, although frankly admitting that he knows little about it.

Monday, October 5, 1959

11:00–11:30 AM   Holaday here to give me his report in his role as chairman of the Civilian–Military Liaison Committee to NASA. The main conclusions are that there are too many vehicles being developed and that Vega should probably be canceled. Also a recommendation to put more emphasis on Saturn over Nova. When I questioned him on that, Bill conceded he meant something like Saturn rather than the particular design now being developed, which he doesn't think is good. He then offered the information that in his role as adviser on missiles to the secretary of defense, he will recommend that the maximum effort should be put into Atlas; that the Titans be canceled after the fifth squadron; that all Minutemen be mobile and that maximum effort be put into Polaris, of which only two subs are planned for fiscal year 1961 by the Navy. The first two recommendations, of course, are exactly opposite to what Secretary Douglas told me the Air Force would come up with, so there will be a lovely little scrap, maybe. Finally, he stated that in his opinion it is impossible to have a sensible space vehicle program so long as both NASA and DOD are authorized to build them; that we will always have everything in duplicate with attendant waste of talent and funds.

> William Holaday, a civilian in the Office of the Secretary of Defense.

12:30–2:00 PM   McCone to lunch at the Mess. Brother John is all sweetness and light. Clearly we have become bosom friends. I wonder when next the knife will be stuck. He showed me a summary of his conversation in Vienna with [K. V.] Novikoff, a big shot from the Soviet Foreign Affairs Ministry, in which Novikoff was trying to find out from McCone whether we were actually willing to agree to nuclear test cessation and challenged McCone's statement that we were, providing the controls were adequate. An interesting sidelight was the remark of Novikoff that the Soviets are not interested in small tactical nuclear weapons; that Khrushchev personally ordered the work on these weapons stopped; and that he only believes in big weapons. This is precisely what the President told me that Khrushchev told him, so the degree of coordination is excellent, or they are actually telling the truth. McCone then told me what he is going to tell Tuesday at the Principals' meeting; namely that we should suggest to the Soviets on October 27 a technical conference of experts on underground detection to determine whether an effective control system is feasible or not. That if this is not done, or the Soviets are unwilling to join and abide by recommendations, we should on January 1 declare for ourselves freedom of action regarding underground tests but not actually resume them, to avoid, as he put it, "provocative" actions while the President is visiting Russia, etc. That we also should unilaterally declare our intention not to resume atmospheric tests. A surprisingly sensible proposal which really represents his complete about-face from last summer. What is behind it, I don't know. He tried to get me to agree to this plan, but I was reasonably coy, being afraid to fall into a trap. We returned to my office together, where we talked about the significance of intelligence satellites as a unilateral aid to inspection controls. After that we got to talking about the space program, and McCone expressed agreement with what I, perhaps incautiously, stated; namely, that I was dubious about our competition with the Soviets for big payloads in space.

2:30–3:00 PM   [Alexander H.] Flax, Air Force chief scientist on courtesy call. He is very much impressed by the way in which the Air Force fails to support applied research on components, throwing all of its money and attention to big weapons systems. I urged him to focus on this subject, which is quite serious.

3:00–3:30 PM   General Starbird, again with a draft of a press release in the eventuality that somebody questions the one-point safety experiments to be undertaken at Los Alamos. Andy Goodpaster called and said that in his opinion this draft would also not meet the President's approval, and I said the same thing, but Starbird said that softening it further would be tending to be dishonest and this may create bad problems at Los Alamos, where staff members may resign in protest, etc. Reached no agreement.

3:30–4:00 PM   Recorded my Voice of America broadcast in Russian. Had a hard time with pronunciation and had to repeat everything several times.

4:30–6:00 PM   Beckler, Keeny, Rathjens, and later Fisk, reviewing events of the day, specifically McCone's statements, and then Rathjens summarizing his impressions of the Strategic Systems Symposium at the Pentagon, which he attended all last week upon the invitation of York's office. The symposium was apparently a flop, because it was all held on a very qualitative level without any technical discussions to speak of. George said that everybody there simply wanted much bigger budgets and so most of the discussion was as to what else to do, rather than how to make the best use of what is available. The general conclusion of some 40 people present was that in the next two or three years a well-executed surprise attack by the Soviets may leave us without an effective retaliatory force, but he thinks this conclusion is colored by the desire for bigger budgets. He found the presentation and remarks by Barlow especially poorly thought through and useless, which is something to remember, since Barlow has been recommended to me as the real genius at Rand.

> George W. Rathjens, on the PSAC staff, was concerned with strategic weapons and some other aspects of our involvement in military affairs. E. J. Barlow, electronics engineer, was on the staff of Rand Corporation.

Tuesday, October 6, 1959

Fisk for breakfast at the Mess, at which we went over the nominees for PSAC. His first choices are pretty much the same as mine. We then talked about the selections of last year and he thinks that the whole thing was horribly handled and we are now paying for it. He considered that my plan of talking with prospective nominees privately and telling each of them what membership entails in the way of work is the way to do it so there will be no misunderstandings afterwards.

Morning spent at NASA listening to presentations of their program, except for a visit to my office in mid-morning, while the rest were having a coffee break. During this visit, Dave Beckler told me that he had a phone call from John Finney (of the *New York Times*), who said that he had learned on good authority that I am conducting a special study of Saturn with the intention of canceling the project, and that this was a shocking and almost treasonable action. Quite clearly I am now reaping the fruits of Jesse Mitchell's simplicity in going to ARPA and talking about our interest in Saturn and inquiring about possible reduction of the program, to all of which Sutton referred on his visit last week. This must have led to the conclusion that we are making a special study. This was undoubtedly passed on by ARPA to ABMA and either there or directly leaked to Finney. Beckler reassured Finney that no such study was being made to the best of his knowledge, which, however, apparently didn't reassure Finney, who was very emotional about the matter.

2:30–5:00 PM Meeting of the Principals. Dillon presented the State plan for tactics, which is essentially the plan I recommended to Farley: namely, to go to Geneva with an offer of a comprehensive treaty in return for many on-site inspections, etc., and then decide on a course of action, i.e., negotiate for such a treaty or not, depending on Soviet response. He was very deliberately fuzzing the situation, not wishing to raise the issue as to whether we are still willing to have a comprehensive treaty or not. Thereupon Gates, and after him Irwin, stated vigorously the DOD policy which is: no comprehensive treaty; instructions to Wadsworth to disengage us from this commitment; and resumption of nuclear tests as of January 1, except for atmospheric tests, which we would renounce unilaterally. McCone made his proposal which, amazingly, was almost what he outlined to me yesterday, except that now there was more emphasis on the certainty that the result of a technical conference of experts would be a finding that monitoring is impossible and therefore comprehensive agreement is impossible. Irwin and Loper renewed their insistence on the Defense proposal and went too far describing the situation, as if in the summer of 1958 we believed that inspection was 100 percent effective and now it was 0 percent. I thereupon pointed out that this wasn't so and tried to assist Dillon, noting that a decision on the required degree of controls depended on many factors, etc. Dillon all the time was trying to fuzz the situation over so that it would appear that there was an agreement with Defense, although clearly there wasn't. Everybody referred to the decision of the President on July 23, that we shouldn't negotiate a comprehensive treaty any longer, but Dillon was very vague thereafter. He was clearly trying to get freedom of maneuver. Gray stated flatly that although everybody was speaking of agreement, in reality there was a sharp disagreement, but even that challenge was never taken up. I got the impression that absolutely nothing was settled and personally I am not going to be very vocal in the whole matter, except to correct misstatements if they are made, because the outcome is pretty clear regardless of what I say.

5:00–5:30 PM Waterman in my office regarding an amendment that the Far Eastern Desk in the State Department wants to put into the treaty on Antarctica, which has now been agreed to by everybody including the Soviets. This amendment would exclude from Antarctica those nations which are not recognized diplomatically [Chicoms, i.e., Communist China] and will clearly make the treaty unacceptable to the Soviets. State is supposed to be divided on the matter and it will be raised at the NSC. Agreed with Waterman that I will ask Gray to invite him to the meeting; that he will object to this amendment, and that I will support him if necessary.

A brief meeting with Fisk and Purcell, who sat throughout the day listening to NASA presentations. Their opinion is that the man-in-space project is good; that there are a number of other good projects, but that the over-all program is very diffuse and poorly thought through. In places it is much too ambitious.

Wednesday, October 7, 1959

9 :00 AM–3 :00 PM   Spent at NASA briefings. The vehicle program looked sensible enough, although it was obvious that nobody wanted to use the relatively cheap Scout, and so we will probably have this vehicle developed, but not used. This is really a shame since so much valuable space research could be done, even under the restrictive conditions of its fairly low payload and inaccurate guidance. The briefings on other subjects, such as auxiliary power, nuclear and otherwise, etc., leave me with the conviction that NASA is trying to cover the waterfront and is terribly eager to get into hardware, where hardware is not called for at present. The whole effect is that when they don't have good ideas, they build expensive equipment. I am sure that my dear friend Keith will be unhappy when I tell him that, but I will.

During the NASA briefings talked with Bill Holaday. He said that he was horrified by the scatter of our efforts in space. Mentioned that actually 22 different vehicles are under development, although the number of official projects is much smaller. He said he was getting up the courage to write a blistering letter to the secretary of defense and to the administrator of NASA about it. I egged him on as much as possible.

Thursday, October 8, 1959

9 :00–9 :45 AM   York's office, for some preliminary discussion of the Army R&D budget, York delivering most of the speeches. He re-emphasized that the Service budgets show a reduction of basic research for fiscal 1961 over that for fiscal 1960, and that his office will not approve of this. Otherwise mostly emphasized a variety of items on which he wished to get information when the meeting reconvenes. The original plan was to have the meeting last all morning, but the 10 o'clock meeting with General Persons made it necessary to postpone it.

10 :00–11 :00 AM   Meeting of Glennan, Horner, Dryden, Gates, York, Admiral Clark, with Persons, Goodpaster, Merriam, and myself. We discussed the fate of ABMA and of the big booster program. Gates emphasized that the Services, especially

the Army but also the Air Force, were getting alarmed over the idea of losing ABMA to NASA. The attitude of the Navy, he wasn't sure of. The Army and Air Force secretaries seemed to prefer the transfer of ABMA to the Air Force. Thus he felt there will be a terrible internal conflict and he will have to go to the President with a split opinion from DOD and JCS. Glennan then developed administrative reasons in favor of the transfer of ABMA to NASA, pointing out that the law establishing NASA specifically required it to engage in the business of developing space vehicles. There was some inconclusive discussion about the existence or nonexistence of military requirements for Saturn, but it was conceded that in any case the Services could produce those at the drop of a hat. I presented my arguments for the urgent necessity to have all superbooster vehicles developed under one management, as otherwise we will repeat the mistake of having duplication of effort as in Thor and Jupiter, Atlas and Titan, Atlas Agena B and Atlas Vega. I emphasized that no amount of coordination will prevent our having two systems under development if two agencies are involved, and therefore recommended the transfer of ABMA to NASA. York pointed out that single management would also be established for all super-boosters if ABMA was transferred to the Air Force, but Glennan noted that in this case NASA would be without any program of its own in this all-important field, and therefore clearly wouldn't fulfill its responsibilities under the law. There was then a long discussion of tactics to be adopted. Both Gates and York con-ceded that personally they favored the transfer of ABMA to NASA. Gates wanted to staff the memorandum of agreement in the Defense Department so it would not appear as a NASA docu-ment, but the meeting noted that in this case 5,000 copies of it would be made and all the press in the country would know of the plan before it was consummated. Gates then agreed to hold it in his own office. He stated that the memorandum being dis-cussed was in principle completely agreeable to him and only language changes were necessary. Persons said that he antici-pated bad political trouble, but that would occur no matter what was done, and he thought we could weather it. He proposed to talk tonight to the President in a preliminary fashion about the proposal and let Gates know of the outcome in the morning. If the President favored the transfer to NASA, Gates could then ask the JCS to prepare an opinion, so as not to appear to be doing it behind their backs.

Vice Adm. John E. Clark in the Office of the Secretary of Defense; Robert E. Merriam, deputy assistant to the president for interdepartmental affairs.

2:00–5:00 PM Navy budget briefing in the Pentagon. This was the first of the briefings on the Service R&D budgets scheduled by York, and we were there as observers. After a silly speech by Admiral Bennett blaming York for the fact that the Navy allocates only 10 percent of its budget to R&D, even though T&E has been now included in it to the tune of 20 percent of the total RDTE, we got down to business. It is clear that in fiscal 1961 the Navy R&D budget is badly squeezed. It is substantially less than in 1960. York and his staff handled the meeting exceedingly well. Clearly they knew what they were talking about and were not befuddled by the complexities of the budget, involving the new appropriation authority, spending level, funds transferred from previous years, etc., etc. Also in regard to individual items, they were well on top of the situation. York instructed the Navy to increase basic research to the 1960 level, and to increase oceanography substantially over 1960. He then indicated approval of increases of certain items over the so-called A budget, which is the basic budget presented by the Navy. At the end of the meeting he made inquiries about certain long-range things promising revolutionary improvements, such as hydrofoil for ships, boundary layer control, etc. My conclusion after this session is that it will be very difficult for the PSAC to really get its teeth into the budget. That clearly requires much patience and time.

Rawson R. Bennett, rear admiral, commander of the Office of Naval Research.

Friday, October 9, 1959

9:00–11:00 AM Navy antisubmarine warfare (ASW) briefing to Harvey Brooks's PSAC panel, which gave me a clearer picture of Navy's budget and plans. Naturally, the Navy spent a substantial length of time telling how pathetically underprivileged they are budgetwise and some of it is true, but some of their arguments lack force, as for instance, the insistence that the number of ships in the navy must remain unchanged. This would require about 40 new ships a year, whereas they are given money for only about 20. Clearly the ships we are building now are so superior to those of World War II days that the maintenance of a navy with a fixed number of ships would represent not a maintenance, but a rapid increase of naval power. Chick Hayward is impressive. He is really on top of everything that the Navy is doing and is a pleasant person to deal with.

Harvey Brooks, professor of applied physics, Harvard University.

1:30–4:40 PM   Air Force budget briefing to York, which started with a vigorous statement by a major general whose name I don't remember, explaining why the Air Force cannot meet the instructions of the secretary of defense to submit A and B budgets, but must insist on submitting a still bigger budget (about $1 billion more) which they call Air Force "minimum essentials" budget. This budget is an incredible can of worms, and clearly a number of proposed reductions from previous years' plans are arranged to have the maximum political impact so that even if the President approves the "essentials" budget, Congress will override him. Another interesting feature is that the RDTE budget remains essentially the same, whether they talk about the A or the B or the Air Force essentials budget, except for one item, namely aircraft nuclear propulsion, which is drastically reduced in the A budget. I have gained very little feeling for what the RDTE budget is, because of the Air Force trick of spending procurement money on R&D projects such as Atlas or Titan and some other weapons systems. An extreme case is the B-70 bomber, which is all on procurement money even though the first operational B-70 will appear sometime after 1965. Also, the coding of many projects leaves one completely in the dark as to what their purpose is.

Saturday, October 10, 1959

9:30 AM   An assistant to a vice-president (Missiles Division) of the Chrysler Corporation, whom I knew in earlier days, spent about an hour and a half. He reminded me that he was the chief of the Rocket Branch of the Ordnance Department when the

Germans, including von Braun, were brought to this country, and that he was largely instrumental in setting up Army work on rockets. He then said he was naturally very concerned about ABMA; that it represented a national asset; that the aircraft industry disliked the German group and would like to see it dispersed; he has convinced Chrysler that this shouldn't be done. He then described a proposition he will present which calls for operating ABMA as a subsidiary of Chrysler, but of course at a cost in government contracts of $100–250 million annually. I assured him that nobody in a position of authority has any idea of dispersing ABMA, that $250 million a year is perhaps a little more than can be afforded, but that it would certainly be kept together. I also said that while I had something to do with technical decisions, a purely administrative matter such as transfer of ABMA to another authority was somewhat beyond my sphere of direct interest. I assured him though that I was sympathetic to the preservation of ABMA. Talked in general about the problem of too many agencies trying to do too much in the space and missile fields. I am not sure what his impression of our conversation was, as I naturally had to be somewhat cautious and some of my remarks might have sounded evasive.

Had York for lunch and until about 3 PM. He read over my "To Race or Not to Race" paper on the space program and to my surprise took no violent exception and, in fact, even improved the English in places and added some facts and figures. The rest of the time we spent in a rather desultory conversation about military budgets.

Monday, October 12, 1959

Glennan came and spent about an hour talking about the scope and purpose of the NASA program. I feel there is not a complete meeting of our minds, since he feels that what NASA is doing now is about the minimum and is pretty high grade; whereas, my feeling is that the effort is diffused, that too much money is being poured into premature hardware projects for advanced space missions that are many years off, and perhaps not enough into the effort to get maximum information with available means in the near future. Glennan stated that if the NASA budget was cut much below the presently proposed figure of $0.78 billion, he would be quite uninterested in taking ABMA, as the presence of ABMA in NASA would require reorganization of the whole NASA and couldn't be effective unless there was an adequate budget. The meeting was very friendly throughout, however.

2:30 PM NEG meeting, Twining presiding. Admiral Sides presented the net evaluation of long-range missiles programs. Conclusions are not in any way different from NIE's, i.e., the Soviets are ahead of us in propulsion, but that is all; that as far as IOC is concerned, we are almost even with them, although their missiles are probably more reliable in terms of launching, but not as accurate. The future situation was mentioned only in terms of NIEs, i.e., the Soviet capability rather than the actual missile forces they will have. The meeting then agreed with my suggestion that the space programs be proposed to the President as an appropriate subject for comparative evaluation. Twining also approved the plan to make a copy of the (PSAC) computer panel report available to CIA and NSA. I must confess that Allen Dulles made an exceptionally poor impression this time. His questions were those of a man who is not totally in command of technical facts and also of a man whose mind is far away.

By the time the meeting ended, about 4:30, the ARPA budget briefing, which started at 3:00 PM, was also over. Those who attended, especially Ed Purcell, said it was exceedingly perfunctory and pointless. Purcell and I dropped into York's office, and the latter explained that he deliberately didn't go into the technical aspects, but simply approved the A budget because the problems were so complicated that he needed time to study them, but couldn't do it now. He felt confident that later on he will be able to shift money within the total budget.

> NEG: Net (or Comparative) Evaluation Group set up under the Joint Chiefs of Staff on the instructions of the President.

> Vice Adm. John H. ("Savvy") Sides, director of the Weapons Systems Evaluation Group (WSEG) and of NEG.

6:30–11:00 PM Reception and dinner of the White House Correspondents Association at the Sheraton Park Hotel. President Eisenhower and Mexican President [Adolpho Lopez] Mateos were present—1100 males in tuxedos and the only women in sight the entertainers. The president of the association, who acted as spokesman, did some gentle kidding of Eisenhower about birthday presents, etc., but absolutely nothing rough. The entertainment after dinner was excellent, with by far the best juggler I have ever seen, a magnificent imitation of Mark Twain in old age by the fellow who makes a specialty of it, and a black woman singer, weighing 300 lbs. and all of it in voice. The other musical numbers were mediocre and the amplifiers were turned up so high that at times the sound was absolutely deafening. I sat between a Mr. [N. S.] Finney of the *Buffalo Evening News*, my host, and Mr. [R. P.] Brandt of the *St. Louis Post–Dispatch*. The latter said very little. Finney talked a great deal, and I gathered that he is very much on Oppenheimer's side and still feels very bitter about that affair. Next to Finney sat Marquis Childs, with whom I talked only briefly, but perhaps his remarks were most significant. He said he had lunch with John McCone just that day and McCone explained to him why no control over nuclear tests or over production of fissionable materials could ever be effective. Childs then said that McCone urged him to get acquainted with me, as I had done much thinking about these problems and Childs would be interested to know of my ideas. I played it cautiously and suggested to Childs that in evaluating the significance of controls, one had to start from a broad basis, namely the need to improve our national security and that dangers of evasion under controls had to be compared with dangers resulting from an uncontrolled arms race.

Marquis Childs, syndicated journalist.

Tuesday, October 13, 1959

2:00–5:15 PM   Army R&D budget briefing to York, which was especially uninstructive because the conversation was disorganized and used too many code names. Furthermore, Gen. [Robert J.] Wood, who represented the Army, was clearly not in command of the figures and much time was spent in just trying to reach a meeting of the minds with York as to what they were talking about. As in other Service briefings, it became clear that the A budget represents a dramatic reduction from last year's budget and will bring serious slowdowns to most areas. York frequently instructed that basic research must be maintained at the FY '60 levels and that the Materials Sciences Program has to be adequately financed.

He proposed to add so much money for CBW that Dick Morse

objected, noting that the budget can be expanded too rapidly and then would not be effective. The meeting became somewhat more lively when we came to the Pershing and Nike Zeus. The A budget for the Pershing would stretch out the R&D by three years, delaying operational missiles from late 1962 to 1965, whereas the B budget would preserve the original schedule, although it would take out of the program some fancier design features such as extra-long range, obtainable with titanium motors. In the course of this discussion, General Wood gave figures which made it clear that operational missiles will cost about $2.5 million each, including ground support equipment (GSE) but not the warhead. York asked for my comments at the end, and I questioned General Wood's assertion that the Pershing had great advantages over tactical aircraft and that it could be used in all weather and could be used in places like Laos. I suggested that in bad weather one doesn't know where the targets several hundred miles distant are located, particularly if one doesn't have tactical aircraft, and if one has the aircraft, one might as well just drop nuclear bombs. Also suggested that considering the cost of the Pershing, it might be more economical to fire Minuteman missiles from the ZI to Laos, rather than transporting the Pershings to Laos first. This created quite a confusion in the Army ranks, but clearly stimulated York to some thinking. Regarding modern weapons, we seem to be pricing ourselves out of the means to conduct warfare. We then discussed[7] briefly the Nike Zeus. Very clearly the difference between the A and B budgets involves to a large extent the items which are not urgent unless Nike Zeus is scheduled for production, i.e., unless a further $1.2 billion is provided. That substantial sum would be required because of a lovely inter-Service argument. The test installation for Nike Zeus needs targets, but the Air Force is unwilling to have the ICBMs it launches from the Pacific Missile Range used as Army targets. So the Army proposes to buy a lot of Atlases and launch them for this purpose, as well as to build a special launch pad on Johnson Island to fire the shorter-range army Jupiters from there. The discussions ended with York deciding to approve the A budget and nothing else at present.

Pershing, a ground-to-ground mobile missile of up to 700-mile range, being developed by the Army.

⌁ Wednesday, October 14, 1959

A staffer from BOB came over to show a draft of a letter from Staats to McElroy, in which BOB notes that total activities in the electronics and communications areas add to several billion dollars and suggests a special meeting to clear up that part of the budget. My comments were only that the list of projects attached cut across virtually all activities of the DOD and that I doubted the wisdom of calling them all electronics. In much the same way one could insist that since metals are used in virtually everything the DOD develops and stocks, there should be a budget review of metals. He saw the point and agreed that the list will be modified to include only communications and the information-receiving and processing operations.

12:45–2:00 PM Lunch with Elmer Staats at the Cosmos Club, talking about the budget. Discussion concentrated on NASA. Staats told me that Glennan had informed him that he had changed his mind and is now unwilling to accept ABMA into NASA unless the President approves essentially his full $0.78 billion budget for 1961. It seems we are back where we were two weeks ago, because Staats flatly said he is unwilling to make such a commitment and doubts very much whether such a budget will be acceptable to the President. The process of government is very similar to a ring-around-the-rosy. I tried to indicate to Staats what parts of the NASA budget could be reduced, specifically emphasizing hardware for long-range projects, but conceded that these could never add up to a quarter-billion dollars, and that if the President cut the budget that far, either the man-in-space project or the purchase and launching of vehicles for scientific purposes would have to be decimated. I noted also that that would mean an open decision not to compete with the Soviets; that the President would have to state so, and that this would have a frightful political effect. Elmer looked quite unhappy, but unmoved. We then talked briefly about military budgets.

2:30–3:15 PM Two faculty members of the University of Missouri. They made the appointments by phone a few days ago and said that they wanted to discuss the problem of improving the lot of "average" college scientists who are neglected by the government under the present practices, don't engage in research, and thus lose their competence. The meeting started with their restating this problem, but then, under the pretext of having discovered a solution, they outlined their university's proposal that NSF give them three-quarters of a million dollars for the purchase of expensive equipment to be used in the life sciences area, such as electron microscopes, centrifuges, etc., etc. What they wanted was my backing because NSF was unsympathetic. It was clearly a trick to get me interested in the general problem and then get my support for a local project and so I dealt with them a little roughly, pointing out that just having instruments is not enough to start research, that one needs the desire to do research, intellectual atmosphere, adequate time free of teaching, etc., I explained why their project was not likely to be approved by the NSF and suggested that they break it up into component parts and justify each major piece of equipment by a worthy research project, instead of just asking for all the equipment and assuring NSF vaguely that they will do good research with it. My remarks apparently didn't make them very enthusiastic, and so at the end I suggested that they meet with our new panel under Glenn Seaborg and discuss the problem in more detail.

⌒⌒⌒ *Thursday, October 15, 1959*

9:00–10:30 AM Special NSC meeting to hear the Net (or Comparative) Evaluation Group's report by Admiral Sides, which was a repetition of the rehearsal the other day in the JCS conference room at the Pentagon. A lively discussion ensued after the report. Vice-president Nixon raised the question as to whether the smaller size of our ICBMs was due to economy; whereupon, I got up and explained that the question of economy played no role in the recommendations of the von Neumann committee; that the size was an attempt to match the predicted warhead performance with considerations of development time for the missiles, which would be longer, the bigger the missile. The President suggested that the three Services write histories of their activities in the long-range missile field which might be useful in the spring, when congressional attacks begin. I ventured to suggest that if those three reports were put together, the mixture would be combustible and probably go up in smoke, which amused the President no end. I then suggested that a single writer be selected not tied to any of the Services. The

President agreed and instructed Gates to do so and have the report ready in three months. I was very pleased that at this meeting the President referred to me as ''George'' for the first time, rather than ''Doctor.'' Clearly I have made the grade in his opinion, since I noticed that he reserves first names only for people he feels are ''members of the team.'' Then Gray reported that the group recommended evaluation of space activities as the next topic and asked me to outline the proposal, which I did. The President liked it, but suggested that NASA be involved and that the activity be separated from the NEG and the report be presented to a joint meeting of the NSC and the Space Council. It was so agreed. I subsequently learned that I am to be made chairman of an interagency committee to supervise the actual work.

11:00 AM–12:15 PM   Air Force and Navy briefings to Perkins McGuire, Herb York, and myself, which were a most violent attack on the Army proposal to have the Army manage all military global communications under its Signal Corps. The presentations were quite partisan, to put it mildly. The Navy revealed its own counterproposal, putting management under the JCS, which would simply perpetuate the present lack of coordination and augment it by inserting additional committees between the secretary of defense and the operating agencies, so certainly nobody would be able to find out what is going on. Rather awful!

12:15–2:00 PM   Wallace Brode for lunch, which went rather well. Brode started by saying he was wondering whether he should terminate his association with the State Department, saying that in any case he would not return to the Bureau of Standards, since he felt his position there would be somewhat awkward. I said I couldn't tell him exactly what to do, but I firmly believed that positions like his and mine should be temporary and rotating, so that fresh points of view of the scientific community could be presented and the holder of the job could not be identified with the Washington bureaucracy. Brode then asserted that he made strenuous efforts not to be associated with bureaucracy. I think my remarks might have had some effect. We then talked about the Federal Council Committee on Foreign Affairs, and I urged strongly that Brode, instead of kicking like a mule, take hold of the job, since this was the only way he could get DOD in line; that State was too weak to manage it alone, and that only by appealing to the White House, the mechanism for which was available in the Federal Council, could he accomplish it. I may have had some effect, although he is generally unhappy and bitter.

4:00–5:00 PM   Teller. Reasonably friendly conversation, in which both of us stated our views rather explicitly. Edward is opposed to any kind of arms limitations because inevitably they will weaken us. First the Communist system must renounce its intention to dominate the world. Hence his unwillingness to accept total nuclear test cessation, although he is willing to accept a ban on atmospheric tests. Asked me whether I would support the funding of AEC laboratories for further weapons research if tests were not allowed. Assured him I would do so. Asked me what types of weapons developments were possible, and I suggested nonlethal C and B agents and research on conventional ordnance, especially the advanced things. He then complained about his difficulties in getting the AEC to declassify certain projects and tried to enlist my aid. The proposal sounded very reasonable, but I pointed out that this would be definitely out of channels and that he would have to operate through the AEC. I stated that while I would not favor complete nuclear test cessation as an isolated political act, I believed in it as a part of a more general effort toward relaxing the cold war tensions. Edward is convinced that tremendous further advances can be made in nuclear weapons, and I told him I was not.

6:30–7:30 PM   Secretary Herter's reception for the delegations of twelve nations which are negotiating a treaty of neutralization of Antarctica and its use for peaceful purposes only. As luck would have it, the man standing immediately behind me in the line moving toward Herter was V. V. Kuznetsov, the head of the USSR delegation here and in Geneva last fall, with whom I then had bitter personal argument because he was once a student of my uncle, W. Kistiakowsky, and very much resented my being on the U.S.A. side. This time he greeted me in a friendly fashion and there followed a rather lengthy conversation in which he only once chided me for my poor Russian and on several occasions translated English words into Russian when I got stuck. There was not much substance to the conversation; he asked me what I was doing and whether I was anxious to return to Geneva for another surprise attack conference, which I said I was not. He expressed considerable optimism about the Antarctica treaty agreement, but on my inquiry about his UN proposal for total disarmament, he said that although he was slightly more optimistic than a year ago, he didn't think that significant progress was being achieved. After we shook hands with Herter, he moved away and obviously considered our conversation finished. Later in the reception I talked to a couple of members of the Russian delegation with technical backgrounds, who seemed to be quite cordial. I might add that before the NSC meeting, I spoke to Herter briefly about the amendment to the Antarctic treaty which the Far East desk of the State Department is pushing, and which Waterman and myself and others are opposed to. Herter assured me that the issue was rather dead; our delegation will not be instructed to push it, but suggested that I nonetheless speak to the senior U.S. delegate during the reception, which unfortunately I was not able to do.

V. V. Kuznetsov, deputy minister of foreign affairs of the Soviet Union.

∽ Friday, October 16, 1959

10:30–11:00 AM   Dr. Nolan to discuss Carmichael's proposal that the National Academy of Sciences, not only Bronk, but the actual membership, be drawn into the discussion of research priorities. I expressed enthusiasm since I have several times before talked to Bronk about getting Academy membership thus involved and got no response from him. We agreed that Nolan and Carmichael will talk to Bronk without me so it wouldn't appear that I was the instigator. We also discussed the question of policy on research facilities and I urged on Nolan that the policy statements about ''in-house'' facilities be kept separate from those about nonprofit institutions, as an attempt to set a common order of priorities would inevitably fail because of differences in the policies of different government departments. He conceded that the standing committee was trying to get up one paper and it already has gotten into trouble. He agreed that separate statements would be politically more feasible.

12:00 M–12:45 PM   Stan Ulam from Los Alamos with a plea that I push the Rover project hard, as a demonstration of nuclear-powered rocket flight and a spectacular that would impress the world. In his opinion, the project has been exceptionally successful and the demonstration wouldn't cost much. I thereupon launched into a lengthy explanation of budget problems and he gave up, but didn't seem to mind, being a good friend. He then suggested that Project Orion of General Atomics, the proposed atom-bomb-powered space flight, be declassified and that the President invite the USSR to do the project jointly. I said that it was an interesting idea and I will follow it up. Stan told me how the two-stage, that is, thermonuclear, weapons came into being. He said that he went with a concept to Teller. In the conversation that ensued the idea was modified and a joint paper followed. It is amusing that yesterday in the course of our conversation Teller flatly said that the idea is his. So even after all these years there is no meeting of the minds on the subject of inventorship.

> S. M. Ulam, applied mathematics, on the staff of the Los Alamos Laboratory.

4:00–4:30 PM   Miss Thomas, a very forceful woman who wouldn't take my no for an answer and insisted on taking a tape recording of my life history for a biography to be included in the book she is writing on leaders of the space age. We got about to age 16 when Caryl Haskins arrived and I shooed Miss Thomas out, which made her exceedingly annoyed, so I had to promise I would see her on her next visit to Washington and complete my life story. At the rate we are going, it will take a couple of days.

> Shirley Thomas, writer; Caryl Haskins, geneticist, president of the Carnegie Institution, Washington, D.C.

⌒ Monday, October 19, 1959

PSAC meeting from 9:30 AM on. The Army, Navy, and Air Force briefings on their R&D budgets, of which that of the Army was the least effectively presented and that of the Air Force by far the best, from the point of view of presentation, although most debatable as to content.

7:30–10:00 PM   PSAC evening session in which we discussed mainly the space policy; the PSAC reaction being that we shouldn't withdraw from competition with the Soviets, but should be very careful in selecting the areas of competition, and that we should emphasize space sciences.

∽ *Tuesday, October 20, 1959*

7:30–8:30 AM Breakfast with Persons, Goodpaster, York, and Glennan, to discuss once more the fate of ABMA. Actually not much new has happened as it is now clear that everybody in this group is in favor of its transfer to NASA.

9:00 AM PSAC meeting. The first hour was Dr. George Harrar's presentation of his study made for the International Cooperation Administration on the role of science and technology in technical aid to Africa south of the Sahara, the report which I read on Sunday. It was a good presentation, and the acting director of ICA joined in the discussion and assured those present that ICA was taking the report very seriously and was going to adapt as much of it as possible in its operations.

J. George Harrar, vice-president, Rockefeller Foundation.

Rest of morning spent on NASA briefing to PSAC, followed by a rather vigorous discussion in which Fisk, Purcell, and others criticized NASA for too much hardware and not enough science, and Glennan defended the plans.

12:00 M–3:00 PM Executive session of the PSAC with York, in which he explained his approach to the military budget. We agreed on a general joint plan of action which leaves us the freedom to recommend changes in the military budgets even though York may have provisionally accepted them. York emphasized he would welcome the changes being made now, although many may have to await reprogramming later in the year.

3:00–4:30 PM ARPA briefing, much of which I missed, as I had to go to General Persons's office to discuss release of a public statement by the President in case he approved transfer of ABMA.

6:30–10:00 PM Drinks in my apartment and dinner with Dunn, Mettler, and Chewning from STL [Space Technology Laboratory.] We discussed first the delays and failures in the Titan ICBM program, and I learned that STL considers the Martin management of Titan to have completely broken down. Schriever will read the riot act to Martin's president, threatening him with loss of fee, requiring reorganization and acceptance of a large STL team which will manage the Denver operations on a temporary basis. We then came to the problem of STL itself, at which time (and three Gibsons) Dunn got very emotional. Before the end of the evening he was almost in tears. I somewhat facetiously suggested a novel approach, namely that STL buy respectability by associating itself with something like Mills College to become a girls' graduate school as well as an arm of the Air Force.

> Louis G. Dunn and Ruben Mettler were the top management team of STL; W. Chewning was their Washington representative.

ᐁ Wednesday, October 21, 1959

9:00–10:00 AM Meeting with the President to discuss ABMA transfer, which he approved. Neither McElroy, who did much of the talking, nor the President was very clear, and so the meeting took more time than necessary. Toward the end Glennan read portions of a letter he is sending to the President, in which he implies that he couldn't accept ABMA unless his budget was essentially what he proposed, $0.78 billion plus the cost of the Saturn, less $50–70 million in economies resulting from integration. The President evaded a clear answer, but said that in his opinion there were three major areas which had priority in space: the meeting of legitimate military needs is first; then comes the development of superboosters to get ahead of the Soviets eventually; and third is the scientific work. The President expressed his lack of desire for many "little shots around the moon" which wouldn't bring us much glory and would even embarrass us at times and cost a lot of money. That may put sort of a crimp into Glennan's program.

10:00–10:30 AM Working with Glennan and York on public statement to be made by the President from Augusta this evening.

10:30–11:00 AM Ewing, to discuss the sad prospects of a reinforced oceanography program, about which I gave him little encouragement, although I assured him that I was working on it. Then he asked me to help him regarding a special experiment involving the AMR [Atlantic Missile Range] splash net, to see whether an acoustic precision navigation couldn't be developed that would be much cheaper than transit satellites. I did some telephoning and next day learned that the Air Force will support the experiments so there was my good deed for the day.

W. Maurice Ewing, professor of geology, Columbia University.

Left at noon by train for Wilmington, where I spent some time on a casual visit to the Hercules Research Laboratories and an even shorter visit to DuPont's. Then two receptions, dinner, and finally my speech, which went reasonably well. At least I got a good hand and a lot of discussion afterwards. I ran into a devoted supporter of von Braun, and we had a fairly heated argument. The evening ended with a beer party of Harvard graduates, many of whom had been my students or had at least taken my lectures. Very pleasant evening. Clearly I am a very big frog now in the little Harvard pool.

∞⌒ *Thursday, October 22, 1959*

At the Limited War Panel to hear some rather fantastic statements by General LeMay: for instance, the way to prepare for limited wars is not to have them and anybody who gets us into them should be put in prison (as if it was up to us to decide, at times). He then said that he certainly wasn't going to provide close air support to the Army. The Air Force was there to destroy enemy installations in the rear and the infantry was there to protect the front. If the infantry couldn't do its job then there was no sense and a waste of money in having it. It is clear that the Army will have no tactical air support unless roles and missions are changed. The panel was so shaken by his remarks that there was even little discussion.

Lunch with Jim McRae, at which he agreed to become the chairman of the missiles panel; so that, thank God, is settled. As a matter of fact, this was very fortunate, since at the very end of the day York phoned me with a message from McElroy, asking that I evaluate Titan vs. Atlas. I agreed and said that my answer would be no answer at present, but a sound evaluation in three to six months, which York thought would be acceptable.

2:30–3:00 PM Bill Elliott, in his role as state department advisor, to show me a draft of a memorandum to the secretary of state, which was about as unintelligent technically as anything I have seen, short of some articles by newspaper commentators. I took a long time explaining solid rocket propellants, what is wrong with Saturn B, what is wrong with Nike Zeus, etc. This may have had some good effect.

> William Yandell Elliott, professor of history and political science, Harvard University.

Then back to the Limited War Panel to hear the end of the Navy presentation and a fairly spirited discussion, in which Chick Hayward referred—as he now does every time—to Kistiakowsky's big holes for the Titans as being the biggest waste of taxpayers' money, and I suggested that nuclear-powered carriers were even bigger rat holes. This seems to have become a standard pleasantry between us, but is done in good humor.

Then I heard a very good presentation by the Marine Corps. It is really quite an organization and is obviously not swimming in funds. The impression I got was that lack of modern attack ships and of special aircraft are really the two greatest shortcomings of the Marine Corps.

Dinner at the White House Mess with Glennan, Greenewalt, and his panel, which is to evaluate the NASA program on a high level. This week seems to consist of nothing but 16-hour workdays.

> Crawford H. Greenewalt, chairman of the board of the DuPont Company.

Friday, October 23, 1959

Divided my time between the Limited War Panel and the Greenewalt committee, which was discussing the competitive aspects of the NASA program vis-à-vis the Soviets. It was clear by the time I left at 11:30 that the majority of the group felt much as I did, namely that all-out competition with the Soviets would only play into their hands. The only standout was Stanton.

> Frank Stanton, president of the Columbia Broadcasting Corporation.

At lunch with the NASA group I sat between Hugh Dryden and James Perkins, both of whom insisted that we were making a mistake by not including in the PSAC public figures such as Don Price or John Gardner, who would provide breadth of experience and social sciences knowledge, since PSAC deals so frequently with problems transcending narrow technical confines.

James A. Perkins, president of Cornell University.

After lunch, and until 3 PM I spent with the NASA, at which Greenewalt summarized the conclusions I referred to earlier. Glennan seemed to be quite willing to accept them and there was much discussion that a change in our attitude should be heralded by a strong presidential speech followed by a carefully developed program, aiming at public education within and outside the U.S.A. Of course, the important thing is to avoid giving the appearance of sour grapes. It was agreed that the group will reconvene in a month's time, at which time Greenewalt will try to get Vice-president Nixon to participate, so as to provide the practical political knowledge as an input for the discussions.

3:00–4:30 PM   Limited War Panel, largely listening to General O. P. Weyland, who sounded very different from General LeMay, although even Weyland stated that in his opinion it was possible to develop aircraft useful both in general and in limited wars, and that he would prefer such weapons systems to those designed only for limited wars.

General O. P. Weyland, commander of the Tactical Air Command.

Monday, October 26, 1959

9:30 AM–12:30 PM   Informal Space Council meeting to review the agenda. Significant event was a discussion of NASA's Vega space vehicle vs. DOD's Atlas Agena B vehicle, which are very similar. Glennan stated that NASA would not accept the Holaday recommendation to cancel Vega, but he would be willing to make a detailed engineering survey after the six vehicles now ordered have been used up; and if such an engineering survey recommends, he will not order any more and will change over to Agena B. Sutton criticized this, pointing out that by employing the more powerful Centaur vehicle, which however was less flexible in that it made no provision for restarting of the engine, NASA could fulfill all its objectives without Vega.

At the end of the meeting we got into a heated discussion of the planning board paper on space policy, in which I found myself on the side of the BOB, while Harr was on the extreme opposite, supporting NASA, which is anxious to include references to our leadership in space, etc. After the meeting I had lunch with Harr and we talked some more and, I think, not without some effect. At least I now understand his position and he understands mine, and they actually aren't so very far apart, except for the evaluation of the importance of political factors.

I find the present arrangement of my chairing the informal Space Council meetings thoroughly unsatisfactory since Phillips briefs me sketchily on the agenda after it is all set up, and I find myself being a complete figurehead with the real play being between Phillips and Glennan. I said, therefore, to Glennan that either I participate more actively in the planning of the agenda and understand it, or he'd better take over the chairmanship, which he doesn't want to do. Part of the trouble is that Phillips regards himself as a Glennan man.

2:45–3:00 PM   Glennan and I to see the President and brief him on what is to come. The President obviously very angry about the duplication of Agena B and Vega and made references to subordinates disobeying orders in connection with this duplication, since last spring he had issued firm orders on the scope of the space vehicle program which did not provide for Agena B. The meeting itself uneventful. It is clear that the President is much concerned about the record of his administration in this whole space business, since it is now the third time that he has referred back to the 1953–1955 period when our missile program was set, and he questions the motives that led us to the selection of comparatively small ICBMs. Was it economy? Both York and I assured him that it wasn't a question of economy, but speed of development which led the von Neumann committee to recommend the small missiles, since they were adequate for the military task.

Had a phone call from Greenewalt, in which he said he thought the meeting with Glennan last week was very successful and useful, but that he now felt even more strongly about the foolishness of overdoing our space program and shouldn't we cut down on the number of vehicles being developed? I understand from Glennan, however, that the Space Sciences Board of the

Academy [NAS], which met last Friday and Saturday, urged expanding the program. A wonderful state of confusion.

5:00–5:30 PM Haskins of NSC staff to coach me in my response to Mr. Mansfield of Senator [Henry M.] Jackson's committee, who will visit tomorrow to question me on the mechanisms of formation of national policy. The meeting was useful, as I now feel much freer to talk than I thought constraints would allow.

Tuesday, October 27, 1959

8:30–9:15 AM Bradbury and Foster. The most recent calculations indicate that the safety of the nuclear bombs is actually very much better than suspected and there is now full confidence that by small changes in design they can be made completely safe. We then had a long discussion about the question of relevant safety experiments to be made at Los Alamos, from which it is clear that General Starbird was not completely candid in his transmission of my views to Los Alamos and vice versa. For instance, he reported that it was my decision that Los Alamos should make no experiments involving fissionable materials but only chemical explosives; whereas I even didn't know of this decision, which was obviously taken by Starbird or somebody in AEC.

> Norris Bradbury, director of the Los Alamos weapons laboratory.

9:30–10:30 AM Spoke at the staff meeting of the Voice of America. About 80 to 100 people present, some from USIA. Henry Loomis (director of VOA) asked me to speak on space and I delivered a fairly standard spiel, emphasizing the distinction between military capability and space activities. I then went on to make my comments that we are playing into Soviet hands by emphasizing our competition in space for immediate achievements since for the next few years they could always best us. I urged toning down this competitive aspect, emphasizing our scientific achievements and also our technological and scientific superiority on a broad front. There was some questioning afterwards, not unfriendly but skeptical. I recall that Raymond Swing asked me whether I recommend not competing with the

Soviets and I said that my recommendation was quite different: that we should compete but not talk about it.

Raymond Gram Swing, radio commentator on foreign affairs.

11:00 AM–12:15 PM   Mr. Kenneth Mansfield from Senator Jackson's government operations subcommittee. Much less tense an experience than I expected. Mansfield had no questions about the operations of this office or of the PSAC but only general inquiries about the role of scientists in government, e.g., should there be more scientific input in the State Department, with which I agreed, suggesting that a position remotely resembling that of York in Defense could be a very good answer. I then strongly supported York and his office and generally emphasized that I knew nothing about organizations. I believed in men, and couldn't provide much assistance on organizational matters. I felt that it was essential to have in government scientists from private life on a temporary basis, just as there are lawyers and businessmen on a temporary basis, but that this should not discount the great importance of Civil Service scientists. It appears that I will have a luncheon with Senator Jackson in January, which I graciously accepted.

1:30–4:00 PM   FCST. It was a smooth meeting and I think it went quite well, although as usual, nothing world-shaking was accomplished. Quite clearly the question of federal policy for research facilities will be a tough nut to crack, as the agencies have such opposing views and not much progress has been made so far. Jim Wakelin presented his comparison of what the oceanography committee program recommended and what is being accomplished by federal agencies in their budget proposals. Only the Navy has met the committee's recommendations and that through inclusion of a $40 million ship which is really designed for ocean survey and not for oceanography. Other agencies are way below the recommendations, especially Interior and Commerce. At this discussion Bennett (Interior) noted that the Bureau of Commercial Fisheries has included in its proposals certain items such as a market study for improving utilization of fish and the study of oysters in polluted estuarial waters, which he didn't think was oceanography but passed on to learn what the FCST would do. It turned out that Wakelin's committee didn't challenge any of it and, hence, the council felt that a substantive review of the programs was desirable. This I agreed to undertake with the intention that when the program is cleaned up it will be presented to the cabinet. Williams defended the latest AEC action in not meeting the recommendations of the committee on high-energy accelerators. In particular he emphasized that the Stanford accelerator was a source of great embarrassment to the AEC because two people on the accelerator staff at Stanford are in a very clear conflict-of-interest situation because of their involvement with a major supplier, and Congress is not willing to accept this. President Sterling of Stanford is not willing to see the situation in its true light and now has had several talkings-to from the AEC people. Because of this situation, Williams urged the council to take no action on the deficiency of the AEC program and specifically not to report on it to the President. With this the council agreed.

4:00–6:30 PM   Attending the PSAC Arms Limitation Panel, which listened to presentations on the problem of effectiveness of monitoring the production of fissionable materials. Had a phone call from Glennan, who is very much upset because the President has now twice stated publicly that the Saturn project must be accelerated. Keith pointed out that he (the President) couldn't say that and then disapprove an increase in the budget. This is very unfortunate, as he pointed out, and I agreed with him. We prepared a statement last week for public release by the President from Augusta [Georgia], which had no reference to accelerating Saturn, but merely noted the accomplishments of ABMA. Apparently that was thrown out and somebody wrote a different release. Keith said that von Braun is already reminding him of this statement of the President.

> Later I was told that Major John Eisenhower (U.S. Army) was instrumental in changing the statement.

⌁ Wednesday, October 28, 1959

7:30–8:30 AM   Pre-press breakfast, almost all of which was devoted to a discussion of the steel strike. The fuss that the staff members make about the exact wording of the memoranda is really quite amusing considering that the President afterwards only glances at the written texts and says something entirely different anyway. I had my time too and discussed the matter of the public release in Augusta and the fact that the President's other statement to the press created the impression that von Braun was being given a carte blanche so far as Saturn was concerned. I learned from Glennan that von Braun is already quoting these statements to NASA. Hagerty maintained that this was not the intent or the spirit of the statements, but authorized me to prepare a corrected one. This I did with the help of Dick Horner after the breakfast.

9:45–10:30 AM With the President. Among the suggested questions and answers, I showed him one on Saturn, the spirit of which was to indicate that NASA will review the project and integrate it with Nova, and that what was needed now was development of a superbooster, and explained to the President the reason for this statement. He was really quite worried about the effect the releases have had and without any hesitation agreed to make this statement. Hagerty then queried the President as to what he would say if a reporter asked him about his reaction to the USSR giving Russian names to various features on the backside of the moon. I suggested that the President might point out that the front side of the moon has Latin names, but we are not worried about it belonging to the Romans. This seemed to amuse the President no end. He kept chuckling about it for about five minutes.

1:00–2:30 PM Jim Conant for lunch. Had very pleasant time chatting about various subjects, but found him unenthusiastic about participation in our work. He is quite wrapped up in his study of secondary schools. An interesting item is that he seemed doubtful about our appeals to strengthen science teaching and produce more scientists. He argued that before we made recommendations on how to accomplish it, we should justify to the nation the purpose itself.

> James B. Conant, former president of Harvard University and an old friend.

3:00 PM Bill Penney from England, who talked about British studies on monitoring of production of fissionable materials and then made a very significant statement, that the outcome of the British elections was such an overwhelming victory of Macmillan that the latter is no longer worried about the outcome of nuclear test cessation negotiations and would probably be perfectly willing to go along with the U.S.A. if we proposed resumption of tests. (Next day Spurgeon Keeny told me that in a private conversation with him, Penney emphasized the same point and implied that the representations by Ormsby-Gore, who stated that England would not accept resumption of tests, were for diplomatic purposes only). All of this naturally will make termination of tests much more difficult.

> Dr. William Penney—now Lord Penney—English nuclear physicist, was in Los Alamos during the war and later became the head of the British Atomic Authority.

Had dinner with Beckler, Rathjens, Kreidler, Bacher, McRae, and Harold Brown, and returned to office at 8:00 PM for session with Rathjens, Beckler, Kreidler, and Keeny to discuss plans for the Friday night and Saturday session of panel chairmen. The trouble is that virtually none of the panels is ready with its report, and we shall certainly have a difficult time trying to formulate a final paper. We agreed that George Rathjens will take his preliminary draft and revise the part on strategic weapons to condense it somewhat. He will then try to cover the same subject in one or two different ways, e.g., without taking up the weapons one by one but concentrating on more philosophical aspects of the problem as a whole. These samples will then be considered by the chairmen and the best format chosen for the whole paper.

⌒ *Thursday, October 29, 1959*

12:30–2:00 PM   Federal Radiation Council luncheon in Secretary Flemming's office. Very mild and uneventful meeting, largely devoted to a presentation by Taylor, the chairman of the Committee on Radiation Safety Standards, on the biological effects of ionizing radiation and the history of standards established in the U.S.A.

2:15–3:45 PM   With York, Charyk, and Rubel, to discuss ICBMs. The meeting was arranged on request of McElroy, who wants to have me concur before approving the Atlas-Titan plan. I agreed to the so-called K plan, which projects 13 Atlas squadrons and 14 Titans, of which the last 6 are to be with storable propellants and with in-silo launch. I also supported 3 Polaris submarines to be started every year and emphasized that when we are planning for 1965 and beyond, the mobile systems should be given strong emphasis because of probable improvements in Soviet guidance accuracy. Rubel wrote a very good paper on strategic weapons, the only part of which I took exception to being one on the B-70. That is being approved in the paper and I said I couldn't do so.

John Rubel, deputy to York in DDRE.

4:30–5:00 PM Dr. David Inglis, representing FAS (Federation of American Scientists) with an impassioned but not very coherent plea to devote my efforts to prevention of nuclear tests resumption.

David R. Inglis, physicist, Argonne National Laboratory.

5:00–5:30 PM Dean [J. P.] Elder [of the Harvard Graduate School] who asked for my intervention with NSF to change the rules of predoctoral fellowships to permit a small amount of teaching for additional pay. I suggested a compromise which might be acceptable to the NSF board, that such teaching be permitted only in the second and later years of fellowships and not to first-year men. Elder thought this would be okay.

⌒ Friday, October 30, 1959

Spent much time with Continental Air Defense Panel, now chaired by Mannie Piore; by and large it was a rehash of the topics which were considered by the slightly different panel this summer, when Wiesner was the chairman. What I heard of the NORAD presentation made me feel that they have only the vaguest of ideas as to how to employ their proposed hardened control centers, and particularly the COC, the command center at Colorado Springs. The presentation by the MITRE staff was very disappointing because instead of analyzing the significance of the so-called DOD Master Plan, they accepted it as given and merely made proposals for additions that would make it better.

11:00 AM–12:30 PM Meeting with Gates, York, Dulles, Scoville, and Glennan, to plan for the comparative study of the USSR vs. U.S.A. space activities, of which I am the chairman by NSC action. The terms of reference I proposed were accepted with only minor changes. We then discussed possible chairmen of the working group, and for a while it looked as if we would get into a completely bad situation because Gates wanted a "public figure" to head the proceedings. Finally we got him back to technical people and agreed on the list: 1, John Williams; 2, Guy Suits; and 3, Hendrik Bode. (I had already contacted the latter by phone and had tentative assurance that he would undertake the job.) Late on Friday I went to see John Williams to offer him the job and got from him a flat refusal. He says he is tremendously overworked by John McCone, and is just barely able to keep his head above water. Nobody in AEC can take over if he gets involved in other activities.

> Hendrik Bode, vice-president for military systems of the Bell Telephone Laboratories.

7:30–11:00 PM Discussions with panel chairmen. The beginning went rather badly in the sense of a considerable waste of time. Din Land launched into a long discourse as to the need for general improvement in the way government does R&D, which had very little connection with our immediate problem, but we may get a special PSAC meeting set up later this winter to study this problem among others. Then we proceeded with discussion of strategic weapons and other topics, and it was difficult to hold the discussion to the subject matter, so that we only completed half the agenda for that evening, although we stayed an hour longer than planned. At the end of the meeting, I made a plea that everybody exercise self-control on Saturday and stick tightly to the subject, as otherwise we would never get through the agenda.

> Edwin H. Land, president of the Polaroid Corporation and a member of PSAC.

Saturday, October 31, 1959

9:00 AM–5:00 PM Things went very much better, as everybody heeded my plea of last night. The discussion was brisk, except when we came to communications and intelligence. Most of the presentations in these areas dealt with general failings of the DOD that couldn't well be incorporated into our paper. We completed discussions on time and should be able on the basis of the conclusions reached to write a staff paper that will not be vacuous, but will contain a very satisfactory set of recommendations.

~ Monday, November 2, 1959

10:00–10:30 AM   Henry Bent, in charge of fellowships for HEW, under the Defense Education Act. [I have known Henry, who is dean of the graduate school at the University of Missouri, from my early years at Harvard.] He described the difficulty of deciding on sound policy for the fellowships, and said that he is getting diametrically opposed views from supposedly competent people such as Waterman, Bronk, and Trytton. I asked whether the PSAC could be helpful and assured him I will ask the Seaborg panel on graduate education to set up a meeting with him and study this problem.

> M. H. Trytton, director of the Office of Scientific Personnel, National Research Council of NAS.

11:00 AM–12:00 PM   [Edward Louis] Keenan, of OCDM, describing its Committee on Specialized Personnel. He talked about various problems the committee is studying concerning the utilization and misapplication of technical personnel by contractors working for the government on R&D projects, which causes increases in budgets and hence in corporation fees. He said that they had a large meeting of technical company executives where the overwhelming majority conceded privately that this was the key problem, but were not willing to go on the record. I offered to raise the problem in public speeches if he provides me with factual data, which he agreed to do, and we parted on assurances of mutual interest and eagerness for collaboration.

Phoned Guy Suits and was told by him that he wouldn't be available at the earliest until next week, and even then would still have to consult with company officials about conflict of interest. I thereupon called York, who agreed that we should now ask Bode, since it was York who put Bode third on the list at the Friday meeting. I then offered the chairmanship to Hendrik, who with only minor hesitancy agreed to do it. The rest of the time I dictated from the notes I took during the Friday and Saturday panel chairmen's meetings. It added to about fifteen pages.

In the evening had a pleasant visit at home from Frank Long. I am much impressed to see how his breadth of view of government problems has broadened in the two years that he has been a member of the Air Force Science Advisory Board and the DOD missiles committee. I am convinced that we must enmesh him in our activities. After quite a few martinis, we went to the Cosmos Club, where the rest of the OSD Ballistic Missiles Advisory Committee was dining, and stayed in their private room until eleven. I was mercilessly pumped by the committee members on matters of high policy, but believe I came out fairly well on the whole, and haven't leaked out any state secrets. To draw attention from myself, I called attention to the question of the wisest space policy, i.e., all-out competition or what will you have, and got completely divergent views. But it was all good fun.

Franklin A. Long, professor of chemistry, Cornell University.

ᑐᑐᑐᑐ *Tuesday, November 3, 1959*

Singularly pointless day. Total accomplishment microscopic.

8:30–9:30 AM   Attended the White House staff meeting, Meyer Kestnbaum reporting on his visit to Russia. The USSR certainly has impressed him very much. He is convinced that within their capabilities they are doing a very good economic job and have good utilization of manpower. His main conclusion is that Khrushchev and company have made so many promises of consumers' goods, shorter working hours, etc., etc., to the populace that short of restoration of the terror as a means of control, they will have to fulfill at least some of the promises, and this makes it essential for them to slow down or stop the cold war. He thinks, therefore, that their efforts to do so for the next ten or even twenty-five years are quite sincere. But even if this is done, they will remain our opponents rather than friends. He believes that with our relaxed attitude we face a real grave threat of finding ourselves second best after a while.

Meyer Kestnbaum, special assistant to the President, and an industrialist.

11:00 AM–12:15 PM   Attended the cornerstone-laying ceremony for the new CIA building. Weather was lovely, the drive beautiful. The cavalcade of cars following the President very impressive. We stopped on the way to cut the ribbon across the new parkway. At the ceremony we had two numbers by the band, two prayers, and two speeches. All pleasantly short and soulful.

12:30–1:30 PM   Lunch with Wakelin and Ewing, discussing mainly the ship building plans of the oceanography program. There is considerable doubt even in Wakelin's mind as to the need for the Navy's 4,000-ton, $30-million ocean survey ship. Ewing is generally opposed to big ships for these purposes and questions the concept of ocean survey so long as navigation is as inaccurate as at present. There was also some discussion of the continuing feud between the Coast and Geodetic Survey and the Navy Hydrographic Office. Ewing's opinion is that the former is much better than the latter.

4:00–5:00 PM   Attended the meeting of a PSAC panel chaired by Ewing, which was reviewing the scientific worth and coherence of the federal oceanography program for 1961. Members feel that while the program contains no items clearly in conflict with the outline of the National Academy of Sciences' Committee on Oceanography, it is poorly coordinated and thought through in some areas. For instance, the Bureau of Commercial Fisheries, although it desperately needs facilities and states so itself, provides virtually no money for facilities and only one million dollars for ship construction, but asks for a substantial increase in research budget. The research proposed is of very low grade and doesn't touch on the basic problems of marine biology. I asked the panel first to comment in general terms on the quality of the program and then to take dubious items one by one and criticize them. Not to worry about politics resulting from criticism; I will take care of them by proper handling of their document.

CHAPTER V

Back to Geneva and the Budget

The Geneva conference on the nuclear test ban had reconvened on October 27. At the outset our Ambassador Wadsworth stated that during the recess the United States had carefully reviewed the unresolved issues and concluded that the chief obstacle to a treaty was the problem of effective safeguards and controls against evasions. The conference then discussed several aspects of the treaty, including the composition and authority of the control commission, the choice and functions of the administrator and his staff, etc. The issue of the veto, that is, the requirement insisted upon by the Soviet Union that certain key decisions of the control commission had to be unanimous, was also taken up. In the course of prior sessions some compromises on several aspects of the veto had been reached, and this process was continued in the fall of 1959. On December 14, Ambassador Tsarapkin introduced a new "package" proposal which, however, failed to resolve some key disagreements including the procedures for on-site inspections.

The next pages will indicate the difficulties we were having in Washington to assemble a strong group to discuss with the Soviet experts the results of the underground test explosions in the Hardtack series of 1958 and the conclusions of the subsequent Berkner panel. The group was finally put together and left for Geneva. However, until November 24 the Soviet Union rejected participation in another technical study of seismic detection of underground explosions, but changed its mind and on November 25 the Technical Working Group II did meet for the first time. It held 21 private meetings which did not lead to a meeting of the minds. Their report submitted to the conference on December 18 noted agreement on ways to improve seismic detection, but not on the results of the 1958 Hardtack tests. The Soviet Union insisted that all explosions down to a few kilotons of TNT equivalent could be seismically identified but the United States group denied this.

The working group adjourned on December 19 and the conference recessed the same day.

And now back to the budget troubles. In the fall of 1958 several PSAC panels were concerned with various aspects of the military and space programs, and advised Killian of their judgments for use at budgetary meetings which he attended.

The reader has undoubtedly already noted that, with the President's blessing, we developed a more ambitious plan for 1959—the preparation of an all-inclusive memo on the proposed military R&D budget for the use by the President and the secretary of defense.

The words all-inclusive are, of course, a misstatement. The memo, which was in the final stages of preparation as this chapter begins, dealt only with large weapons systems projects. Our attempts to understand and evaluate what else was being done or proposed by DOD were not successful because of the volume of information and lack of time to digest it.

I recall a half-day presentation requested by me from the Air Force Research and Development Command on their half-billion-dollar applied research budget. It was mostly cut up into pieces under a half-million dollars each to eliminate the need of seeking approval from the Office of the Secretary of Defense through DDRE. Frankly, it overwhelmed me. I still recall becoming indignant on discovering that the cost of exclusively paper studies in industrial establishments on "Strategic Defense of Cis-Lunar Space" and similar topics amounted to more dollars than all the funds available to the NSF for the support of research in chemistry. I tried to raise hell about this with York but the results were negligible.

Much of the fall was (and probably still is) dedicated by the Bureau of the Budget to full-dress reviews of the budget by the director. I attended one or two of these meetings and some members of our staff were invited to others. We found these meetings not very useful or satisfying because very little time was given to analysis of the substance of the budgetary items considered, but our technically conditioned minds yearned for just such discussions.

The President's decisions were being sought when BOB and an agency head could not reach a compromise. This chapter gives a lengthy account of a major budgetary session with the President, who kept our memo in front of him during the meeting.

Wednesday, November 4, 1959

8:30–9:00 AM   With Goodpaster, largely to check on the record of the July 23 meeting of the Principals with the President to find out whether the President had firmly decided on the policy for only a limited test ban. Goodpaster's check indicates that the President approved only a tactical paper then presented by the State Department, and since this plan has been swept away by the course of events there appears to be no definite decision on the record.

11:15 AM–12:00 M   With the President. He told me of his response to a question at today's press conference regarding resumption of nuclear tests, which led to my remark that no nuclear weapons, if they contain nuclear materials inserted, can be absolutely safe against accidents. We then talked about space, and he agreed to read a prepared statement on our objectives in the space program, which Keith and I would prepare. I then mentioned that we were working on the military budget, and the President spoke at length about the diffuseness of the program and attempting to do too much. He felt that concentration on a secure deterrent force and the ability to act effectively in very limited engagements should be the focal points of our military program. I spoke in general terms of the Federal Council, noting that some progress was being made, but it was very slow. The President asked whether there was deliberate obstruction or just the typical inability of the government to agree or move fast. I assured him it was the latter. I mentioned the facilities program, but the President felt there was no use building research facilities unless there were first-class people available, and questioned whether in fact there were such available.

12:35–2:00 PM   Ed McMillan and John Williams for lunch. The former complained about inadequate funds to modernize the

Bevatron and generally to keep facilities operating at maximum effectiveness. He spoke of Alvarez engaging in a major program to accelerate evaluation of data from his bubble chamber, which enabled me to point out that many physicists felt that Alvarez is hugging these data instead of being generous, as any scientist should, and is not providing other groups, without the Bevatron facilities, with opportunities to study the films. Ed conceded that Alvarez was difficult, but assured me that some improvement was being made. John Williams was very positive that no changes in the AEC budget toward a larger allocation of funds to high-energy physics was possible, and so the meeting was not very effective.

> Edwin M. McMillan, director of the Lawrence Radiation Laboratory, Berkeley, California; Luis W. Alvarez, professor of physics, University of California, Berkeley.

2:00–4:00 PM Hendrik Bode's Comparative Space Evaluation Panel organizational meeting. I gave a brief introduction, whereupon there was a long discussion which finally crystallized into who does what and what will be the procedure. My impression is that good progress was made.

> This panel was an interagency group (NSC and Space Council) who were to evaluate the relative accomplishments of the United States and of the USSR in outer space.

Afterwards I had a protracted argument with Beckler regarding what I should do about the staffing of the second Geneva conference of experts. Dave insisted that regardless of what the conference was for, I should bend every effort to get the best people; whereas I maintained that only if our objectives for the meeting of experts in Geneva to evaluate new seismic data are positive in the sense of attempting to reach an agreement, would I try to get the best personnel. If all that State is trying to do is to prove the impossibility of monitoring, Harold Brown will do perfectly well.

Thursday, November 5, 1959

After the NSC meeting this morning I remained behind with Herter and McCone to discuss staffing of Geneva conference. I emphasized the point I made to Beckler, the night before, and although I didn't get a clear-cut agreement, Herter did say that he would much prefer an imperfect control system that would provide some deterrent against cheating to no treaty at all. With this I heartily agreed and McCone didn't object. McCone mentioned resumption of tests and stated that in opposition to his staff and Defense, he was willing to postpone resumption of underground tests well beyond January 1, but that the one-point safety experiments should be started. I didn't disagree, nor did Herter. Both Herter and McCone stated that we needed the highest technical level representation in Geneva to convince the world that we were honest and also to force the British to send Bill Penney to Geneva. We went over possible candidates, of whom Berkner and Fisk were rejected by McCone, and decided on Killian as first choice. At the end we went to see the President, who asked Herter to urge Killian to go and to tell him that he, the President, would excuse him from any other Washington commitments during the period he is in Geneva. As we were leaving I said to McCone that I wanted to talk to him about the Materials Sciences research program, to which he responded that he too wanted to talk to me about it and also about high-energy physics.

After this I left for Florham Park, New Jersey, the site of the new Esso Research and Engineering Center, where I had a heavy but unsatisfactory lunch and delivered my speech, which moderately needled Esso for failure to engage in basic research. My speech was preceded by that of the governor of New Jersey, whose speechwriter obviously didn't know anything about controlled thermonuclear reactions. He promised to make deuterium! The governor left in the middle of my speech. From conversations after the speech, while we were touring the facilities, it was clear that my needling was noticed and welcomed by a number of people. President Murphree of Esso grinned and said they will try to do better in the future. So that wasn't a complete waste of effort. I returned to Washington at about 6:00 PM.

~~~ *Friday, November 6, 1959*

9:00–10:30 AM   Cabinet meeting. The first hour was devoted to a summary by Herter, Cabot Lodge, McCone, and Benson, of their impressions about the Khrushchev visit and their visits to the USSR. The general summary is that Khrushchev is anxiously presenting the appearance that great progress was achieved at Camp David toward relaxation of tensions and that we must test him by insisting on deeds. Khrushchev's attitude toward Red China is questionable. His very recent speech to the Communist Party Congress in Moscow sounds almost apologetic, as if he were defending himself against accusations of being too soft. McCone emphasized the tremendous progress of the USSR in the field of atomic energy and their ability to do technical jobs faster than we do. A somewhat discordant note was produced by Benson, who had found nothing good in the USSR on his visit to see agriculture. He ended his speech on a very emotional note, describing his visit to a crowded church in Moscow and said this was the only favorable point in his visit in Russia.

> Ezra Taft Benson, secretary of agriculture and an elder of the Church of the Latter Day Saints.

1:00–3:00 PM   The Edison Foundation luncheon at the Statler Hotel. Sat at the head table and listened to an incredibly long speech by Alan Waterman. Our joint speech writer at her worst and the text totally undigested by Waterman. Long quotation from a speech of mine, and so I thought that she was quoting herself, but afterward discovered she chose a speech written for me by somebody else. I must never deliver speeches written by professional writers without spending several hours on each of them.

3:30–4:00 PM   Dick Leghorn, to bemoan his troubles with Charlie Coolidge. We discussed the question of having the Coolidge group, our panel, and perhaps Phil Farley, making recommendations (to whom? to the Government Organization Committee?) about a continuing organization of the study of arms limitation.

ᴏᴄ⁓ Saturday, November 7, 1959

In the morning had a phone call from Herter, who said that Killian refused the Geneva job. I suggested that Brown might be undesirable and Herter without hesitation said he felt so too, that Brown couldn't do an objective job. We then sort of agreed on Bacher, and Herter will approach the President for personal intervention with Bacher.

ᴏᴄ⁓ Monday, November 9, 1959

9:30 ᴀᴍ–6:00 ᴘᴍ   PSAC meeting, which went rather well. It ended on a slightly burlesque tone, with a report by Lloyd Berkner about the findings of the NAS Space Sciences Board. Berkner really went overboard selling space, stating for instance that if Project Orion were developed it would cost only a few cents per pound of payload to put it into space (whereas even a very optimistic estimate of the cost of small atom bombs, only $100,000 apiece, shows that the cost would be about $50 per pound).

During the meeting Britton Chance told me that he definitely wanted to resign. I expressed regret, but left it there.

> Britton Chance, professor of biophysics, University of Pennsylvania, and a member of PSAC.

We reconvened at 7:30 ᴘᴍ with York, to review the military budget paper whose last draft we had worked on all through Saturday. The draft was received well and even York had very little to object to. In fact at one point he suggested that some people might think that he and we were in collusion.

This session ended at 10:30 ᴘᴍ, at which time I asked Bacher, Fisk, Bethe, and Killian to stay behind. I should note that during the day Secretary Herter asked Bacher to become the chairman of the U.S.A. delegation to the experts' Geneva meeting on seismic detection, which is to begin in a week's time or so, on Russian acceptance. We spent about an hour with Bacher, most of the talking being done by him and me, although Jim Killian also urged him to accept. Bacher's attitude was quite negative; he contended that there was no policy of any kind, that the session would be a holding operation, and that he saw no reason for being involved in it. At one time Hans Bethe argued that since there was no policy, it might be best to have Harold Brown head the delegation, so he would state the extreme position that it was impossible to monitor the tests. This, according to Bethe, would force the Russians to really get involved and might clarify the atmosphere. The meeting ended about 11:30 PM, with nothing agreed to, but Bacher consenting to think about it overnight.

Tuesday, November 10, 1959

8:15 AM   Picked up Bacher at Cosmos Club, and on the way to EOB he said he had decided not to accept the job because of "complicated" grounds, some of which were personal and very strong. I asked him to inform Secretary Herter and emphasize to Herter his misgivings that lack of policy would make it impossible to get any "independent" scientists to participate.

9:00 AM–1:30 PM   PSAC meeting continued. During it I was proffered a compliment by Din Land, who said the PSAC meeting on Monday was the best he ever attended.

Captain [A. P.] Aurand [the President's naval aide] came with a proposition, which I found slightly incongruous but followed up on it, to approach Din Land with the proposal that the Polaroid Corporation offer to the President up to a hundred Polaroid cameras as gifts in the Middle East and other areas, where the exchange of gifts is a custom. I did approach Din Land, and he agreed readily enough to discuss it with the company management.

4:00–5:30 PM   Visit from McCone. Very friendly and we covered a lot of topics. He is much concerned about Russian accomplishments and feels that we will rapidly fall behind. He is convinced we must have a drastic reorganization of government and put most of the R&D work in the area of the physical sciences, including that by NSF, AEC, NASA, and a good deal of DOD, into a new department. He wanted even PSAC included in it. He asked me to make cautious inquiries as to the opinion of scientists regarding the Argonne National Laboratory. McCone is afraid that Argonne is not terribly good intellectually and lacks leadership. We spoke then about the nuclear test cessation conference and I explained to him the difficulties of getting independent scientists to a conference which had no clear-cut purposes and represented a holding operation. We spoke then about the Stanford SLAC project, about which McCone is very unenthusiastic on a number of grounds. He thinks Stanford has pulled some fast tricks in response to AEC's request for more detailed independent evaluation, by hiring of engineers and seismologists who are connected with probable subcontractors that would be engaged in the construction of the linear accelerator. He is also annoyed about the conflict of interest with regard to the Varian Associates Corporation through the stock held in that company by key Stanford personnel. I pointed out that after all the klystrons needn't be bought from Varian, but could be, for instance, equally well purchased from Eimac Corporation. I reminded him that the President was committed so far that we shouldn't embarrass him and must get him off the hook in case the Stanford project didn't go ahead and suggested we return to the matter a little later after I made some inquiries. I emphasized my concern about the lack of adequate facilities at the existing high-energy machines or those now under construction. He agreed with my concern, pointing out that the second largest machine in the USSR had 17 essentially independent laboratories attached to the machine, whereas we tend to build machines only, without any laboratories. He conceded that he will have to look at the AEC budget again, trying to get more money for the purpose. I then brought up the Materials Sciences Program, explained how critical it was to demonstrate the importance of the Federal Council for Science and Technology, and asked him to cooperate. He noted that he was on the record as so flatly opposed to giving buildings to universities that he couldn't reverse his position without a clear national policy decision. I then explained how it was impossible to get a national policy decision in time for the '61 budget and suggested he accept the recommendations of the materials subcommittee, but retain title for AEC to buildings at Berkeley and the University of Illinois. This suggestion he liked, he said, and was going to see whether it couldn't be put into effect.

ᲝᲦᲮ᳐ *Wednesday, November 11, 1959*

8:30–10:00 AM NSC meeting. After the usual intelligence summary by Dulles began an extensive discussion of NATO, introduced by reports of Merchant and Burgess. They both emphasized that our Allies were not meeting the objectives for military strength agreed to in NATO documents; that they were not building up their land forces, but that some of them were trying to develop their own "deterrent." They advised the council, however, that our withdrawing of large bodies of troops from Europe could have very bad political effects on NATO. There followed remarks by McElroy, Gray, Anderson, and others. The President took an active part and stressed his conviction that it was high time the Allies assumed their proper share of the burden and relieved us of providing the major part of the "shield" and said he felt that a "trip wire," a thin line of U.S. troops just to establish our presence there, was all that should be needed and that our job was to have the "sword," namely, the strategic deterrent forces. At one time Gordon Gray summarized the findings of the planning board, describing the growing threat from the USSR. The President quite suddenly stopped his usual doodling, raised a hand, and said: "Please enter a minority report of one." The clear implication was that he considered the threat decreasing. He then reminded everybody that when he went to Europe in 1951, he had explicit instructions from President Truman that stationing large U.S. forces in Europe was an emergency measure; that he so informed all NATO governments, and that an emergency which lasted eight years was not an emergency any more. Stans, as usual, gave horrifying figures on the cost of maintaining our forces in Europe—$2.5 to 4 billion—but then was forced to concede, when McElroy challenged him, that moving these forces to ZI wouldn't save anything like that amount. His estimate is a saving of $100 million per division. Much of this discussion was concerned with the need to conserve dollars, because of the very unfavorable balance of payments which now exists. The discussion, as was understood from the beginning, did not end in any firm decisions, but later that day, when I was at lunch with McElroy, he commented that this was one of the best discussions that he ever took part in at NSC and wondered why we couldn't have more of them.

W. Randolph Burgess, United States Ambassador to NATO;
Robert B. Anderson, secretary of the treasury.

10:00–11:30 AM   Cabinet meeting, which was not interesting. When it ended, I spoke briefly with Herter about the need to discuss the personnel and policy for the Geneva experts' meeting. Thereupon, Herter and McCone went to see the President. An hour later, which time I spent mostly with Spurgeon Keeny talking about the Geneva problem, I had a phone call from Herter saying that the President has given specific policy instructions, which are that the experts should make every effort to determine the form of monitoring system which would be satisfactory for detection of underground nuclear tests. Herter was very elated and so am I. This is a tremendous change from the attitude which prevailed recently that really all we can do is to seek disengagement from the comprehensive treaty concept and try for an atmospheric ban. I guess "boring from within," which I have been doing the last couple of months, has finally borne some fruit. Let's see how long this decision will stand, since Defense will unquestionably hit the roof when it learns of it. I suggested to Herter a scheme which had occurred to me in discussion with Spurgeon, namely to ask Fisk to be the chairman for only three weeks, and to have Panofsky as vice-chairman during that time with no real duties, so he could continue working at CERN, and then have Panofsky take over. Herter bought this without hesitation. I had then a phone call from McCone, essentially confirming the understanding that Herter got from the President, although with some shadings, which were to be expected in view of McCone's general attitude. He proposed to persuade Rabi, with whom he was going to have lunch, to be the chairman. I warned him that Rabi wouldn't accept because of a heart condition. Shortly after that Rabi visited me, and I tried the idea on him and got a very definite no.

Phoned Goodpaster to get from him a record of the meeting with the President. What I learned confirmed Herter's understanding, so now I can move. After lunch McCone phoned and said Rabi was not available, whereupon I telephoned Fisk and almost got him to agree to my plan. He was afraid Herter may not have full confidence in him because of the outcome of the summer of 1958 conference, and also said that the president of AT&T must release him. Phoned Herter, who said that he had maximum confidence in Fisk and that he would phone him immediately and would also ask the president of AT&T for a release.

CERN: European Center for Nuclear Research, on the outskirts of Geneva; Wolfgang K. H. ("Pief") Panofsky, professor of physics at Stanford University.

Lunch with McElroy, Gates, and York, to each of whom I handed a copy of the PSAC magnum opus on the military budget. Lunch went very well, only Gates, however, reading the report in detail. Gates was positively enthusiastic, noting repeatedly that what we recommended was just what he was fighting for. McElroy talked in more general terms. The impression I got was that they were agreeing with most of our recommendations. They proposed to cut B-70 to just the airframe and engines, no avionics, accept our recommendations on Nike Zeus, etc. We will have to see how it will all work out when they meet with the Services to approve their budgets. At the end of the luncheon, I invited McElroy and the others to attend our meeting with the President, which I had arranged with Stephens on Tuesday morning to take place on Monday, November 23. McElroy said there was no point in having that meeting since he was taking the military budget to the President in Augusta on Monday the 16th. I said, well, we couldn't get time with the President until the 23rd because of his imminent departure. McElroy answered that in that case I should certainly come to Augusta and that he would phone Goodpaster about it. Sometime in the afternoon, I called Goodpaster about this, and he said he should have thought himself about asking me to come, and that certainly I should. During this same afternoon, I did a lot of telephoning about the Polaroid cameras. Land called me to say that this was most agreeable to the company and that it will provide up to a hundred full kits, which include flash bulbs, etc., etc. By the end of the day it turned out that the White House couldn't use more than forty kits, but those the President expressed eagerness to accept, and Goodpaster read to me over the phone a letter they will address to every king and princeling receiving the camera, which would explain that the gift was being made by Polaroid and the President jointly, and would express the hopes of Dr. Land and the President that the recipient would enjoy its use. So I guess Din Land will get a lot of free publicity and advertising for his generosity.

Goodpaster expressed the desire to have an advance copy of the PSAC report, and Irene, my secretary, proceeded on the heroic task of retyping a clean copy, which she finished at 7 PM and I assembled and proofread.

Thomas E. Stephens, the President's appointments secretary.

Thursday, November 12, 1959

7:30–8:30 AM Breakfast with Goodpaster, Morgan, Gray, and the usual NASA delegation—Glennan, Dryden, and Horner. Glennan read a draft of a proposed letter to the President from himself, in which he recommends that the President: 1. Ask Congress for new legislation by which the Space Council (as well as the liaison committee) would be abolished; have NASA given complete responsibility over space vehicle boosters, and a military applications division established in NASA. 2. Make a major address to the nation, in which he would announce this legislation and explain our purposes in space. 3. Approve plans for a spectacular demonstration of the accuracy of our missiles to be made by having public observers stationed in the splash net on the Atlantic Missile Range and several shots fired on the same day. We discussed the proposed legislation and made some suggestions. Goodpaster recommended substantial changes in the first portion of the letter, since as he said, to tell the President that unless he acts, Congress will, would not dispose the President favorably toward the writer. I noted that to expect that we could make several shots on the same day was a little optimistic, because of troubles with the range which have nothing to do with the reliability of the missiles. It was suggested the President should show a short movie film, showing re-entry of the nose cone. Such movies exist and are spectacular. He could also cite the accuracy of our missiles which York had assured me could be declassified.

Shortly after breakfast had a phone call from Fisk, who to my extreme delight informed me he accepted the Herter invitation, and we immediately proceeded to discuss the staff, etc. Jim told me he telephoned Hans Bethe and Bethe has agreed to serve, which is terrific, considering the negative attitude Hans took on Monday night. I certainly hope to be able to hold the line in Washington while these people are in Geneva. If not, my name will be mud and possibly deservedly.

12:30–1:30 PM   Lunch with Admirals Burke, Hayward, Russell, etc. Pleasant and by now rather usual banter about big holes (for the Titans) that I am fostering on the country, with my response that if the Navy will stop building big carriers and cancel the big radio astronomy "dish," I might plug the holes. Also a lot of good-natured kidding about Harvard football prowess (Harvard-Princeton score last Saturday: 14–0).

> The big radio astronomy dish being built as a highly classified project (opposed by PSAC) was abandoned as a technical failure after an expenditure of more than $150 million. It was Admiral Hayward's pet.

1:30–2:00 PM   Special intelligence briefing on Soviet naval activities. They certainly have a tremendous fleet built since the war. There was also a very interesting account of the ways in which our Navy gets intimate information on the Soviet naval activities, but that is so hush-hush, I can't put it on paper. Someday it will make a very exciting news story.

> It did in 1975.

2:15–2:45 PM   Called on Undersecretary of Commerce Ray, to show him the report of the Ewing panel on oceanography which we set up to analyze the agencies' fiscal '61 plans for oceanography, compared with the recommendations of the Academy's Committee on Oceanography. The Ewing report is quite critical of the effort in the Commerce Department. Had a pleasant meeting and Ray assured me he will look into the possibilities of improving the situation, especially as regards the very meager research budget of the Weather Bureau. He was less optimistic about ship construction.

A short visit from Glennan, who showed me his revised letter and revised wording of the proposed bill, both of which were a substantial improvement. He said he will go to Augusta next Tuesday or Wednesday to talk to the President about it. Then Undersecretary of the Interior Bennett came and I gave him too a copy of the Ewing report. In contrast to Ray, Bennett was pleased pink. He said he was delighted that at last the Bureau of Commercial Fisheries will learn a lesson and prepare a decent program. He said that he was going to put on the maximum pressure to force them to do so and increase allocations for better facilities and ship construction at the expense of ridiculous and meaningless "research" proposals they stuck in.

∾ *Friday, November 13, 1959*

An utterly mad day, with enormous activity, but nothing of importance. Jim Fisk arrived in the morning and until almost noon he, Spurgeon, and I worked on detailed plans for the conference and telephoned a number of people, such as Panofsky, Press, and Turkevich. A little after twelve, Phil Farley arrived and we four had lunch and talked more about the importance of spelling out on paper the national policy on conference objectives and getting good terms of reference and instructions.

> Frank Press, professor of geophysics, California Institute of Technology; Antony L. Turkevich, professor of physics, University of Chicago.

In the meantime, I started talking with staff people from the Bureau of the Budget on military budgets but had to ask them to leave. They came back at 4:30 PM, when we had a lengthy discussion. On the whole, BOB comments and proposals aren't too bad except that here and there they make rather ridiculous statements which obviously weaken the force of their paper. Naturally they cut out more than should be cut and increased nowhere, as they should have done.

Right after lunch, Don Paarlberg came to ask us to organize a study of the Pure Food and Drug Administration's activity in preventing contamination of foods by insecticides and other toxic materials. He noted this is getting to be an acute political matter—as exemplified by cranberries. He gave me his memos to Jerry Morgan, the first of which antedates cranberries. Thank God! I tentatively agreed to do it as a part of our study of life sciences.

> Don Paarlberg, special assistant to the President for economic affairs.

Just before the end of the day got a telegram from Press, declining the invitation. In the meantime, Panofsky and Turkevich accepted. Phoned Lee DuBridge and urged him to change Press's mind. Did it with considerable relish, remembering what happened to me last spring.

> Lee A. DuBridge, president of the California Institute of Technology. The relish was because of remembering that in the spring of 1959 when Dr. Killian, on behalf of the President, invited me

to succeed him, I went to the president of Harvard University, Nathan M. Pusey, to ask for a leave of absence in the event I accepted. Mr. Pusey refused, pointing out my extensive absences from Harvard since November 1957. Rather than resign from Harvard at my age (58) I declined Killian's invitation. Only a couple of days later Pusey called me to say that President Eisenhower had phoned him and that I should take a leave. A little later I called on President Eisenhower, who welcomed me to his staff and said that we have a strange man as president of Harvard.

∝⸱ Saturday, November 14, 1959

12:00 M–2:00 PM   Meeting of Principals in Secretary Herter's office. Herter passed around a "Memorandum of Conversation" with the President, of McCone and himself. The memorandum stated that McCone informed the President that according to my report to him, scientists were unwilling to serve on the delegation to the second experts' conference in Geneva because the U.S. was attempting to disengage itself from a comprehensive treaty and they wouldn't be doing anything useful. After reading this memorandum, I noted that scientists didn't necessarily object to the policy of disengagement, but they objected to the lack of policy which existed at that time and were afraid that they would have to drag their feet in Geneva while policy was being made up in Washington. Herter then presented a statement of policy on the basis of the President's instructions, which was very clear and satisfactory to me. Thereupon followed an hour and a half thoroughly wasted while Jack Irwin was nit-picking this policy statement. After all this time we ended with a text which, although completely rewritten, said exactly the same as the original. Irwin is really a completely impossible person, and I was amused to see how horrified and shocked he was by the Memorandum of Conversation. He said: "Do you mean to say that our policy is not any more to disengage?" and clamped his teeth. Gates, on the other hand, was quite relaxed about the matter after McCone assured him that he had protected the interests of DOD.

After the settlement of the policy paper, we spent half an hour arguing about the proposed speech by Lodge in the UN, to which Gates and McCone took violent exception. Herter very tactfully explained that unless Lodge were to support and vote for a resolution which would urge us not to resume testing during negotiations, it was quite certain that a much tougher Indian resolution would be adopted. This would have the General Assembly urge us not to resume testing until the treaty for the cessation of tests was signed and also urge us not to do any R&D on weapons in the meantime, so we would have to close the laboratories if we were to comply. Herter really handled the matter very well, I thought. Finally we came to the discussion of undertaking the one-point safety tests and got nowhere as usual. The meeting adjourned and we agreed to meet again Tuesday.

After the meeting, I went on McCone's invitation to have lunch with him and Dulles at the Metropolitan Club. It was a cordial atmosphere, but I suggested to McCone that he check on the new calculations regarding the one-point safety of certain nuclear weapons, and he agreed to telephone Floberg, who was that day at Los Alamos, and instruct him to get the data given to me during a visit by John Foster of Livermore and Carson Mark a couple of weeks ago.

∽ *Sunday, November 15, 1959*

Had a phone call from General Starbird, saying they had a letter from Carson Mark and they wanted me to see it. As I had just returned from the office, I rejected the idea of going back, and invited them to come by my apartment; they arrived shortly thereafter. I read the long letter from Carson Mark to Floberg, in which he stated that now Los Alamos and Livermore calculations were in agreement; that the conclusion was that the slightly modified weapons design now in production was probably safe against accidental explosion, but that the margin of safety was not great enough for the laboratories to guarantee it. It was clearly a sort of double-talk letter. Since my apartment isn't de-bugged, Starbird suggested we go for a walk, which we did, and had a long discussion on the bridge over the Rock Creek Park, during which I explained to them that the President is firmly opposed to tying the one-point safety tests to the general problem of underground nuclear weapons tests; that they had to accept this position and try to see how they could formulate a press release acceptable to the President, and then start the safety tests. I also suggested that such a release take the bull by the horns and explain in considerable detail to the people what is involved and in what way the proposed experiments differ from weapons tests.

～ Monday, November 16, 1959

6:00 AM   Took off for Augusta from MATS terminal, feeling not very cheerful, in the plane of the secretary of defense. On board were McElroy, Gates, General Twining, Gordon Gray, Stans, General Persons, John Sprague (acting DOD comptroller), Mr. Schaub (BOB), and Gen. [Carey A.] Randall (McElroy's aide). Herb York, who was coming, has a bad ear infection and was told to stay at home, so McElroy told me I had to represent York's point of view as well as our own and try to do it fairly, but he had confidence I would do it.

8:30 AM   Meeting started in the big room of the clubhouse of the Augusta National Golf Club. A lovely room with a burning fire, coffee aplenty, and a long table in the middle. One side of it was occupied by the President and Persons, Gray, Stans, and myself as well as Goodpaster flanking him. On the other side, McElroy was facing the President, flanked by his people. The meeting began with McElroy saying that in order to keep the budget down, he proposed to withdraw some U.S. forces from NATO, and realizing that withdrawal of whole divisions would create a grave political issue, he proposed to cut the Air Force complement from 46 to 32 squadrons, and said that perhaps the French unilateral action in refusing to store our nuclear warheads would justify this step. Some discussion followed which was ended by the President saying that the forces could not be pulled out now from NATO even though it would save some dollars, but that NATO should be notified diplomatically of our plans to reduce the forces. The President then raised the issue of the Sixth Fleet and why it was in the Mediterranean all the time. He said why not pull out at least part of the fleet; in the coming missile age the value of this fleet in a big war is very doubtful. Gates said that the Sixth Fleet is in the same legal position relative to NATO as our other forces in Europe. The President questioned the value of carrier bombers in the future and pointed out that we have quit fighting with the bow and arrow and it was about time we switched to missiles from bombers. Gates said he was for withdrawing the Sixth Fleet but was afraid of reaction in Greece and Turkey. The President commented that he couldn't see any reason why we should assume the complete burden of the cold war; certainly our allies in NATO had resources to assume a substantial contribution, but we should start with a political preparation of NATO and convince them that we have as much right to reduce our forces as Great Britain, which recently reduced her forces greatly. He then started reminiscing about [John Foster] Dulles, who used to grow violent when the suggestion was made to reduce forces in Europe. Dulles expected the dissolution of NATO as a result. McElroy suggested that at that time this could easily have happened. The President said that our forces had gone to Europe only as an emergency measure, but it was too late now to withdraw them in time to affect the 1961 budget. We should do the withdrawal gradually. The President then returned to the Sixth Fleet and asked why we needed two carriers; that one would be plenty. Gates explained that a one-carrier operation is not sound because when a carrier is refueling there is no platform for aircraft to land. Twining said that as soon as we moved the Sixth Fleet from the Mediterranean, the Russians will move in. The President asked: "With what?" They can certainly move their ships into the Mediterranean now; if they don't do so, it isn't because our fleet is there, but is because they don't gain anything by doing so. The President said he wanted to have the Sixth Fleet moved as soon as possible, and to explain to the Allies that the greatest help we can give them is having a secure deterrent, and that we have it. He said the trouble is the Allies want not only our deterrent but also every other conceivable military force. Gates noted that in the past we always took a stand within NATO which would make it very difficult for us to inform them that we are planning to reduce our forces. He reminded that in the past we always urged forces to be increased.

This led to a general discussion of the size of land forces and brought the President to a very sharp remark about Gen. Maxwell Taylor's article in *Look* magazine. Looking at Twining, and white with anger, the President said: "Will you tell the Chiefs that they can't scare me by such inexcusable indiscretions as that of Taylor. I shall still continue doing what I believe right, and I consider that his behavior is totally inexcusable." McElroy accepted the President's decision, but noted it will mean increasing the '61 budget over his own plans and urged speedy political action to get favorable environment for decreasing at least the '62 budget. The President said that the State Department should be instructed to prepare the ground for the reduction of forces— not to make a study of how to prepare the ground. Stans noted that in the past the State Department unilaterally made commitments for our participation in NATO forces and that the budget implications were never considered. He suggested that in the future this matter be submitted to the NSC before commitments were made that tied our hands for at least two years ahead. The President readily agreed and went on to say that the deterrent is the important thing, but it is difficult to explain this to the Allies. He thinks that he may be able to convince Macmillan and Adenauer of the correctness of this point of view, but that de Gaulle was hopeless and a complete "psychopath" and will refuse to accept the reduction of U.S. forces in Europe even though the present situation would lead to an eventual weakening of our deterrent forces.

McElroy then brought up the subject of Army Reserves and the National Guard, suggesting that reduction was needed. The President asked what the JCS thought of it, and Twining said the Army objects because reduction would mess up their training plans. The President asked: "Why doesn't the Army start thinking realistically?" He said that in an all-out war the reserves will not be able to do anything useful. They won't be able to move because of destruction of communications, transportation, etc. He said he wanted to talk to Lemnitzer about it and to send him to Augusta right away. McElroy noted it will be difficult to get reduction through Congress. They have defeated such proposals two years running now. There followed a general discussion as to whether to ask for a reduction to 630 or 500 thousand men. The President noted that the only forces that we could count on were those in being, but it was hopeless to get much reduction from Congress, and we should ask for reduction to 630, expecting little response. Also we should announce that the JCS have begun detailed study of the need of reserves.

> Gen. Lyman E. Lemnitzer, Army Chief of Staff, later chairman of the Joint Chiefs of Staff.

McElroy turned to Air Defense and stated what the planned force reductions of the Navy and the Air Force will be and said that in order to accomplish this, the Navy will reduce either the ship or aircraft part of early warning in oceans. The President made a long statement in which he questioned the value of BMEWS because of the threat of nuclear submarines, but did not press the subject and then agreed in general with the reduction of the early warning by aircraft. He thought that Mr. Khrushchev was probably sincere in his conversations at Camp David, when he questioned the value of aircraft and navy surface ships. The trouble is we are trying to protect against everything imaginable and he instructed McElroy to talk to the Canadian officials about reduction in strength of the mid-Canada antiaircraft early warning line. ''Those people, God damn it,'' he said, ''don't know what the progress of technology means. They're still thinking in World War I terms.''

McElroy then came to air alert, explaining the OSD proposal which would give the Air Force the capability of maintaining Strategic Air Command on air alert for a limited length of time upon strategic warning of Russian threat. The plan involved buying spare engines and parts, but not adding to the numerical strength of SAC. The President conceded the need to make the Air Force more resistant to surprise attack, but asked why shouldn't we abandon BMEWS if we should have an air alert? Stans objected to the way in which individual items were dealt with, without regard to their effect on the total defense budget. Thereupon the President gave him a rather sharp lecture, pointing out that the reason for having a civilian secretary of defense was to get away from the Service judgements of the three Chiefs of Staff. ''I am the only army general,'' he said, ''to have disassociated myself from Army thinking, and I have been called a traitor for this more than once, but I believe that the secretary of defense is as concerned about maintaining a balanced, sensible budget as I am, and I hope you will bear that in mind. You and I can't form judgements on the military merits of special items like air defense, BMEWS, missile base hardening, etc. In these matters we should accept the secretary's judgement.''

> ''Strategic'' as opposed to a ''tactical'' or imminent warning was supposed to provide up to several days' or even weeks' notice.

McElroy turned to missiles and explained the K plan, which would lead to 13 Atlas and 14 Titan squadrons and provide for a development program to have storable propellants and in-silo launch for the last 6 Titan squadrons. There followed a long discussion on the merits of Titan, the problem with the Martin Company management, etc., in which I got heavily involved. The

President then said he approved this program, but not to make the announcement that the Titan program will be increased beyond 11 squadrons until the storable propellant development has been proven out. Followed some discussion in which I was involved regarding scheduling advantages of Titan over Atlas, etc., and the President instructed DOD to go ahead with the K plan except for the extra Titan squadrons. He asked, why not simply ask for a lump sum for missiles without specifying what we are going to do, but McElroy explained this wouldn't work in congressional committees. That brought the President to ask us whether he shouldn't talk to the JCS and Service secretaries, and it was agreed that he should. The chiefs will be invited to Augusta this week and the President will see the Service secretaries early next week. McElroy then came to the B-58, explaining the program and saying that the Air Force wanted 3 wings, 116 aircraft, costing one billion more on top of two billion already spent, but that this would provide low-level attack capability which we didn't have now. He preferred to have the program cancelled, but was worried about the Convair plant in Fort Worth, which would have to shut down, and this would cause grave local unemployment. The President asked why we couldn't build Atlases in the Fort Worth plant that was building the B-58, but I explained why this wouldn't be economical and the President accepted the explanation. He then asked how the "Hound Dog" air-to-ground missile was moving. He was told the B-52 will start being equipped with Hound Dog next summer. Stans pointed out that according to the WSEG report, the B-52 with a Hound Dog is better than a B-58. Twining noted the importance of a B-58 as the first supersonic bomber and noted, of course, that the B-70 was a logical extension of this progress. The President's reaction was instant and vigorous. He said he wouldn't spend one cent on the B-70; that by the time it would be available, the missile age would be on us and he wasn't going to buy obsolete equipment at the cost of billions of dollars. McElroy asked me to present York's and our opinion on it, and I did. Thereupon McElroy argued the advantages of the B-70 as, first, a bomber, second for reconnaissance, third as civilian transport, and fourth as military transport. He felt we must develop this aircraft as the Free World leader. The President again got rather steamed up and said that present civilian transports were plenty fast, and that he certainly couldn't see why "we" should develop Mach 3 civilian transport under the guise of supporting the defense effort, but actually giving the development free to aircraft companies. He could see absolutely no military value in the B-70.

There followed a long discussion as to what will be the nature of destruction in the next war. The President flatly stated that in his opinion there will be general destruction, mainly of cities and industrial centers. No one would be foolish enough to waste thermonuclear weapons on missile sites. He ended by a statement that the B-70 is simply not needed! McElroy continued to defend the B-70 weakly, at least as a training need, as a reconnaissance and troop transport, and Twining said we must have at least one supersonic aircraft. The President replied by saying that you will have the B-58, which meant that the B-58 was approved. McElroy then came to the Navy problems and said that they propose to start three more Polaris submarines in '61. The President asked when the program would be cut off. When McElroy answered that this wasn't decided, the President started laughing and suggested that cutoffs were good things to have. He felt the Navy should be concentrating on limiting itself to fulfilling its basic mission—control of the seas—implying, as I understood it, under conditions of limited wars.

Then followed a discussion of carriers. The President said he wanted a conventional carrier, but that Congress rammed through a nuclear one. McElroy said that Navy plans call for 12 modern carriers, so that 4 would always be deployed, and since three years have passed since the last one was authorized, this was the right time to start another. Stans said he wanted to postpone the carrier one more year. The President said he thought it was silly to have so many carriers deployed. The "other fellow" is laughing at us for spending so much money on pointless armaments. Discussion followed as to whether Congress would add the carrier on its own initiative even if none were requested. This ended by the President remarking that he is tired of spending billions just to increase by a small percentage the performance of current weapons systems. He would like to postpone the carrier. He has lost faith in carriers, but would approve a carrier if it were a conventional one. Smaller ships were preferable. Stans said that the actions this morning wouldn't reduce the total budget substantially and pointed out that when the complete budget was added up there was danger of a deficit. He urged that a number of small projects, such as Dynasoar, SAGE, superhardened air defense centers, etc., be cut out. The President urged McElroy to try and reduce the cost of these items. McElroy noted that for the first time a really thorough and sound examination had been made of the Service budgets by York and the PSAC, and complimented us. I said we had excellent cooperation from York and that our opinions were very close. The President stated that to fight "that demagogue" (Khrushchev) we have to keep the budget balanced.

He then went on describing the difficulties he had in the postwar period about getting money from Congress for the military and suggested that the secretary of defense, in making up his budget, should think of what might happen five years hence if Congress adopts the '49 attitude. He said McElroy should cut out the fat and reminded all of us of our precarious financial position in the world.

Tuesday, November 17, 1959

Meeting of Principals at which for once Dulles said something. Evidently it was the result of a conversation I had with him when I drove with him from a preceding meeting, in which I pleaded with him to help the State Department, which was the underdog vis-à-vis Defense, and he said well he didn't like to speak on policy matters because his was a technical role. Just before leaving the Principals' meeting, which he did earlier than the others, he gave a little speech in which he endorsed the State Department terms of reference for our delegation to the Technical Working Group II in Geneva. During the long discussion of these Herter emphasized that Fisk was to be the judge as to whether such terms of reference give him freedom to carry out the policy instructions which we got on Saturday from the President. Fisk pointed out that he could certainly carry out these policy instructions; more explicit terms of reference would make it easier but he was willing to operate under the terms proposed. This is what really took the wind out of McCone. When we finished this discussion, Fisk left and we came to the one-point safety tests, with McCone stating that in view of the new assessments (that is, the letter which I described in my notes on what took place Sunday), the urgency of carrying out one-point safety tests was very much reduced. This was now a matter which no longer had to be resolved before January 1, but could be postponed another month or so. On Monday, when we were returning from Augusta, I talked to Gates about one-point safety tests and explained what took place on Sunday (Starbird's visit) and then remarked that the President was anxious not to have the general issue of resumption of nuclear tests tied to the one-point safety experiments and that I was trying to carry out his wish. At which time Gates grinned and said: "Well, of course, we are trying to do just the reverse." He is really a nice guy. I said to him then that I knew they were, but of course couldn't accuse them openly of it.

∾⁓ *Wednesday, November 18, 1959*

10:00–11:15 ᴀᴍ Had a visit from Messrs. Sedov, Dobronravov, and Krasovsky, accompanied by A. V. Grosse, and Don Ling from Bell Laboratories was in my office. Our visitors all spoke Russian, although I know (from an earlier meeting in Cambridge) that Sedov speaks English and it was obvious that Dobronravov understood English quite well. I spoke in Russian on casual matters, but used English for more sensitive statements, which were then translated by Grosse. The visit began with a presentation to me by Sedov of a book, which contained photographs of the other side of the moon, and of a few other publications by the USSR Academy on artificial satellites. We then chatted for a while about science in general and exchanged ideas about the organization of science, the visitors maintaining that without effective planning and direction, as they have through the Academy mechanism, it is impossible to advance because of the great complexity of modern science. We contended that while this will be a more effective method to achieve local successes, our method of freedom of choice, minimum of direction, and support of science by several sources, assured us of progress on the broad front. They inquired about my function, and I explained that I had in a way a triple function, presidential adviser, chairman of the PSAC which consisted of scientists from outside the government, and chairman of the Federal Council for Science and Technology, representing government organizations. They asked then what was the position of the Academy and I explained that it is not a government organization and has in general little influence on federal R&D policies. We then agreed that there is need for better mutual understanding and international cooperation of scientists. They emphasized how important that is.

Then we got on a rather hot subject as they brought up the matter of disarmament and I agreed that it was highly desirable, but pointed out that there were very fundamental difficulties in the way because they had their military establishment hidden in secrecy whereas ours was rather open. This got their goat, particularly when Grosse started pointing out that we made no secrets of the space rockets we are using, whereas we didn't even know what theirs were or the location of launching sites. Sedov said, he saw here no connection with scientific aspects of the work, and it was a silly curiosity to insist on knowing what the rockets looked like and where they were fired from, and commented that they disclosed everything scientifically important, whereas we did less. This argument went on for some time. Professor Krasovsky got into the fray with a rather vehement statement that we are hiding much more than they—for instance, why didn't we disclose the secret of the nuclear weapons after the end of the war. He was obviously very hot under the collar and simply didn't listen to Grosse and me when we pointed out that the Baruch plan offered the Soviets an opportunity to share in nuclear developments. This plan was totally dismissed by them. Krasovsky then went on to point out that he hasn't been able even to buy the fast photographic emulsions here for his astrophysical work and said that that kind of secrecy was far worse, because it interfered with scientific work, than the secrets of how the rockets worked, which had no bearing on science. The discussion of who had more secrecy was obviously getting nowhere, so I changed the subject and suggested that the thing to do was to try to reach better understanding through scientific cooperation because it was so much more international than other activities.

Sedov then went into a long discourse on the need to reach understanding in the military field, but stressed the difficulty of doing so and the necessity of doing it gradually because, as he said, in neither country can a single man make sweeping and revolutionary decisions. He said: "You probably think that Khrushchev can make them, but that isn't so. In fact, even Stalin couldn't make really sweeping decisions on his own."

The issue of secrecy obviously annoyed them, because until the end they kept returning to the subject and, for instance, at the very end Sedov noted that although we displayed the Explorer

VI satellite in Moscow, we didn't disclose until quite recently that it carried photographic equipment.

At the end I showed them the picture taken from the Atlas nose cone of the clouds over the Atlantic and said this type of meteorological observation will be very important for humanity. This also annoyed them and they suggested that the photographs of the back side of the moon might be far more important for the future of humanity, and that in any case instead of fooling around with little rockets, one should go to big payloads and install automated and subsequently manned stations in space, as this is the only scientifically sensible way to progress. That was the parting remark, and we said goodbyes.

As we were arguing about secrecy on both sides, I said that as a chemist I had a very legitimate curiosity about what rocket propellants they are using, but Sedov only grinned, and when Dr. Grosse said that we made no secret of our space program, Sedov immediately rejoined that they didn't either, and that we could learn from open publications precisely what they are planning to do.

> Leonid I. Sedov, Academician, one of the leaders of the Soviet space program; V. V. Dobronravov, Academician, also a leader in the space program; N. N. Krasovsky, professor of astrophysics, Moscow University; A. V. Grosse, professor of chemistry, Temple University.
>
> Donald P. Ling, Bell Telephone Laboratories.

3:00 PM   Visit from Luca Dainelli accompanied by Emilio Fiorio. Dainelli is now chief of the UN branch of the Italian Ministry of Foreign Affairs and has been at the UN from the beginning of the General Assembly session. He told me that things are moving very badly; that the Western nations have absolutely no initiative in regard to disarmament and are letting the Soviets get all the credit and keep the initiative; that the U.S. is exceptionally bad (and I am not surprised, considering what Charlie Coolidge is able to accomplish). Kuznetsov at the UN has been able to put through a resolution which will convert the ten-nation disarmament committee (to meet early in 1960) into a committee to discuss the Khrushchev proposal on total disarmament, which is obviously futile, but will give great opportunities for propaganda by the Soviets, instead of working on first steps. The real object of Dainelli's visit was to urge me to influence the State Department to take the initiative in establishing cooperation among the Western nations prior to the convening of the ten-nation disarmament committee. He felt that last year's Geneva session on surprise attack (at which I met him for the first time) was a magnificent example of what can be accomplished by unity of action of Western delegations. He hoped that the same could be accomplished now, but that everybody was floundering and didn't know what to prepare. He hoped that if the State Department was not ready to outline the U.S. positions, it would at least be able to suggest to the Allies the areas which should be studied by them. I tried to explain to Luca that I wasn't personally involved; that it was a State Department activity, but that I would do what I could. We passed the time of day and discussed general matters, and at departure he urged me strongly to visit Italy.

~~~ Thursday, November 19, 1959

9:30–10:00 AM   Visit from Glennan, who told me that the President agreed with the principle of changing the NASA law along the lines suggested by Glennan and conceded that he may have to make a major speech on our space program. Then Glennan bitterly complained of the Budget Bureau proposal to cut his budget to $530 million. We went over the proposed cuts and I assured Glennan that I would vigorously support him on asking reinstatement of certain of the cuts, but not of all.

11:45 AM–2:15 PM   Attending lunch of Saints and Sinners, General Persons being the guest of honor and the "fall guy." The proceeding started with what was intended presumably as a side show, an incredibly lewd strip tease. Certainly it is not a mild humor that is exercised here, and the cracks about Persons, the President, etc., were sharp at times. The main import of the skits was that the President never does any work and that Persons, under his pleasant exterior, is as tough as Sherman Adams.

2:30–4:00 PM   In Glennan's office with him and other NASA people discussing national space policy paper. Was sleepy and anyway I can't get excited about minor semantic matters, so I agreed pretty much with everything Glennan wanted, but urged strongly not to accept the DOD-JCS version of the section on foreign policy because it would totally prevent the possibility of international agreements.

4:30–5:00 PM George Sutton of ARPA, with whom I talked about the future of ARPA. He conceded that it should concentrate on advanced research and do more of it under its own management, but maintained that ARPA should be independent of York, reporting directly to the secretary of defense, because of internal political reasons. He pointed out that York agreed with this concept, so I said if York agreed, I had no objections.

Friday, November 20, 1959

9:00 AM–12:15 PM Attending a meeting of the President's Board of Consultants on Foreign Intelligence. We first listened to a report by a man who for two years now has been trying to define a set of intelligence indicators for a coming surprise attack on the U.S.A. and feels much more time will pass before he will complete the job, although the present set of indicators is rapidly becoming obsolete because of missiles. Next we heard a report from General [J. H.] Walsh, the Air Force intelligence chief, speaking on grandiose electronic classified projects [now defunct]. This was a very clever presentation, emphasizing that these projects are really quite straightforward and already partly functioning—like the big radars—and minimizing the magnitude of the remaining job to achieve ultimate specifications. I would say it was a thoroughly dishonest presentation. When the Air Force delegation left, we had a brief meeting of the committee with just me and Jerry Wiesner present, and I spoke freely about the real problems involved, namely that the Air Force wants fantastic, complicated electronic systems in the naive belief that computers can do a man's thinking. I emphasized that machines can only help men, and recommended against endorsing Air Force plans. Jerry supported me. We then had a report from an assistant director of NSA, who described what was being accomplished in COMINT and ELINT. He also complained bitterly about people in high places carelessly revealing facts which could have been learned only through COMINT, pointing out that after the vice-president's (Nixon's) casual public remark that the Soviets had three unsuccessful tries before they launched Lunik I, the Russians increased greatly the sophistication of codes used in messages dealing with missiles.

3:15–3:45 PM Visit from Ambassador [Walter] Dowling (who is going to Germany) on the suggestion of Jim Conant, who is a close friend of his. He asked my opinion about missiles, conventional forces, the proper role for Germany, etc., etc., and I gave him a rather frank outline of our beliefs on all of these matters, emphasizing of course, that I was not a chief of staff, but only a scientist. He then asked me for the relation of space to international and national security and politics, and I again gave him a frank opinion, namely the need to stop the rather silly competition with the Soviets for bigger and more spectacular payloads, because we can't win anyway in the next few years, and will just make fools of ourselves; and that no possible advantages except in the intelligence and communications field can accrue from space exploits in the next years. He seemed to enjoy the visit, we parted best of friends, and he extended to me an invitation to visit Germany. I said, just think up a good reason.

Had a visit from Harold Brown, who is unhappy about the policy instruction for our delegation in Geneva and asked for my clarification of it. I said it seemed to me to be as clear as any policy statement could be: the technical people should be objective, present capabilities and limitations, and not be influenced by political considerations. We talked generally about these matters. He was violently opposed to cessation of tests unless there is a high threshold.

> What Brown was referring to was a treaty that would forbid nuclear weapons tests only above a specified explosive yield. Thus the current SALT negotiations involve a threshold of 150 kilotons—about ten times the yield of the Hiroshima bomb—which would permit weapons development to continue almost without hindrance.

∿ Saturday, November 21, 1959

The day started with a visit from the President's assistant naval aide, who gave me a document prepared by Admiral [R. B.] Pirie, which is obviously the Navy's counterattack to our condemnation of supercarriers. The usual arguments, nothing new. I probably was unwise in pointing out to the visitor the weaknesses in the Navy document and our reasons for recommending against building large carriers.

10:45 AM Private conversation with Herter. I started by telling him of my conversation with Brode, in which I urged him to resign, and also that Bronk had a similar conversation and that now Brode was anxious to have an appointment with Herter at which he might tender his resignation. I explained that while we thought that rotation in this job was desirable, the main thing was we didn't think Brode was the best man for the job. Herter spoke very explicitly and stated that as far as he was concerned, Brode was not useful, that he could never get advice from him on scientific problems and that Brode was only building up his own office. Having heard Herter, I conceded that Brode was difficult to work with from our point of view. Herter said he will certainly call in Brode and discuss it. I asked Herter then about Koepfli as a possible successor, saying that I was afraid that he was too much associated with Dean Acheson, and thus would be undesirable. Herter said there had been violent dislike of Koepfli by Hoover, who forced him out, but that it had nothing to do with State Department matters, being related to something that happened in California. That he, Herter, knew Koepfli and liked him. We ended agreeing that as soon as I get the word from him (Herter) I will get together with some other people and present him a list of choices to succeed Brode. I tried then to take leave, but he asked me to stay and raised two other matters. One is that we will probably have to go further in compromising with the Russians on the terms of reference for the second Geneva conference of experts than was agreed to in the Principals' meeting, and that McCone is raising hell about it. Herter then said that he saw McCone as personally really desiring an agreement, but being under terrific pressure from his staff and Defense to block it. I then discovered that Herter feels about Jack Irwin exactly as I do, and he told me in confidence that Tom Gates has the same opinion. This is just ducky! He then said he had on his desk a memorandum from his staff to be transmitted to the President, urging that we vigorously push toward our supremacy in space over the Russians, and he said he was very troubled because he couldn't get sound technical advice on this subject. I said if he would allow me, I would speak for five minutes on the subject, and did, pointing out that no matter what we do, they could always beat us with spectaculars for the next few years, and that politically it might be advantageous to take a more aloof attitude, deny competition and emphasize cooperation. Herter seemed interested in the thought.

Joseph B. Koepfli, former science advisor to the secretary of state, professor of chemistry at the California Institute of Technology; Herbert Hoover, Jr., former undersecretary of state; industrialist.

11:10 AM–12 M   Appointment with Glennan at NASA. We first went over their budget, or rather I listened to a lot of yelping and whimpering. They feel that the most they can cut is $50 million as against $250 million that Stans wants to cut, and assert that going below $730 million will wreck their program. I assured them that I would support reinstatement of certain items and left it vague on others. Specifically, I will support the Nova engine program. I suggested that Dynasoar and the follow-on, the Mercury, that is, the man-in-space program, should be unified, and that I wouldn't be willing to support both, but they argued that the purposes were quite different and both should be supported. Their arguments were plausible enough so that I don't know where I stand now. The last half of the meeting we spent going over the agenda of the Monday morning informal Space Council meeting, which will have a very impressive representation: McElroy, Twining, etc. They agreed that I will imitate the President in the way he handles the formal meetings, in that I will just say a few words, introduce Glennan, and then let him carry the ball.

Lunch at the Mess with Secretary Flemming. I pulled a bad boner, I guess, because I was so tired that, when Flemming came to my table, I thought for several minutes he was Stans, since I had been thinking about Glennan going to see Stans, and so I spoke to Flemming as if he were Stans, which is damn embarrassing. He probably thinks I am totally nuts. After that we had a pleasant conversation in which he welcomed a study of the cranberry herbicide problem by PSAC and also on his own initiative suggested that we should evaluate NIH. This is quite something and really makes it possible to be effective. He thanked me for our work for Henry Bent and the fellowships, and altogether was very cordial. Said he will come to the Federal Council meeting which I asked him to attend for the item on out-of-house facilities financing.

I went to the reception at the Soviet Embassy for Emelyanov and company, which was not too bad. Talked to the boss of the legal department of the Soviet Ministry of Foreign Affairs, who is apparently also professor of international law at the Moscow University, and with the head of Tass here in Washington.

Quite a squirt, who devoted a great deal of time to explaining how Tass is the only news agency that is objective and never distorts facts.

∼ *Monday, November 23, 1959*

9:30 AM–12:45 PM   Informal Space Council meeting. Much time was taken up by the two items, the document on national policy on space and the 10-year program proposal of NASA. It was interesting that during the first discussion a perfectly frank admission was made at last that we are in space to an extent not justified by the scientific importance of the subject, but because we cannot afford to let the Soviets continue their very successful political efforts "proving" that they are ahead of us in every way merely by putting bigger payloads into space. Dillon was particularly emphatic that we cannot carry out successful foreign policy if we are always second in space. Expressions of horror were abundant when the possibility was mentioned that the Soviets may be the first with a man in space, which, by the way, is what I guess they will be. Everybody expects a frightful upheaval in our planning and a very violent reaction in Congress, but nobody knows how to be prepared for it. Glennan then presented the 10-year projection which, as I noted, was not quite realistic because it assumed that after 1965 no new rocket booster vehicles will be developed. By assuming that we stop developments with the Nova, he was able to present the budget as stationary after about 1965.

> My prognosis was essentially correct. The Soviet Union was the first to put a man in orbit (Yuri Gagarin), in the spring of 1961. The political reaction in the United States was strong and came on the heels of the major embarrassment to the administration over the Bay of Pigs fiasco. President Kennedy then embarked, without consulting his science advisors, as he publicly stated, on Apollo, a crash lunar-landing program that cost the United States altogether over $30 billion and whose social value is therefore debatable. The Apollo is now followed by the Space Shuttle, a largely political pork barrel involving the expenditure of many billions.
>
> It might also be noted that a well-established government technique for soothing budget-conscious listeners is to present long-

range plans with no starts of development projects well before
the end of the time span considered so that the budgets cease to
rise. Thus, Glennan's presentation was neither the first nor the
last one of this kind.

1:00–2:30 PM   Lunch with Herter, Coolidge, and Coolidge's
associates, to discuss disarmament. Coolidge is not an effective
spokesman, I am afraid, and didn't give Herter much meat in
the proposals which he outlined, so that Herter urged him on to
greater efforts. We spent some time discussing various control
measures related to nuclear weapons. I raised the question of
ICBM controls and Coolidge spoke about the first steps which
might be a freezing and subsequently a thinning out of forces in
a zone in Europe. Herter emphasized that the proposal was still
too vague to be useful to the President, and that Coolidge had to
come through with more details, etc. The meeting was interesting, but possibly not very productive. Quite obviously Herter is
much interested in the problem of disarmament and at one point
said that in his opinion this would be the central topic of our
foreign policy for years to come. He also said the President
approved setting up a continuing organization within the State
Department for the study of arms controls. That looks as if our
(PSAC) intention to influence the design of the organization to
work on this subject has come to naught.

4:30–5:00 PM   Meeting with Staats concerning the oceanography budget, to discuss the augmentations which are needed to
meet even partial objectives of the NASCO oceanographic program. It is very difficult to get to these BOB people the sense of
what is intended; one almost talks a different language. I took
the opportunity when Staats mentioned NSF to give a most emotional although short speech on the need of giving more money to
it. Hope that it will do some good.

> NASCO: The Committee on Oceanography of the National Acad-
> emy of Sciences prepared a report which was critical of the
> existing federal activities in oceanography as fragmented, uncoor-
> dinated, and underfinanced. It outlined a long-range plan which
> the Federal Council for Science and Technology tried to put into
> effect. Secretary Wakelin's presentation at an earlier cabinet meet-
> ing was a part of these planning activities.

Tuesday, November 24, 1959

8:30–9:00 AM White House staff meeting, Persons and Hagerty talking about the President's trip and other such matters.

9:00–11:00 AM In office working on a memo to Persons regarding the NSF budget, which was the strongest memo I have yet written in Washington. Then spent about half an hour with Hendrik Bode's comparative evaluation panel on space developments. Ling did a magnificent job writing a summary and Hendrik added some important thoughts. It looks to me as if this report may be a real milestone in our understanding of the space situation.

11:00 AM–12:00 M M. Jules Moch, General Pierre Genevey, M. Michel Legendre, and M. Pierre Pellen, who visited me to discuss disarmament. Clearly, even though I emphasized that I was not involved and could only speak as a private individual, they thought that what I said carried a great deal of weight. Their own plan in the UN is to emphasize abolition of missiles rather than of nuclear warheads. They quickly conceded that the Russians would insist on the abolition of SAC bombers, and when I pointed out that thereafter the U.S. would have no choice but to withdraw its forces from Europe and become in effect a Fortress America, not giving a damn about the rest of the world since it would have no big stick left, they were much nonplussed. This point obviously made a deep impression, which was interesting. So I then indicated that the only thing that made sense to me was a balanced program of arms limitations of long-range nuclear weapons delivery systems, as well as conventional forces. That got the French very sensitive because they immediately think of zones, and so I took the bull by the horns and said that that is exactly what should be done because it is the only thing which could be discussed usefully with the Russians. The whole discussion might have been of some small use, although it is hard to say. The French really are stubborn.

> The French visitors, of whom Jules Moch was a former prime minister, whereas Genevey and Legendre I knew from the Geneva surprise-attack conference in 1958, were participating in the United Nations General Assembly disarmament discussions in New York.

12:00 M–12:30 PM With Jerry Persons, to whom I gave my memo urging that the NSF budget be increased at least $15

million over the $182 million allowed by BOB. It was a strong
memo but Persons took it well. He said he agreed with me and
that he will try to get it through. "Stans is the best budget
maker that I know of, but he just doesn't understand these
things, and as you noted in your memo, Waterman is just not
enough of an SOB to get what he wants in Washington." He
said if necessary he will take the memo to the President for final
resolution.

Persons then said our paper on the military budget was very
good; that he read it twice and thought that this year a good
portion of it could be put into effect and next year the rest. He
then said that General White came to him and asked his support
for a much reduced B-70 program, involving only $75 million
and one or two planes eventually, and what did I think of it. I
said I would go along with it. The impressive thing is that later I
had a phone call from Secretary Douglas, who described this
modified B-70 program and asked whether I would approve it.
Quite obviously our paper has not been written in vain and is not
without influence.

> James H. Douglas, Jr., secretary of the air force and in 1960
> deputy secretary of defense, a lawyer in Chicago.

1:30–4:00 PM   Federal Council meeting started with a review
of the oceanography program. Clearly we are not going to meet
the objectives of the NASCO or Wakelin report, but we will have
a substantially augmented program that is a good beginning and
could even be called a coordinated program. This is the first
concrete success that can be credited to the Federal Council.
Staats suggested that to mark it, the budget message should
specifically refer to the coordinated oceanography program and
describe it, and the council agreed, so we will be immortalized.
Secretary Flemming joined us then and we took up the question
of research facilities in nonprofit institutions. At issue was a
policy paper which would lead to new legislation authorizing all
federal agencies to make grants for the purpose. There was a
long discussion in which Flemming took an effective part and
only Bennett of Interior really objected. The paper, with minor
language changes, has been approved by the council and this
looks like another success, providing the proposed plan will clear
the agencies and be enacted by Congress. We then discussed the
problem of institutional grants by NIH and NSF. Staats
strongly objected, but the discussion in which I took a vigorous

part, probably improperly since I am the chairman, may have influenced him somewhat. Finally, we had a somewhat ineffectual discussion of research priorities. Now all the agencies except DOD have submitted them, but York flatly says that DOD is interested in everything related to research except certain aspects of archaeology, and it would be much simpler for him to write a paper on what they are not interested in. He is quite disarming about it, but obviously uncooperative, so the whole project may collapse.

4:30–5:15 PM   In Barton Hall (CIA). First a briefing on a Soviet project, which General Walsh described to the Board of Consultants last Friday in very frightening terms, implying that it was a weapons system more dangerous than any ICBM. Having listened to the detailed evidence, it is clear to me that all it is is a supersonic cruise missile, nothing very extreme. Walsh was dramatically exaggerating to win approval by the board of some grandiose Air Force projects.

⌀ Wednesday, November 25, 1959

9:00–11:00 AM   NSC meeting, which started with a summary by the DOD of the military program as of the next fiscal year. Among the objectives stated was one of having highly mobile, small and specialized forces available for local conflicts which were also to be used in a general war. When the presentation was over, the President said that he couldn't figure out how that could be accomplished and would JCS please write him a memorandum explaining how the small, highly mobile, etc., forces would be valuable in a general war. Twining was obviously squirming. Then McElroy, and after him John Sprague, the acting DOD comptroller, presented the budget which added to slightly over $41 billion, essentially the same as last year. The features of interest in connection with our budget paper are that the B-70 is to be continued at $74 million, the B-58 to be built up to 3 wings, and Atlas and Titan to 27 squadrons. As on previous occasions when the Air Force program was presented, the President showed his distinct displeasure by suggesting that too much money was being spent on aircraft if the Air Force was confident in the success of the missiles, and if it wasn't, then too much money was being spent on missiles. He made a remark about having been lectured to several times about how wonderful these missiles will be, a remark which I feel was a little too personal but it didn't register too much with the others. McElroy, in his presentation, threw us a little bouquet by saying that: "Kistiakowsky and PSAC were of tremendous help to my office in planning the budget." Perhaps that was just using us as a shield, although it did sound like a bouquet. Gordon Gray called the President's attention to the changed statement of national policy, which emphasized the use of nonnuclear weapons, and the President asked Twining what was the situation in this respect. He got a completely noncommittal reply that of course the national policy still laid great stress on nuclear weapons and, therefore, those were "a little ahead of the nonnuclear weapons," but the latter would catch up soon.

1:00–2:15 PM Lunch at French Ambassador Alphand's house in honor of M. Moch. Very French. Most of the conversation was in French, even though Phil Farley and his wife were obviously not in command of the language. I turned on my maximum charm, including hand-kissing, etc., and suspect I made a favorable impression. Rather enjoyed myself. At lunch I sat next to Mme. Alphand, who is a stunning looking woman, very young for her admitted 39 years of age.

2:30–3:30 PM Mr. George Bunker, chairman of the board, and Mr. William Bergen, president of the Martin Company, visited me at my invitation, which I extended to Bunker this morning. This was done on the request of Benny Schriever, who yesterday evening explained that he was getting desperate because he couldn't make the Martin Company management realize the seriousness of the situation and would I do something about it personally. I used the soft approach with the Martin people, explaining that I was personally much concerned about the Titan, having been its supporter from the very beginning; that I was substantially responsible for having the Titan program approved a year ago when DOD wanted to have it canceled, but that I was now reaching a point where I couldn't give it my support any more and that it would probably be canceled shortly. Both of my visitors got very emotional and claimed that if only they could be left alone [by the Ramo-Wooldridge Co.'s management team] they would have everything straightened out; that the difficulties were not serious, just a little lack of discipline which is now being corrected and that in the next few weeks they expected several flight tests and would have the program back on the rails. I kept up the gentle pressure, emphasizing that this was their last chance. We parted on speaking terms, but obviously they are not in a state of happiness. The conversation may have helped Benny.

> Bernard A. Schriever, commanding general of the Air Research and Development Command.

4:00–4:30 PM   Phil Farley, who explained the background for the planning of an arms limitation organization in the State Department. He showed me a copy of a memorandum from the President to Herter in October, which said that the President wanted a permanent organization for the study of arms limitations, and that probably this should be within the State Department because of the unfortunate experience with such an organization in the White House Office. He also showed the reply to it and a proposed organization chart, which consists in beefing up Farley's office for operational responsibilities and setting up on a permanent basis a planning group somewhat like that now working for Coolidge. We discussed the matter and Farley conceded, after my challenge, that the State Department would have difficulties getting money for this operation from Congress, getting cooperation from the military, and getting good men to work. We parted with the understanding that Killian's PSAC panel will continue working on its own organizational plan, since the matter is not completely closed.

> The reference in the President's memo was to Harold Stassen, who as the presidential assistant for disarmament, with cabinet rank, displeased John Foster Dulles, then secretary of state, and was promptly eliminated in 1955.

Monday, November 30, 1959

At Brookhaven with McCone, Floberg, and the AEC staff, listening to presentations by a panel of all the AEC laboratory directors about their needs in high-energy physics. The meeting got off to a bad start because McCone was delayed by weather and this snafued the program. I have taken detailed notes during the meeting, but they can be simply summarized by saying that as time progresses every builder or operator of a large accelerator machine discovers that he needs much more in the way of facilities and expensive equipment than he thought earlier. Some of it is clearly due to the change in the character of experimentation: the easy things have been done and the tough ones require more laboratory equipment. The result is that all the laboratories are clamoring for budget allocations almost comparable to the original costs of the machines. The estimates of a year ago by the PSAC/GAC panel that to meet the needs of high-energy physics the budget must rise to about $135 million a year five years hence, are completely unrealistic. To put into effect the program suggested by the present panel would cost more than $200 million a year. McCone is quite obviously shaken by all this and intends to do some thorough program review and certainly not to initiate any new machines in the immediate future.

> The Brookhaven National Laboratory on Long Island is the only one of the AEC national laboratories not to engage extensively in classified research.

On the flight home had a long talk with McCone. He is a strange person—so intelligent and well informed in certain areas and so completely uninformed in others. For instance, he started with a rather emotional argument that our missiles were under-designed and couldn't carry big payloads, which was just not good. This may have been the beginning of a tirade to resume testing, but if so, it was squelched by my pointing out that when Titans get operational, more than 8 megatons of warhead yield will be on them, and that even the redesigned Atlas nose cones could have warheads of several megatons yield.

While I was at Brookhaven, I had a very emotional telephone call from Keith Glennan who said BOB insists on cutting $150 million from his budget and this is catastrophic. He announced that in his compromise reply he was cutting out the Nova engine. When I said this was absurd and suicidal for the United States, he asked my permission to quote me to Persons. I so authorized him. When I got home (9 PM, into an uncleaned apartment) I talked to Glennan again, who is still very upset about it all, and he asked me to talk to Persons.

Tuesday, December 1, 1959

8:30 AM   NSC meeting, which started by a presentation by Charlie Coolidge of the preliminary findings of his group. It was anything but an effective presentation revealing essentially negative results. Coolidge emphasized that it was hopeless to present package deals of successive stages of disarmament because they were too complicated to encompass in advance. He recommended instead to present the statement on the ultimate goals and this he handed to the President. It stipulated that we want to see a world governed by law, rather than force; that gradually the UN be given armed forces so strong that no nation or group of nations (which in the meantime have undergone disarmament) could successfully hope to oppose the UN forces; finally that no weapons of mass destruction be retained by individual nations. Coolidge went on to say that the conclusions of his group were that it was impossible to develop monitoring controls for a reduction of nuclear warheads. He then said that they had reached the same conclusions on the all-out war, and that in any case before the Soviets would discuss control of missiles, we would have to surrender our foreign bases. Having reached this essentially negative conclusion on the impossibility, or at least the extreme improbability of disarmament, he then went on to suggest some very small initial steps, the purpose of which was primarily to test the sincerity of the Russians. One of them was the adoption of the so-called Norstad plan, which as I recall involves initially just an agreement not to increase forces in Central Europe and then gradually moves to reduction of these forces, limitations on deployment, etc., and introduces increasing measures of inspection.

After Coolidge's presentation, Herter made a brief remark which sounded as if he was desperate because he had to go abroad and talk very soon about disarmament and Coolidge gave him nothing to talk about. In the course of these remarks, however, Herter in fact confirmed Coolidge's statement by saying that he had discussed missile controls with General Goodpaster as an officer of the Engineer Corps and that Goodpaster was convinced that 100 missiles could be easily hidden from inspection, and if everybody else disarmed, such a force would provide a nation with complete control over the earth. The President spoke up voicing a strong belief that notwithstanding Coolidge's statement, as he put it, we must find ways of control and later elimination of the means of delivery for nuclear warheads. He conceded that nuclear warheads by themselves could not be controlled, but also wouldn't be very dangerous if there were no ways of delivering them on another nation. This gave me an opportunity and so I called to the attention of NSC the fact that almost two years ago I reported to it on the study of a PSAC panel which looked into the feasibility and advantages of seeking a cessation of missile tests, and that the conclusions were negative. I suggested, however, that times have changed; that we now had operational missiles and, therefore, the situation was not as asymmetrical as it was two years ago. I suggested, therefore, that the question of cessation of missile tests, coupled, of course, with controls, could be reopened with some hope that a realistic proposal could be developed which would not be to our disadvantage. I noted that the time during which a start on missile controls was feasible was rapidly running out and that a few years hence, when comparatively compact solid-propellant ICBM's became available, they will be as difficult to control as nuclear warheads. I conceded that Goodpaster was right about the concealment of 100 missiles, but suggested that if these missiles had to use LOX and launching facilities of the kind that are involved in present missiles, it would be impossible to launch a salvo of operational missiles.

At the end of the NSC meeting, McElroy was presented with the Medal of Freedom by the President, with a fine citation and some warm handshaking. Gates is succeeding. I went over to Gates who was sort of standing alone, and said there was a silver lining to every cloud and that I was looking forward to working with him. He clearly seemed to appreciate it, and it didn't cost me anything to say. Actually, I do like him, and find working with him rather easy. He is a much better thinker and has the facts better at his command than McElroy. He is also much more interested in the internal Defense Department problems.

Shortly after the NSC meeting, I met with Persons, Morgan, and Goodpaster, on the NASA budget. The situation at present is that NASA has been going down with its budget total and has finally reduced it by $130 million. This, however, has been done by proposing to cancel the Nova engine project as well as the Vega vehicle and taking bad cuts in some important activities. Stans wants to cut another $130 million and it is this prospect which is getting Glennan so emotional. We talked inconclusively about the whole issue and finally Persons told me to talk to the NASA people and acquire a better understanding of the details of their budget. This I did later in the morning, meeting with Dryden and Horner, since Glennan was out of town. The meeting was very exasperating, but also quite an eye opener. There is no doubt that Dryden, who is supposed to be a scientist, is completely controlled by two considerations: politics, and empire building of the old NACA organization. All proposals for cuts I made he rejected as politically impossible and when I suggested that $80 million for construction and equipment of NASA facilities was too much in a budget in which the Nova engine had to be dropped, he got quite angry. Horner, however, pointed out that if Nova were cancelled, we would lose many years restoring the project. We discussed Saturn B and Dryden rejected all suggestions for reorienting it so as to combine it with Nova. This is an interesting new development, since before the ABMA had been transferred from the Army to NASA, he and Glennan assured me that such a reorientation was feasible and would be undertaken by NASA. It was on the basis of this assurance that I endorsed to the President the plan of transferring ABMA. The meeting ended by my suggesting that they did not have a space program, but only one to feed the many hungry NASA mouths, which didn't set well with Dryden. When they left I wrote a memo to Persons and used that same description of the budget. Later when I saw him, he seemed to rather relish my description. It may become a small classic.

Met Mr. [Howard] Simons, a former Nieman fellow at Harvard and now a free-lance writer, who agreed to be my next speechwriter. Hope it will work out.

Then had a visit from BOB people to discuss AEC military reactor projects. We agreed that BOB will request that in the ANP program only one project be continued but that that one be given adequate funds. BOB wanted to cancel the Pluto, but I think I persuaded them not to cancel that one. Rover will be carried on a slightly reduced funding, rather than being expanded into an all-out effort.

ANP: The aircraft nuclear propulsion project included the development of two competitive nuclear power plants for aircraft. Eventually first one and then the other were cancelled, after more than $1 billion had been spent.

Pluto was to be a nuclear power plant for a supersonic long-range missile (SLAM) and was being developed by the Livermore laboratory. Rover was to be the nuclear rocket engine and was being developed by the Los Alamos laboratory. All three were later canceled after a total expenditure of over a billion dollars.

NACA : National Advisory Committee on Aeronautics, which
was established in 1915 during World War I and through its
development centers was responsible for most of the technical
innovations in American aircraft before World War II.

Hugh L. Dryden, a physicist from the National Bureau of
Standards, became the director of NACA in 1947 and remained in
this key post until 1958. One of the consequences of Sputnik I
was the creation of NASA : the National Aeronautics and Space
Administration, which at birth in 1958 was but NACA, minus the
committee, plus an administrator, T. Keith Glennan, who came
from the presidency of the Case Institute of Technology. Hugh
Dryden became the deputy administrator, holding the loyalty
of senior NACA personnel who took over the top managerial
positions in the rapidly growing NASA establishment. The expanding NASA activities competed with the space plans of the
Air Force. They also meant rapidly growing budgets, from the
small beginnings of the NACA budget in 1958. As the Budget
Bureau was and is intensely allergic to any budget growth, the
conflicts and Glennan's tribulations here described were unavoidable.

At 4 :00 PM had a visit from Dick Morse. He is also in an emotional state. It appears that Secretary Brucker is getting really
furious about Nike Zeus since nobody has officially told him that
Nike Zeus will not be put into production and will be carried on
only as an R&D project, although unofficially he has been informed. Now Brucker feels that this is going to be his political
end, and he is building up steam to have a press conference and
blast everybody. Dick Morse, having in principle accepted our
recommendation not to start production of Nike Zeus, now finds
himself in an untenable position. He showed me his memo for the
record of the conversation he had with me regarding Nike Zeus,
and told me he had given this memo to Brucker to read, but not
to retain. When I read the memo, I hit the roof myself, because
he attributed to me a number of statements which I did not make
and, moreover, in each case represented me as asserting that this
was the detailed opinion of the President; actually during our
conversation I did not refer to the President at all. If Brucker
has photocopied that memo and shows it to the press, not only is
my goose cooked, but it will be a scandal of the first magnitude. I
urged Dick very strongly to rewrite the memo and told him that
if that memo were kept on the record, I would refuse ever to talk
to him privately again. Later that afternoon he telephoned me
with the strangest story of it all. He said that the memo he

showed me wasn't the one he intended to show and that the one he showed was not seen by Brucker or anybody else. The one he intended to show and the one seen by Brucker was very much milder.

4:30 PM Pre-press conference in Jerry Persons's office. Totally uneventful, except for a joke which was enjoyed by everybody present, including Mrs. Wheaton. The joke came about because everybody expected at tomorrow's press conference some questions about birth control, in conjunction with the Catholic bishops' maneuvers to eliminate Kennedy's chances as a candidate. The joke is: "A young lady was going to be married and went to see her physician about birth control. The physician said, well, just eat an apple. She asked before or after, and he said, no, instead."

Ann W. Wheaton, assistant press secretary to the President.

5:30 PM Charles Mohr of *Time-Life*, who is to accompany the the President on his trip and write an article on him as the man of the year, came on Hagerty's suggestion and stayed almost an hour and a half. The main theme of questioning was related to space. Is it true that the President has lost all interest in space? Is it true that his scientific advisors, including myself, discount space, etc., etc.? I may have said too much in return, but emphasized that these were my personal opinions. I tried to present a balanced picture of space activities and other activities in science and technology as they affect national welfare through intellectual, economic, and political progress. I noted that only yesterday the high-energy nuclear physicists at the Brookhaven meeting accused me of selling out science for space, and suggested to him that I might after all turn out to be rather fair and objective. He took it all down and I don't know what use he will make of it.

Had supper with Killian and Rabi at the Metropolitan Club. Killian was just back from a session with Secretary Herter, as one of his advisors. Apparently my remarks at the NSC meeting this morning were not fully understood and were causing puzzlement, which is very unfortunate. At any rate, what Killian suggested that I meant to say, was exactly what I actually did say, so I guess this has been straightened out. I understand that Gordon Gray will ask me to present an extended set of remarks on the same subject at a forthcoming NSC meeting. During dinner we went over various PSAC activities. Killian and I discovered that we both suffer from the same kind of insomnia while in Washington, i.e., waking up about 5:00 AM. Killian left to catch a plane and Rabi and I continued the conversation. Before Killian left, we talked about my successor, but I didn't get much help because they simply emphasized that this job is not one from which it is possible to resign without having a first-class replacement. Rabi even talked about my having to continue long into the next administration, but Killian came to my rescue and pointed out that that wasn't so. The only man considered by them as a possibility is Harvey Brooks. I like him very much too, but would have a strong feeling of guilt taking him away from Harvard and a very promising academic career. I would have to find an older type instead.

> Harvey Brooks, dean of applied physics and engineering, Harvard University, and member of PSAC.

⌒⌒⌒ Wednesday, December 2, 1959

7:30–8:30 AM   Pre-press conference breakfast. Uneventful. Birth control again a major topic of conversation. Also steel strike and the President's trip. I got no assignments to write.

> What was involved, as I recall, was public agitation that the United States offer its help regarding birth control techniques to the underdeveloped nations which would request such aid, as a part of its general economic assistance. I might note that at an earlier date I had suggested to the President that PSAC as a part of its future involvement in life sciences consider also the technical aspects of birth control and was told not to do so. In 1962 then, as the chairman of the newly created Committee on Science and Public Policy of the National Academy of Sciences, I initiated a project on this subject which resulted in a report, "World Population Growth." This, it seems, was influential with the Kennedy administration and it generated extensive and favorable press comments.

9:45–10:30 AM   Briefing the President, which was quite amusing when the subject of birth control came up. The President got quite steamed up when the proposed text was read to him, and he refused to reproduce it to the press. He insisted that birth control was not a subject in which the federal government was or should be involved and that was what he was going to say, staff members notwithstanding. Besides, he emphasized, it is our basic policy not to interfere in other nations' affairs. If another nation wants to introduce birth control, we certainly can't apply pressure not to do so. The question came up about nuclear test resumption after January 1, and Hagerty mentioned that McCone announced at a press conference that we may prolong the moratorium but only on a "week-to-week basis." The President was obviously annoyed by this statement and said he wouldn't accept it. When the steel strike issue came up, the President got angry and used strong language in referring to steel management which just sent out its "last offer," long before the legal deadline and thus created the impression that negotiations have been broken off.

10:30–11:15 AM   Another meeting with Persons, Goodpaster, and Morgan, regarding the NASA budget. They read my memo of yesterday, and I pointed out the items in the budget which are empire building, for instance, almost $3 million for additional facilities to be used in conjunction with work on nuclear reactors, which has only the remotest connection with space and should be done by AEC. They generally agreed with me and the plan of action is to offer to Glennan somewhat less money than the last NASA proposal, but substantially more than Stans wants to give, or to accept Glennan's lowest figure; but in either case insist that such acceptance is conditional on NASA revising the program, throwing out all gold-plating and financing real projects. It was decided that this couldn't be done quickly enough to have the President approve the budget before departure; that I would probably be assigned by the White House to monitor a NASA review of the budget (What a lovely prospect. This is the way to make friends!), and the revised budget, with my endorsement, would then be sent to the President for approval in Rome or somewhere.

1:45–2:35 PM   Meeting with Persons, Stans, Goodpaster, Glennan, Dryden, and Horner, concerning the NASA budget. The situation was briefly summarized and then I was quickly put on the spot by Jerry Persons, who asked me what I thought of the proposed stripped-down NASA budget. So I said, one example of places where money could be saved were additions to the plant and I called attention to nearly $6 million for a building to house the group working on Project Mercury, but which would become available only after the completion of Project

Mercury, and to some other buildings that looked rather plushy. Dryden became exceedingly indignant and insisted that these were absolute necessities, that NASA has an orderly plan for rejuvenation of facilities on a ten-year basis, and not to have good facilities means having overhead and no useful output.

Conversation turned to Saturn B and I pointed out that when private discussions regarding transfer of ABMA were begun, NASA had assured me that they could reorient Saturn so as to combine it with the Nova project into a single superbooster project; that I so assured the President and recommended transfer of ABMA; that their present attitude, which treats these two as entirely separate, was at least embarrassing to me; and finally, that I had exceedingly little enthusiasm for the present Saturn. Goodpaster joined in, emphasizing the President's emphatic desire that there be only a single integrated superbooster program. The NASA boys squirmed and admitted that at the time I referred to they made such assurances but that now they found such reorientation was impossible because von Braun had made so much progress. I asked how much money was already spent on Saturn and they said $70 million. I then asked how much would be spent before we would have a useful booster, and the answer was about $300 million. So I asked whether it was a good policy to throw $300 million good money after $70 million bad money. The answer was that politically Saturn B could not be stopped as it would throw most of ABMA out of jobs. So at least that point was cleared up and there followed a rather heated general discussion, the details of which I don't remember, with the NASA boys insisting that they must have not only the $825 million asked for by them, but another $25 million to include Nova in their program, and Stans insisting that about $700 million total would be enough. At this moment I had to leave hastily to catch a plane, which had to be held a few minutes for me at the airport.

⌒⌒ *Thursday, December 3, 1959*

Arrived in Washington just in time to go to the NSC meeting, most of which was devoted to a presentation on our mutual security program, by Dillon who spoke about economic assistance, and Jack Irwin about military assistance. The most interesting part of the discussion occurred when the secretary of the treasury argued that we should drastically cut foreign aid, and the President very vigorously replied that supporting free people was a much better way to maintain peace than buying missiles and navy ships. He also firmly stated that he would not agree to our unilaterally telling those countries which have become financially capable of maintaining their own military forces that we will cease giving them military assistance [another proposal of Secretary Anderson]. The President said that this was not the way to keep friends and that cuts should be done gradually and only after conversations with our Allies. There was also much discussion as to whether the State Department figures on the magnitude of the Soviet and our foreign aid programs were a fair comparison. McCone, for some reason not clear to me, insisted that all the technical people working abroad for private U.S. concerns should be added to the government employees working in mutual aid, which of course would change drastically the ratio of the numbers involved.

After the NSC meeting, Jim McRae came in and told me he would accept membership in PSAC, and we discussed that and the make-up of the new missiles panel, and then had lunch together and talked about PSAC problems. In the afternoon I had a 45-minute visit from Harold Urey, who started by saying that he has been told repeatedly that I am against the space program, and am trying to stop it. Followed an at times slightly heated conversation, in which I emphasized the importance of a balanced over-all program in science and technology and also a sensible space program, and not just a silly attempt to overtake the Soviets without the means to do so. He left somewhat mollified, but not convinced, and will probably remember some of the things I said and use them against me.

> Harold C. Urey, professor of chemistry, University of California, San Diego.

Had a phone call from Glennan, who said that after the meeting Persons suggested that he accept $825 million as the budget, if he could somehow keep Nova going on this money, and that he agreed. Persons called and told me that he took this figure to the President, who approved it, so the matter is settled.

I forgot to note that after the NSC meeting I had two interesting conversations. One was with Secretary Herter, who expressed his extreme distress at the "accomplishments" of Coolidge and his group. He said that as Coolidge was briefing NSC, it seemed at times that he was working on rearmament rather than disarmament, and I expressed equal distress. Herter said then that something very vigorous had to be done to remedy the situation and I agreed with him. God knows what that will get me into.

The other conversation was with Maurice Stans, who evidently didn't know of the President's decision about the NASA budget, because he said he could not accept a figure as high as $825 million. He then said that Persons talked to him about the NSF budget and my concern about it; that he was acting on the President's suggestion that NSF could get along with just what it had a year ago and that it was only out of his own generosity that he raised the $160 million to $182 million. Furthermore, that Waterman was accepting $182 million. I then put as much emotion into my expression as I could and pleaded with him that he was condemning the future greatness of the U.S. for a measly $10 million, etc. I seemed to be rather effective, because at the end he agreed to look once more at the figures and maybe add a little.

Had a visit from Leghorn about a number of matters, including his plan for control of missiles, which is still rather nebulous and not terribly original. Sounds very much like what we talked about in Geneva a year ago, but I urged him to write a memo. He also asked my opinion as to whether he and Wiesner could go to Moscow and talk strictly privately with some USSR Academy people about disarmament. I said I saw no reason why not, so long as they spoke strictly as private individuals. He then suggested that such private conversations could be followed by a meeting of representatives of the U.S. and USSR academies to carry substantive discussions and to this I strongly objected since the two academies are in very different positions vis-à-vis their governments. I said I favored discussions arranged by the State Department, like the experts' conference in Geneva in the summer of 1958.

CHAPTER VI

Meetings with the Vice-president

THE MONTH OF DECEMBER brought several meetings with the vice-president, who, as the diary reveals, seemed highly intelligent though calculatingly cordial. My attitude towards Richard Nixon before I joined the White House Office was quite negative because of memories of his ugly congressional campaigns against Jerry Voorhis and Helen Gahagan Douglas, his persecution of Alger Hiss, of whose guilt I was not convinced; and then, after he became the vice-president, his various political activities as the administration's hatchet man. It was this conditioning which led me to be surprised (and dubious about his motivation) at hearing very reasonable statements by Nixon in the infrequent White House meetings he attended. These doubts were reinforced when in August or September on an oral instruction of the President I personally took to the Capitol a copy of the McRae panel report to show (but not to keep) to the vice-president. To my surprise he manifested no interest in the report that assessed the urgency of our resumption of nuclear tests but spent the entire time of my visit talking about himself.

Throughout my term of office Nixon was cordial and on the few occasions of more or less private meetings he persistently explained to me his political beliefs and ideas. From my limited vantage point Nixon seemed to play only a minor role in the activities of the White House but he had friends there and in the top executive echelon. My own attitude to him remained skeptical to the end of my stay in Washington.

Friday, December 4, 1959

12:00–12:45 PM   Visit to Vice-president Nixon's Senate office to report as requested on our activities. He however spoke at length about his views, which I found very sensible and, in fact, enlightened. For instance, that he considered the suspension of nuclear tests of no value by itself, but of great importance as a first step in attempts to reach arms limitation; his conviction that attaining arms limitation is completely indispensable because a nuclear war catastrophe through accident can easily occur in the future; that in his opinion we were stuck with the space program, not of much value by itself, but unavoidable politically. Generally I had a feeling that although I was invited to come in order to provide information and express my views, actually he was trying to impress me and, should I add, successfully. I still wonder, however, whether he would have the strength of his convictions if an occasion arises. I was able to clarify some of the things on which he was uninformed, such as the degree of technical urgency of the resumption of nuclear weapons tests, a subject on which he was briefed recently by Teller (need I add more?). I explained the purpose of Teller's radiation weapons and also told him about the Latter hole problem. So, on the whole this was a satisfactory visit. Another reason the vice-president may have had for asking me to his senate office was that on this particular day a herd of photographers were taking pictures of him during all of his activities, and we were duly photographed together and filmed, while both of us spoke animatedly. This will clearly establish the fact that the vice-president is in close touch with the special assistant to the President for science and technology.

> The film, I learned later, was to be a small part of a movie, "One day in the life of the vice-president," to be shown in 1960. I was not invited again to the office of the vice-president.

2:30 PM   Had a visit from Walker Cisler, who showed me his memo to the secretary of defense on the required managerial setup of the missile test ranges to bring order out of present chaos. His proposal to create a director of scheduling and operations directly under the secretary of defense, with specified authority etc., I found excellent, and hope it will be put into effect. McElroy on his last day in office gave a general approval, but, of course, it will be up to Tom Gates to put it through. It will be a terrible blow to a lot of people because, in effect, both Navy and Air Force will lose control of their ranges, although still providing the money for running the operations. It will also take whatever little control NASA has over its own installation away from them and put it in DOD. I assured Cisler that I would support his recommendations whenever necessary.

Monday, December 7, through Thursday, December 10, 1959

Monday and Tuesday I spent with Jim McRae and George Rathjens at BMD in Los Angeles, listening on Monday to their missile program and on Tuesday to their space program. Much of the time during the missile presentation was devoted to a discussion of what to do about Titan and the Martin Company. We were shown the record of ''mishaps'' which have prevented Martin Company from flight-testing missiles now for eight months. It is a dismal record, but according to Gen. Osmond J. Ritland [the commanding officer] the Martin top management is not yet convinced that something is seriously at fault and refuses to take heroic measures. On Tuesday we had a general briefing on the military satellites, which emphasized that BMD is suffering greatly from contradictory orders which they receive almost on a weekly schedule from Washington and which keep changing their program. BMD acts as if they were at the point of despair. They will clearly object to the PSAC recommendations because they believe that ''readout'' SAMOS is much more promising than ''recovery'' SAMOS and also feel that the polar communications satellite is the system on which to concentrate in that field. In both cases our recommendations go diametrically opposite to these feelings. Jim McRae and I attended a special briefing on the reconstruction of the technical characteristics of the Soviet ICBM. They have done a really outstanding job. I must try and arrange to brief the President someday, since I am sure he will enjoy it. While we were listening to the briefings, the Titan was going through a countdown at Cape Canaveral, but in the middle of the afternoon the test was postponed. (This missile then blew up on the stand on Thursday, during the initial moments of a second attempt at launching.)

> BMD: Ballistic Missiles Division of the Air Force Research and Development Command. With the help of STL of the Ramo-Wooldridge Corporation as its technical deputy, it was managing the long-range missiles and space development projects of the Air Force.
>
> SAMOS: the intelligence-gathering satellite, the data either being read out and transmitted to the ground electronically or recovered as a package of photographic film released from the satellite and retarded by a retrorocket.

Wednesday George Rathjens and I spent at Edwards Air Force Base and had very good time. It is a tremendous facility both for rocket engine tests (investment already exceeds $70 million) and for the original purpose, testing of aircraft. General Carpenter [commanding officer] was friendly and accompanied us the whole day. News travels fast in the Air Force. He already knew that we have recommended against funding the B-70 and expressed sorrow at my blocking the development of supersonic aircraft. I was given a supersonic ride in the F-100 B, which was interesting, but not exciting. Part of the trouble was that I had a bad head cold and the doctor wouldn't let me fly at high altitude. They have promised a ride in the F-104 at full altitude next time I visit Edwards. That is the one that has wings only seven feet long and drops like a brick when the engine quits, according to Ritland and party, as they teased me and tried to scare me on Monday about this flight.

> Gen. James Wilson Carpenter III, USAF general in command of Edwards Air Force Base.

Thursday was a rather exhausting day, as I landed early in Washington and immediately went to prepare my presentation to the NSC on whether there should be another study to determine the desirability of treaty negotiations aiming at a halt to missile flight tests. I made the presentation as impartially as possible, stating that quite likely the answer would be negative again as it was in the first study, which I chaired almost two years ago. I pointed out, however, that circumstances have changed and may call for a reappraisal. Tom Gates reacted violently and said he flatly opposed such a study since his mind was already made up, that he would never agree to cessation of missile tests, and all the study could do would be to leak out the information to the public, since this is what scientists do. This made me see red, but I controlled myself, and Gordon Gray noted that there have never been any leaks from the PSAC. I said that I certainly had in mind not a study like the Gaither panel but a study by a small group and that it has to be an interagency undertaking. Admiral Burke stated that he also was opposed to the study and his mind was already made up against cessation. Herter then stated rather positively that he couldn't understand this attitude; the French were coming up with a proposal for cessation and that we certainly had to have an up-to-date understanding of the problem even if only to reject the French proposal. Nixon also spoke in favor of the study and the study was decided upon.

Afternoon spent with Greenewalt committee at NASA, which was devoted to restatement of the conclusions reached at an earlier meeting. Nothing new.

I had also a private session with Dr. George Beadle, whom I didn't dare invite to join the PSAC because of political uncertainty, but asked him to join for a special session on life sciences in January. He made a fine impression, and I think he would be a good member of the committee.

> George W. Beadle, professor of genetics, California Institute of Technology.

In the evening had a dinner with Vice-president Nixon and party at a restaurant. Already in the morning and again this evening Nixon started calling me "George." I am not sure that I will be enthusiastic about it, because he leaves me with a strange impression. So far I haven't heard him once make a statement which was wrong from my point of view, and most of the time they are very sound observations, but I have a feeling that the real motivation for these remarks is strongly political and he admits so at times.

Friday, December 11, 1959

8:30–10:00 AM   Cabinet meeting, which was exceedingly dull except for a ridiculous item raised by Karl Harr. Harr warned that the movie "On the Beach" is getting tremendous advance publicity and will have premieres in something like sixteen cities throughout the world, including Moscow; that the movie will strongly arouse public feelings against nuclear weapons and fallout; that, therefore, it would not be in the government's interest to support it, and so he urged cabinet members not to attend the premiere in Washington. Karl Harr, from all my contacts with him (for instance, his emotional defense of all-out competition with the USSR in outer space), leaves me with an impression of an extremely limited intellect.

Following the cabinet meeting we had the meeting of the Principals with the vice-president, secretary of the treasury, and attorney general to discuss recommendations to the President, on his return (from the Mediterranean and Mideast) regarding continuation of the nuclear test moratorium. As usual, Gates took the line that the moratorium must come to an end immediately. He was sharply put in his place by Herter, who asked him what impression would be created in the world if the President almost immediately upon his return from a great and very successful visit to eleven nations, which was undertaken to seek peace, should announce resumption of nuclear tests. The vice-president agreed with Herter and then asked me to present my views on the urgency of the resumption of tests. I noted the conclusions of the McRae report and then we had some discussion of the one-point safety tests. Thereupon discussion became general and utterly inconclusive. It is really appalling what a state of indecision we are in.

3:30 PM   Left Washington to be the moderator at the Harvard Law School Forum, devoted to outer space. As this was presented after cocktails and dinner, I was in good form and, I am afraid, quipped a little too much about the speakers and took vigorous part in the debate, which didn't sit so well with one of the speakers, a Mr. George Feldman, now on the U.S. delegation to the UN, and formerly chief counsel for the House Select Committee on Outer Space. Actually, Feldman's remarks showed a lack of consistency, because he discussed favorably the question of the right of peaceful passage as applied to satellites, but then defined it in such a way that it would be impossible for us to use reconnaissance or early warning satellites (he took exception to my suggestion that his interpretation of the right of peaceful passage was equivalent to being blindfolded and handcuffed prior to crossing a neighbor's lawn). He then got very emotional about dangers from Soviet satellites carrying nuclear warheads and wouldn't accept my assertion that this was not a useful weapons system. However, when it was all over, we seemed to be on reasonably good terms.

⤳ Monday, December 14, 1959

An all-day PSAC meeting. We had an active and long discussion during the executive session which resulted in our covering only half the agenda. The afternoon was devoted to the presentation by the Arms Limitation Panel. Most of the committee was favorable to the presentation except Rabi, who grew quite violent and insisted that all we would be doing according to this plan would be increasing the armaments race and that anyway he didn't believe in the concept of secure deterrents which is central to the plan. Bacher told me on Wednesday that he spent a long time on Tuesday evening with Rabi, trying to alter his attitude, but was unsuccessful, and that Rabi is acting almost irrationally on the subject.

⤳ Tuesday, December 15, 1959

PSAC meeting continued. Most interesting was the presentation by Bode and Ling of the findings of the interagency panel on the evaluation of the U.S. and USSR accomplishments in outer space. The presentation was longer than necessary and stimulated so much discussion that the rest of the program had to be sharply curtailed.

The afternoon I spent in York's office with Dick Bissell and a number of others discussing the troubles with the Discoverer program. It is a very distressing situation, but we ended by urging that flight tests of the entire system continue since it seems that no test conditions can be discovered on the ground to simulate the peculiar failures which occur in orbit.

> Discoverer program: an early military intelligence satellite project.

Went to General White's dinner in honor of General Pate. A big affair, at which General White for the first time was friendly, so we ended agreeing to call each other Tom and George. The dinner was about as boring as a formal dinner can be.

⌒ *Wednesday, December 16, 1959*

Day started by an hour with Miss Thomas again, who is persisting in trying to immortalize me by writing my biography, together with those of other "leaders of space." Clearly it is a balanced proposition: she makes money and I make immortality, but I have a strong hunch that her biography will be absolutely awful since she has no knack for asking penetrating questions.

Bob Bacher dropped in to tell me that Rabi is still firmly and vehemently opposed to the general idea of the disarmament plan proposed by the Killian panel yesterday because he is opposed to the concept of stable deterrents and has a notion that disarmament can be negotiated quickly.

10:00–10:30 AM General Starbird and a colonel from AEC, discussing the proposed AEC letter to the President. I expressed misgivings about the proposal that the President make a statement on the one-point safety tests at the same time he will announce extension of the moratorium, because he made no reference to laboratory programs in his prior statements about the moratorium and this change would arouse suspicion. It was agreed that a different tack will be tried: the President will make a very bland statement and only if the reporters get nosy will Los Alamos put out a more detailed statement. Finally, if that is not sufficient, the White House will make a still longer statement. These drafts I am to look at tomorrow.

11:00 AM–12:00 M McCone in my office, alone. This was a very important meeting. I suggested that we might see whether we couldn't agree on a policy recommendation to the President regarding nuclear test cessation. McCone started with a fairly emotional statement accusing scientists of having gotten us into the mess we are in now, politically, and referred to the Puerto Rico meeting of the PSAC as having started it. He mentioned further how bitter Foster Dulles had been about this shortly before his death. I took exception, noting that the initiative was that of Dulles who told the President early in 1958 that he couldn't carry out U.S. foreign policy unless tests were suspended and that the group which concluded that test monitoring was feasible contained only one PSAC member and otherwise included agency representatives. This cooled off McCone and the rest of the discussion was quite friendly. We have agreed on a joint recommendation which is likely to become policy since it falls half way between State and Defense points of view. It is to disregard the potential evasion of monitoring by means of far outer-space tests or underground tests using big Latter holes, and toss the hot potato back to the Soviets by stating that we will accept a suspension of tests down to a threshold yield which will be determined by the number of on-site inspections acceptable to USSR. Thus the fewer inspections they insist on, the bigger the weapon it would remain possible to test as below the threshold. I reached the decision on this plan because I am now convinced that a comprehensive treaty would not be ratified by the Senate since AEC, DOD, and Teller will all testify in opposition. Of course, McCone can easily reverse himself, but if he sticks to his commitment to this plan it will be a substantial success for me after the incredible morass of indecision in which we have been for several months. Compared to the previous McCone attitude, the proposed plan represents a major liberalization.

At the end of the meeting we discussed quickly the high-energy particle accelerators and I suggested that the machines being built by Harvard-M.I.T. and Princeton-Pennsylvania, as well as the Berkeley machine, be declared national establishments rather than those of local institutions, if they are to be given additional federal funds. McCone liked this proposal very much, and when I tried it yesterday on PSAC nobody murdered me.

12:15–12:30 PM   Alan Waterman, with apologies for his inability to assist me in increasing the NSF budget. By now he has thought out a perfectly logical justification for his position, so I was sympathetic and even apologized for not having gotten him more. He was soothing in return. He said that the $3 million extra which I got for him will be valuable.

12:30–1:30 PM   Bronk [president of NAS] for lunch at the

White House Mess. From my past attempts to influence him, I learned that he was firmly opposed to getting the membership of the NAS involved in any federal advisory activities. Now he came to talk to me about the Carmichael-Nolan proposal, which I had pushed hard in the Federal Council, so that the council endorsed it. It calls for the National Academy membership to study and present the most important priorities from the point of view of pure scientific knowledge, and human welfare, and then for the Federal Council to compare these recommendations with the so-called research priorities of the federal agencies concerned, as presented to it. I hope thereby over a period of years to achieve an increasing influence on the part of the Academy, that is, of the cream of our scientific talent, upon government planning and, incidentally, increase the stature of the Academy. Now, during lunch, Bronk, who had already been approached by Carmichael, represented the whole idea as being essentially his own initiative! But naturally I did not deny his authorship because that is the way to get him guiding the project effectively. So we discussed the details, from which it was clear that he is anxious that I do get involved in the operation, for instance by speaking to the Academy, etc. I refused to write letters outlining my long-range plans, because that may get me in Dutch, but volunteered to present them orally.

1 :30–2 :00 PM Admiral Russell to discuss our proposed PSAC three-day meeting at Key West. Apparently the only housing facilities are rather antiquated BOQs, and it may not be even possible to provide sleeping accommodations for the secretaries. We suggested that perhaps we all should stay at a motel but he doubted whether reservations could be made for such a large group.

2 :30–4 :45 PM NSC meeting. Nothing of great interest except a rather vigorous discussion between the vice-president and Dillon, in which Nixon sharply criticized Rubottom for being unwilling to even discuss the so-called Cuban refugee policies, which refusal, in Nixon's opinion, was simply doing nothing. He warned Dillon that politically the Cuban situation was getting very explosive and that even [Nelson] Rockefeller will severely criticize our present policy, which, Nixon said, he has a perfect right to do. Gray said that Rubottom refused to discuss the issue in the planning board; and the attorney general [William P. Rogers] said that his agents, who are mixed-in with the Cuban refugees, know what the refugees want to do, but since they receive no policy guidance from Washington, do nothing useful.

Roy Rubottom, assistant secretary of state for inter-American affairs.

I was never informed about the military training of Cuban refugees by the CIA in Venezuela and this remark of Nixon was the only hint (not then recognized by me) that he was active as a go-between for the White House and the Cuban refugees.

4:45–6:20 PM Space Evaluation Panel meeting with the steering group to which Bode and Ling made their presentation on space. Ling's was outstandingly good and the steering group, comprising Dulles, Glennan, and Secretary Douglas, was very complimentary. So were York and Gray, and the presentation was approved for the NSC-Space Council joint meeting on the 29th.

Spent the evening in Schriever's home. Bennie raised the issue of whether space activities would remain in the hands of the Air Force or would be transferred to NASA in their entirety—a piece of scuttlebutt which he has picked up. He pleaded that the latter course would result in terrible reduction in efficiency. In this conjunction, he made a very disturbing remark, namely that "everybody in the Air Force from the secretary down now thinks that you control the entire military R&D program." I kidded him about it and took it as a joke, but, even though slightly high at the time, I was and am very disturbed. This is the kind of belief that will certainly lead to measures by the military to change the situation; i.e., to leaks of ugly stories to newspapers, etc. Clearly we must be more careful in the future and emphasize more York's decisive role. We also talked about the appalling wastage of technical talent in the aircraft and electronics industries.

Thursday, December 17, 1959

8:30–9:00 AM White House staff meeting to listen to Ling's presentation of the problem of space communications. Don did a good job again and a number of people complimented him and rightly.

9:00–9:15 AM Went over the new draft of McCone's proposed letter to the President concerning resumption of one-point nuclear weapons safety tests and found the new draft completely acceptable.

11:00 AM–12:00 M   Met at IDA headquarters with the "bright young physicists," a group [called Jason] assembled by Charlie Townes to do imaginative thinking about military problems. It is a tremendously bright squad of some 30 people to whom I talked off-the-cuff on the general problems facing us because of too many simultaneous technological revolutions in military hardware, such as nuclearization of warheads, transition to turbojet and rocket propulsion, computerization and automation. I emphasized that new ideas which didn't suffer from complexity and wouldn't involve billions for development were most important and then spoke of the general objectives as being the creation of a secure deterrent and of an effective small force for use in limited engagements. Although feeling slightly hangoverish from last night, I think I did a reasonably good job and my remarks were followed by a very active discussion. This was followed by a luncheon which was just like other luncheons.

Talked by phone with Spurgeon Keeny in Geneva, who told me that things at the Technical Working Group II were going very badly; that no agreement with the Soviets was in sight, and that Fisk was faced with a choice of making unacceptable concessions, or submitting a separate report, or recessing the meeting to reconvene after the New Year. I recommended against the last course of action unless a reasonable agreement was definitely in sight. Also urged Spurgeon to write the report in such a way that an otherwise comprehensive test-ban treaty, but with an underground threshold, would be easier to achieve.

> To remind the reader, the threshold being referred to would be the lower limit of underground nuclear explosions forbidden by a treaty. Explosions below the threshold, being unidentifiable by seismic means, would therefore be permitted by implication.

ᨆ Friday, December 18, 1959

10:00–11:30 AM   Cabinet meeting presided over by the vice-president. The first of two highlights was the discussion by the attorney general of the difficult situation in which the Justice Department finds itself when other government agencies fail to carry out their statutory obligation to report to Justice when substantial allegations are made that a government employee is guilty of a crime. Rogers pointed out that time and again Justice learned of the problem from congressional. hearings and he pleaded, as well as threatened, to change the procedures. In the ensuing discussion dear John McCone got in a lick against the PSAC (what a bastard!) by noting that he is having difficulties with a congressional committee regarding conflict of interest in his General Advisory Committee and suggesting that the PSAC and my office will undoubtedly have far greater difficulties. Rogers noted that this was quite an irrelevant matter since no crimes were involved, and that shut McCone off, but not until after the seed of suspicion was neatly planted.

The other interesting item was a report by Senator Thruston B. Morton on his impressions throughout the country. He was very elated about the Republican victory in the Iowa congressional election and maintained that the change in attitude of the President—coming out fighting—has created an incredible change in the mood of the Republicans. He felt that if the efforts are maintained, the Republicans have a very good chance of not only winning the presidency, but regaining control of the House as well. A very smooth gentleman, who has a clever way of blending optimism with cautionary warnings that efforts are essential.

12:45–1:30 PM Lunch with the deputy secretary of the navy and several assistant secretaries, including Wakelin. Very pleasant occasion, largely punctuated by the Navy people trying to find out from me what was wrong with the Air ·Force R&D program. I obliged to the extent that I thought was wise, condemning the B-70 and much of the long-range air defense plans, but defending missiles and then getting in a few licks about the weaknesses of the Polaris system.

After lunch had a short private meeting with Wakelin, during which he assured me that he has ordered the Navy to take care for the next six months of the ship *Pioneer* owned and operated by the Coast and Geodetic Survey, which was to be laid up because Navy cancelled its contract with Commerce and the latter has no money. This is something about which Secretary Ray phoned me a week or so ago, and I have been trying to get BOB to approve a supplemental budget request for $375,000. So this seems to be taken care of.

3:45–4:15 PM With General Persons, whom I acquainted with the results of my private discussion and agreement with

McCone regarding nuclear test cessation policy. I also discussed the rather catastrophic Titan situation, and gave him the conclusions of the Bode panel on space evaluation to read. Jerry suggested that the White House take over from the Air Force the Martin Company–Titan problem, and force the Martin Company out, but I asked for a little more time, so as not to appear to jump over the head of the secretary of the air force. Persons was very pleased with the McCone-GBK agreement, but I warned him that McCone will probably renege on his word, to which Jerry said he will see to it that McCone doesn't. As regards the Bode report, he said he expected something nearly as bad as it reveals and would like to read the complete report.

6:45–7:30 PM   Attended the vice-president's reception at his residence and found it a surprisingly pleasant affair. It was strictly executive branch, no congressional characters being there. I had a number of pleasant chats with the Morgans, with somebody whose name I don't remember who used to be a White House staff member and is now working for the vice-president, General Lemnitzer, General and Mrs. White (the latter, being English-born, loves to have her hand kissed), et al. As I was ready to leave, the party was really just getting into full swing and obviously wouldn't end by eight, but the vice-president, who, as well as Mrs. Nixon, was circulating among the guests rather than standing and receiving, insisted on showing me his house. The only false note was his taking me around to see all the gifts which he has received from the potentates and other VIPs on his travels. The whole performance was in the style of a museum director. At the end he stepped out of the front door and, calling me George all the time, kept talking to me for about a quarter of an hour on various subjects, including nuclear test cessation. He said he was very surprised that in his latest conversation with McCone the latter has completely reversed his stand and is now very reasonable and recognizes the political realities; namely that we cannot resume tests and that therefore we might just as well reap the benefits we can from an inspection system. I explained (perhaps I shouldn't have done so) that I had a long conversation with McCone and hammered out an agreement on a joint policy proposal and then added that in view of McCone's instability this, of course, didn't mean very much. Nixon raised his eyebrows and laughed, so this may not be a complete surprise to him.

Had phone calls from Beadle and Weinberg, both of whom

agreed to serve on the PSAC committee and I started getting replies from "life sciences" invitees to the organizational meeting on January 4. So far nobody has turned us down.

ᕕᐵᐞ Monday, December 21, 1959

Spent day in office except for a short meeting with Stans, Floberg, and York, discussing ANP. York is alone in trying to continue the work of both contractors. It is also clear that if the GE program is cancelled, York will be under terrific attack, and so none of us drove him very hard. Agreement was reached that York and I will present the matter to the President, since it is largely a political issue, the project being definitely a technical failure.

Had a long session with Spurgeon Keeny, just back from Geneva. He is very unhappy about the failure of the Technical Working Group II to reach agreement on key issues and believes that the Russians were under political instructions to stand pat against an effective monitoring system. In his opinion even a policy based on a threshold limitation of underground tests cannot be developed because of technical uncertainties. I had a phone call from Jim Fisk, who also believes that thresholds are unfeasible, but is inclined to think that we should still go for a comprehensive treaty. I don't think that is politically possible, because of the high probability that the Senate would not ratify such a treaty, after listening to testimony by Gates, McCone, and Teller. So the whole situation is a complete mess and my efforts to reach agreement with McCone have been wasted. What a nice Christmas present!

ᕕᐵᐞ Tuesday, December 22, 1959

Had a phone call from Walker Cisler in Detroit, saying that yesterday he had a thoroughly unpleasant two-hour conversation with York, who refused to accept his recommendation for the management setup over the missile ranges and satellite-tracking facilities (bypassing York's office). Cisler was rather emotional and pleaded for help. I promised it, since his type of management setup can make sense of the incredible chaos and mismanagement which exist now. Late in the afternoon had a phone call from Al Waggoner, saying that this morning in a meeting between Gates, Douglas, and York, with himself present, Gates decided on Cisler's plan over York's protests, and has instructed Waggoner to organize the office, so I don't have to do anything to help Cisler.

Alvin G. Waggoner, engineer, Office of the Secretary of Defense.

1:30–4:00 PM   Federal Council meeting, which was quite uneventful except that dear Wallace Brode got thoroughly under my skin. Foolishly I referred to the National Academy as representing nongovernment scientists, whereupon Brode launched into a rather vehement tirade, saying he wanted to correct my erroneous statement, that nearly 10 percent of the Academy members were government scientists, etc., etc. That man does have a facility for being unpleasant! I didn't argue with him, but conceded my error and still he kept on. Obviously, for some reason this issue is a sore point with him. John Williams handed me McCone's letter regarding the FCST paper on research priorities. It is a combination of silly nit-picking and an announcement that AEC will not make use of the proposed legislation, which is pointless, because the whole policy is permissive only.

After the meeting, Peterson of Agriculture stayed in my office to complain about the shocking actions of Secretary Flemming in the cranberry case, and told how all the farmers, cattle raisers, chicken producers, are up in arms, passing resolutions damning Flemming, etc. He was very pleased to hear that we are organizing a panel to study the general issue of chemicals in food and left with soothed feelings.

Ended the day by talking to Elmer Staats about McCone's letter, who agreed to handle it with a minimum of fuss.

∾ *Wednesday, December 23, 1959*

9:30–10:00 AM   Meeting with Jerry Morgan, Kendall, McPhee, McCabe, and Harlow, to discuss Congressman Overton

Brooks's ''invitation'' to me to testify before the House Committee on Science and Astronautics, in my capacity as chairman of the Federal Council. The conclusion was reached that I could refuse to testify on the space program, which the Brooks letter intimated I would be asked to explain, because the Federal Council is not concerned with this subject, it being vested in the Space Council. As chairman of FCST I could testify on science in general, however painful that may be. Harlow agreed to telephone Ducander, the staff director of the committee, about this and by the end of the day sent me a draft of a reply in which I explained why I couldn't testify on space and invited Brooks for an informal discussion of the subject.

> H. Roemer McPhee, Jr., assistant counsel to the President; Bryce Harlow, deputy assistant to the President for congressional liaison; Edward A. McCabe, assistant to Harlow.

12:30–2:30 PM   John Rubel, in charge of strategic systems in York's office, for lunch and discussion. Talked about Titan and it became apparent that York's office is staying aloof, although keeping informed. Discussed an emergency measure of transferring all Titan test activities to Douglas Aircraft Company, leaving Martin Company in charge only of manufacturing. This appears to be the best extreme measure if it is feasible legally. Later in my office talked about air-launched ballistic missiles and Rubel explained the absurdity of the original project, unwillingness of the Air Force to explain what it was doing, and totally unrealistic so-called operational requirements. He explained that York's office is threatening to withhold funds from the program unless by the end of January the Air Force, which is bypassing BMD and STL and managing this project from the Wright Development Center with Douglas Company as prime contractor, comes through with a realistic development plan and reasonable operational requirements. If that happens, York intends to approve the program and Rubel asked me what would be my attitude. I said we opposed this program partly on the grounds of its unfeasibility, but also on operational grounds. Having made my recommendations, however, I will take no further action if York's office doesn't accept them.

> The air-launched ballistic missile called Skybolt was energetically promoted by the contractor and the Air Force, although having little substance except on paper. This nuclear-tipped missile was to be launched from aircraft flying outside the range of Soviet

interceptors or defensive missiles and so would have obvious tactical advantages (and some less obvious drawbacks). Sometime after he took office President Kennedy was unfortunately sold a bill of goods on Skybolt and in a meeting with Prime Minister Macmillan made a firm commitment to provide Skybolts for the British Vulcan bombers, thus modernizing them. Much cheer was generated by this in London and when soon thereafter the Skybolt had to be canceled as not feasible technically, the reaction caused grave embarrassment to the Macmillan government.

During the day, a colonel from General Starbird's AEC shop was in the office at least half a dozen times with proposals for the exact text of a press release that would be made by Los Alamos in case their one-point safety experiments aroused public curiosity. For once I found myself on the side of the AEC rather than the State Department and rather enjoyed telling Starbird over the phone that I was most unhappy to be on his side, which provoked a loud gasp. This matter, which was in the nature of a tempest in a teapot, was finally resolved to the satisfaction of Dillon, McCone, and myself.

During the day I was in a state of complete bafflement about next week's schedule because of the President's uncertain plans, and only in the afternoon became reasonably sure that I could leave tonight for Boston.

Friend Peterson, who visited me yesterday, obviously immediately leaked to the press that we are planning to study the problem of toxic chemicals in agricultural products and John Finney of the *New York Times* called up the office saying he was told that I had been handed the whole tainted-food problem as a project and was that true. I believe we ended by simply saying "no comment."

Otherwise the day was restful.

CHAPTER VII

Thresholds and Cranberries

THE REPEATED REFERENCES in the following weeks to my own and a PSAC panel's involvement in the problem of toxic chemicals in foodstuffs suggest that it may be useful to sketch the contemporary issue—the cranberry sauce crisis—which created so much public and political commotion.

The cranberry crisis arose out of the presence of trace amounts of the (probably carcinogenic) herbicide aminotriazole (ATZ) in a substantial portion of the 1959 cranberry crop. The control of weeds in cranberry bogs had presented a major problem to growers. It seemed solved in the mid-1950's by the discovery and application of aminotriazole as a herbicide. However, the Food and Drug Administration had refused to establish a tolerance limit for the presence of ATZ in cranberries because of uncertainties about its toxicity. The Department of Agriculture in 1958 approved the agricultural use of ATZ but only for the post-crop season, in order to ensure that it would not be present in the next year's crop.

A substantial portion of the 1957 cranberry crop had been found to be contaminated by ATZ (because it was used before the crop) and was then voluntarily withdrawn from the market by the cranberry-growers' marketing association. The association urged the growers to be careful in their use of ATZ.

Meantime public concern about the carcinogenic action of some chemicals was growing. As early as 1955 the International

Union against Cancer promoted the policy that substances with reversible toxicological actions should be judged differently from those with irreversible, e.g., carcinogenic, actions and urged that no "nontoxic thresholds" for human use be set for the latter. In 1958 an amendment to the Pure Food and Drug Act became law, strengthening the authority of the Food and Drug Administration (although subsequent appropriations did not strengthen the FDA budget accordingly). One clause, the Delaney amendment, which was to come into effect in the spring of 1960, categorically forbade the presence in foodstuffs of any additives shown by experiment to cause cancer in test animals.

Contrary to the hopes of the FDA and USDA the voluntary actions of the cranberry-growers to ensure the absence of ATZ in the 1959 crop did not prove successful, as FDA discovered when early in November they tested several shipments of the berries. Accordingly Secretary Flemming, on the advice of the FDA, announced the discovery of contaminated lots at his regular press converence on November 9 and then conceded that a consumer had no means to tell whether the cranberries on the store shelf were contaminated or not. Public response was widespread and intense. Some supermarket chains (and even some urban administrations) halted all sales of cranberry products. Many restaurants crossed them off their menus. President Eisenhower was rumored not to have had cranberries on his Thanksgiving Day table. Sales dropped to almost nothing. Naturally the response of the cranberry industry was prompt and hostile to Flemming. Governors of cranberry-growing states and their congressmen intervened and characterized Flemming's statement as "a classic example of bureaucracy at its worst." USDA Secretary Ezra Taft Benson criticized Flemming publicly for unjustly hurting American farmers. Vice-president Nixon on a speech-making tour of Wisconsin in search of votes for 1960 ate four helpings of cranberry sauce—"just like the kind mother used to make," according to media reports in a cranberry-growing region. Senator John F. Kennedy, also in search of higher office, drank cranberry juice in the same region and later criticized Flemming, as reported by the media.

The FDA expanded the testing for ATZ until it involved a large technical staff, and Flemming in late November reached an agreement with the cranberry-growers' association to label as safe all their products that had been tested and found free of ATZ. The sales of these products, however, lagged disastrously and many families chose to be without cranberries for Christmas.

The President asked me casually whether he should include cranberry sauce on his Christmas menu and I urged him to do so, in view of the public turmoil. He did.

By mid-December the White House was under strong political pressure to find a means for financially compensating cranberry-growers. The White House was also urged to make sure that future actions of the HEW secretary would not harm other sectors of American agriculture. Specifically, it was feared that Flemming might condemn the use of a sex hormone (diethylstilbestrol) used to accelerate the growth of poultry and cattle.

It was in this context that I had the conversation with Peterson (himself an ex–cranberry-grower, as I recall), and an earlier one with Don Paarlberg, and then proceeded to set up a panel on carcinogenic chemical additives in food, whose main objective was to clarify the administration of the Delaney amendment which created serious problems because of its categorical wording. The PSAC plan to set up a panel on life sciences had no connection with the cranberry sauce crisis, although it proved a lucky coincidence.

The immediate problem confronting me as I returned to Washington after Christmas was the approaching collapse of the nuclear test-ban negotiations. It was the result of a total impasse between the U.S.A. and the Soviet Union on the capabilities of a global seismic monitoring system. It was further exacerbated by the internal conflicts in the administration in Washington regarding the feasibility of identifying seismic signals produced by nuclear explosions of specified magnitude. It was the internal conflict which led me to propose the definition of a threshold in a treaty in terms of seismic signal strength, i.e., a number on the Richter scale, rather than in TNT kiloton equivalents.

Not that this proposal resolved the conflict, but it did make the issues clearer in the Principals' Committee and hence more amenable to compromise. Similarly, in Geneva much of the impasse reached in December involved the relations between the kiloton equivalent of an underground explosion and the strength of the seismic signal generated. If the treaty were to be written so that it specified the threshold in kiloton equivalents, the signatories would not be able to agree, after an observed seismic signal, whether the seismic event producing it was below the kiloton threshold, or was above it and therefore called for an on-site inspection. Drafting the treaty in Richter-scale terms alone removed the ambiguity. Once this issue had been taken off the agenda some constructive negotiations could be hoped for.

As the following will indicate, the selling of this new approach took some time and argument. Its acceptance, however, did indeed reduce the internal conflicts and also led to a major progress toward a treaty that would have been almost comprehensive. Unfortunately, the progress was to be totally negated later by purely extraneous events, namely the U-2 crisis.

Monday, December 28, 1959

11:45 AM–1:00 PM   With Herter, Dillon, Farley, and [Gerard] Smith, discussing the nuclear test cessation problem. When I arrived it was clear from the remarks that the staff proposal—which according to Spurgeon Keeny was to involve a comprehensive treaty but with a stipulation that after three years' time the underground problem would be reopened if in the meantime no satisfactory control methods were established—had been pretty much discarded by the secretary. I then outlined the threshold plan as it gradually evolved from my recent thinking and discussion with McCone. It involves defining threshold not in terms of kilotons but in terms of the size of the seismic signal—that is, of earthquake equivalents. This resolved the problem of decoupling and the Latter hole, in the sense of accepting the risk that the other side will test bigger weapons by decoupling them. Another new aspect is that all tests permitted under the treaty, that is, below the threshold, would be declared in advance and utilized by the control organization for the development of better seismic detection methods. Naturally the threshold would be thought of as subject to gradual reduction, with improvement of the detection methods.

Had lunch with Jim Fisk, who had just arrived, and Spurgeon. Early in the morning Spurgeon dropped in and assured me that he had changed his mind over the weekend and was now in favor of the threshold plan, but at lunch I found Fisk strongly opposed. He thinks it is not negotiable with the Soviets and much too complex and uncertain to be spelled out in a treaty. I asked for alternatives and found him still in favor of a comprehensive treaty, which I feel is politically hopeless to struggle for.

2:00–2:45 PM   Visit from McCone, who is emotionally most

upset about the outrageous political attack by [E. K.] Fedorov in his final statement on Fisk and others. McCone feels a drastic response on the highest level is called for, and hoped for a much more vigorous reply by Fisk than he made in response to Fedorov. We then reviewed the treaty problem and I found him still willing to go for the threshold concept; he accepted my added gimmicks.

> At the end of the sessions of the Technical Working Group II on December 18, Fedorov, the chief of the Soviet group, delivered in public a violent personal attack on the American group and especially James Fisk, its leader, accusing them of unscientific attitudes, subservience to politicians, etc. The report submitted on December 18 by the technical group to the conference brought out basic disagreements between the American and Soviet components of the group.

3 :00–5 :00 PM Meeting of Principals with Fisk and Wadsworth present, as well as Eaton (who will be the chief U.S. representative at the UN ten-nation disarmament committee). The meeting started with Fisk reporting on the session and then proceeded towards a rather relaxed discussion of the threshold concept, which was acceptable to all but Fisk (not emphatically), Wadsworth, and Farley. Wadsworth feels it is completely non-negotiable and he also is deeply convinced that the political ill effects of our resumption of testing would be so severe that a comprehensive treaty is far more preferable. Gates and Twining made statements for the record asking for unrestricted underground testing, but their hearts were obviously not in it. It was pretty much agreed to recommend the threshold-type treaty to the President, provided the threshold concept stands up under critical technical analysis to be undertaken in a great hurry. Wadsworth obviously exceedingly unhappy. We then discussed the question of resumption of tests and agreed on a statement which in effect says that our self-imposed moratorium is over but that we will not resume tests without prior announcement.

F. M. Eaton, a New York lawyer.

Back in the office, had a phone call from Bill McMillan from Rand Corporation, who tried unabashedly to pump me about the decisions just reached at the Principals' meeting. Very strange behavior and I think I will change my mind about making him a consultant to the PSAC. Then spoke to Harold Brown and asked him to undertake quickly a review of the threshold concept.

~~~ *Tuesday, December 29, 1959*

Got up at 4:45 AM (what a life!) to fly to Augusta with the Principals, Fisk and Wadsworth, where we have an appointment with the President at 8:30 AM. Flight uneventful and during it I got approval from Herter and Gates of the proposed terms of reference for the Missile Test Cessation Study (NSC action). Gates keeps saying that the whole thing is a waste of time since we obviously wouldn't touch such an agreement. In the conversation with Herter, I learned that the Wallace Brode situation is most unsatisfactory. Herter called him in and told him that Brode shouldn't assume that his position has permanency or is the beginning of a State Department career, but Brode responded with such enthusiasm about the joys of his office and how he loves it and how he feels that he has made a great success of it that Herter, as he put it, didn't "have the heart to tell him to quit." So there we are. I merely reassured Herter that if he got Brode to quit, I would find him a substitute, but Herter questioned the wisdom of it because of the short time left to this administration (the first visible sign of the "do-nothing attitude," which will certainly overwhelm us in the next few months). The meeting with the President was really quite unsatisfactory. He was obviously tired and impatient with the whole subject. His difficulty in expressing his thoughts was far greater than I have ever encountered before, and the result was that it was almost impossible to conclude what he wished to be done. He was clear only when he started talking about what he felt was the really important task, the leadership of the U.S. and Western Europe in giving economic aid to underdeveloped countries and getting them on their feet economically. He did agree though to a fairly strong condemnation of Fedorov in a press release, approved vaguely Herter's plans for Wadsworth (but the latter said as we were going outside: "I certainly don't know what my instructions are and how I am to conduct the meetings"). He also agreed to the formula for nonresumption of nuclear tests by the United States. All of this took only an hour and then Herter and Dillon stayed alone with the President for another hour or more, discussing the plans for the President's trip to South America. Our return trip was uneventful and I spent the afternoon in my office fighting sleepiness.

> President Eisenhower stated publicly on December 29 that the United States would resume negotiations on the test-ban treaty notwithstanding the unwillingness of the Soviet Union to give serious consideration to the new data on the degree of effectiveness of seismic techniques for the detection of underground nuclear explosions. He deplored as "intemperate and technically unsupportable" the Soviet attack by Fedorov on American participants of the Technical Working Group II. He further stated that although the voluntary American moratorium on nuclear tests would expire on December 31, the United States would "not resume nuclear weapons tests without announcing our intention in advance." He further noted that an "active program of weapons research, development, and laboratory-type experimentation" would be continued. The reference to the latter, of course, was covering the one-point safety experiments at Los Alamos to which several references have already been made here.

∞⌐ Wednesday, December 30, 1959

Flew during the morning to New London to deliver the address (an oration in the style of a senator of 1900 vintage) at the commissioning of the U.S.S.N. *George Washington*. A pleasant and well-run ceremony which went off without a hitch. Reports from the audience were that my delivery was not bad. Thereafter went to the reception which was so well oiled that a couple of admirals had to be essentially carried out on the arms of their aides-de-camp. I also felt better after four large drinks. We then went through the submarine and were finally driven to Boston in a company car since trains were not on schedule.

> *George Washington:* The first Polaris-type nuclear submarine.

Monday, January 4, 1960

Spent the whole day with the PSAC ad hoc Life Sciences group, made up of thirteen people, since Beadle, unfortunately, was unable to attend. The meeting was devoted to identification of problems and areas which might be studied by the PSAC. On the whole the session was not very useful, the overwhelming weight of the responses being in the nature of recommendations that the federal government give more money to colleges, graduate schools, and fellowships, even though in my introduction I urged them explicitly not to make such recommendations. A number of people showed themselves quite unable to separate identification of areas for study from firm recommendations of what should be done. In this respect, Coggeshall, Robbins, and Schmitt were the worst. The most "statesmanlike" attitude was shown by Haskins, Loeb, Harrar, Long, and McNew. Thimann, Gonzales, DuShane, and Weiss were not very helpful. It is clear that panels made up of people like these will require a great deal of personal effort on my part to direct into useful channels. Otherwise they will simply put out reports asking for more money for basic research and the training of scientists. Nobody showed any enthusiasm to get involved in the cranberry problem, but I said I had no choice but to start that study and before the meeting was over got a list of names of those who might participate:

> L. T. Coggeshall, professor of medical sciences, University of Chicago; W. J. Robbins, professor of biology, Columbia University; F. O. Schmitt, professor of biochemistry, M.I.T.; C. P. Haskins, geneticist, president of the Carnegie Institution; R. F. Loeb, professor of medical sciences, Rockefeller University; J. G. Harrar, applied biology, Rockefeller Foundation; C. N. H. Long, professor of biochemistry, Yale University; G. L. McNew, botanist, Boyce Thompson Institute; K. V. Thimann, professor of biology, Harvard University; L. M. Gonzales, professor of medical sciences, University of Puerto Rico; G. DuShane, professor of embryology, Stanford University; P. Weiss, professor of medical sciences, Rockefeller University.

⁀ Tuesday, January 5, 1960

Spent all day in New York at the meeting of the Educational
Policies Commission. Barely managed to get there, feeling like
hell with flu or something and then discovered that Conant,
whom I hoped to see, was too ill to attend. The proposed policy
draft to indicate the purpose of education in American schools is
an appallingly bad piece of paper; vague and almost meaningless
in parts. There were a lot of big shots at the meeting, e.g., the
heads of the Rockefeller, Ford, and Carnegie foundations, a Mr.
Josephs, the chairman of some big bank in New York, etc., but I
can't say that I was much impressed by the wisdom and clarity
of their remarks. Some of it was pretty bad, and I broke a
promise to myself and didn't keep quiet, which probably is just
as well, because otherwise the meeting would probably have rejected the need to reorient schools toward a better math and
science education. Dean Rusk made by far the best impression on
me. What he said was good, even though not world-shaking.

Dean Rusk, president of the Rockefeller Foundation.

⁀ Wednesday, January 6, 1960

9:30–11:00 AM   Cabinet meeting which was largely devoted
to a reading by Moos of the latest draft of the State of the Union
message. I found the document good and nothing in it that
seemed objectionable. Following its reading, there was a long
discussion but strictly in the nature of nit-picking. Then Secretary of Labor [James] Mitchell presented the inside story on the
steel strike settlement. It appears that unbeknown even to the
rest of the cabinet, Mitchell and Nixon for more than two weeks
have been meeting daily with the steel workers' union and steel
managements and feel terribly proud of the agreement they
finally worked out. The thesis which they presented and which
was initially challenged by several cabinet members, especially
Benson, so that Nixon got rather hot under the collar at one time
responding to Benson, is that the settlement is extremely favorable to the anti-inflation policy, considering the circumstances;
that it concedes a lower percentage rise in wages than any recent

agreement concluded with labor, such as the one on the part of ALCOA or the tin can companies. An interesting sidelight on the mood of the American people is that polls conducted by the steel workers' union and the management indicated that 93 percent of the union members would vote for resumption of the strike rather than for acceptance of the so-called last offer of the management. Both Mitchell and Nixon emphasized that had that vote been taken, McDonald and his union would have been in a tremendously strengthened position and would not have accepted a compromise which is only slightly above the midway point between the management and the union proposals.

Malcolm Moos, presidential speechwriter.

The rest of the morning spent in telephone calls and attending the meeting on seismic thresholds which involved Panofsky, Tukey, Brown, Romney, etc. Had lunch with this group and was told that it has concluded that a simple definition of thresholds, understandable, as I urged, even to diplomats (I said that deliberately in the presence of the State Department people who seemed to find it very funny) was quite feasible. After lunch spent a couple of hours more with the group and got a reasonably good understanding of all the recommendations they were making for the Friday Principals' meeting.

Spent about an hour with Panofsky alone, who, to begin with, handed me a long letter from him to me, which is in essence a vigorous protest against his having been used in Geneva in the Technical Group II for political purposes. He feels intensely unhappy and says that Fisk feels likewise; also says that the experts' inability to agree will be used as an excuse for a change in U.S. policy, whereas that change had been decided upon before the Geneva meeting. I gave him an honest account of the situation I inherited from Killian when I stepped into this job and of what I tried to achieve during the fall; conceded that the chances for a comprehensive treaty were small when the conference began, but maintained (and honestly too) that they were not nil. I think this discussion ended by his feeling substantially soothed, although he is not very happy (and neither am I). We went then to the other painful topic—the Stanford linear accelerator, which he is lobbying for—and I learned to my surprise, that this whole business of the neutron-generating machine and letters by physicists to AEC urging its construction in preference to the Stanford accelerator, is a maneuver by Rabi, who is opposed to more high-energy machines in principle, and so has tried to block the Stanford accelerator. What a bastard! He took me in! Panofsky said that Serber, Lee, and others are now extremely unhappy for having written and feel that they have been had by Rabi. Of course, the damage has been done: with these letters in the hands of McCone, who obviously doesn't want the Stanford accelerator to go through, little can be done, as we agreed. I explained about the panel meeting that I am going to call shortly and Pief accepted this as the only sensible course.

> J. W. Tukey, professor of statistics, Princeton University; C. F. Romney, seismologist, USAF Technical Applications Center (AFTAC); Robert Serber, professor of physics, Columbia University; Tsung-Dao Lee, professor of physics, Columbia University.

∽ Thursday, January 7, 1960

The day started with Hi Watters warning me that Secretary Brucker is refusing to accept the decision that Nike Zeus will not go into production and is instructing all army personnel that advocacy of a delay in production would be considered disloyal to him and the army.

10:15–10:30 AM   Mr. [J. L.] Atwood, the president of North American Aviation, a sad-looking gentleman notwithstanding a good rate of compensation, was here to convince me that the B-70 bomber is a terrific thing; that North American Aviation has achieved tremendous breakthroughs in its design; and that it would be tragic for the country if this project were reduced in scope.

10:30–11:15 AM   Glennan to outline his proposed presentation to the President at a meeting which will take place tomorrow, and which I am to attend.

11:15–11:45 AM   Committee on Specialized Personnel, OCDM, which asked me for my views on problems of poor utilization of scientific and engineering personnel. I gave a rather glib statement, which, however, was received well, and left hastily in a big burst of applause (first time that has happened to me in Washington).

12:00 M–1:30 PM   At the Capitol, watching the President deliver his State of the Union message. Quite a show, which I will certainly not forget. The delivery by the President was good until nearly the end, when he began to lose his voice and also stumbled on some difficult words. He got a warm welcome when he was escorted to the speaker's dais, and a few good rounds of applause early in the speech, when he spoke of his desire for peace, but after that the applause was rather feeble. Afterwards Malcolm Moos said that he got about as much applause as he ever received after the first two years in office, and that his delivery was not the best, but certainly not the worst among the speeches Moos had heard him deliver. I was shocked to learn from the message that they expect $4.2 billion surplus in fiscal '61, and have sworn to myself not to be such a naive guy next fall, when Stans and Co. lecture me on the need for economy in research funds.

3:00–5:30 PM NSC meeting. The President was in rare good form, laughing, kidding various people present, etc. Did a lot of talking too. The first item was MATS and cargo aircraft for potential military use. This subject, which really isn't very important, took a lot of time and induced very detailed discussion in which a number of people took part, but Quesada did most of the talking. Next, Tom Gates asked approval of the new missile program, including 27 ICBM squadrons and additional Polaris submarines, and the President gave his approval. Then Dulles, in his usual dull way, reported the conclusions of USIB on the Soviet missile threat. For the first time this estimate was based not on capabilities but on probable plans because there is no evidence that the Soviets are engaged in any crash ICBM program and hence obviously are not using their full capability. This report concludes that mid-1961 is the time of maximum threat because as of then we still won't have hardened missile bases and SAC will be vulnerable too, but the threat is not catastrophic. In fact the missile gap doesn't look to be very serious (I hope this estimate is not a political effort to cut down on trouble with Congress). The President started asking a lot of questions and very soon Tom Gates ran out of knowledge and suggested that George give the answers, so I got up and answered the President's questions for about half an hour. He was very friendly and cordial. Everything got inquired into: Why have Titans? Why with storable propellants? Could we scrap Atlas missiles and put Titans into Atlas bases? Why harden missiles if, as I believe, Soviet missile accuracy is substantially better than USIB estimates? Etc., etc. The meeting ended with a dull account by Karl Harr of the OCB findings regarding Iran, Hong

Kong, and Turkey. Comments on Iran led the President into a long discourse on his meeting with the shah and how he liked the shah and how the shah seemed to be smart and wanted to scrap his foot soldiers and buy instead some modern aircraft.

Friday, January 8, 1960

7:30–8:30 AM Breakfast with Persons, Jerry Morgan, and Dave Kendall joining us. I first talked about the GAO report on the Air Force's missile management; clearly very colored politically and unfair. We agreed that a strategy meeting on the missile issue is needed in the near future and Persons asked me to prepare a summary of the report prior to the next press conference of the President. I subsequently asked Dave Beckler to do that. I then took up the Secretary Brucker problem, namely the emphasis of Dick Morse that Brucker is unwilling to accept the decision that Nike Zeus will not be in production; his instruction to Army that anybody who accepts this decision becomes disloyal to him, etc. Kendall, who is a personal friend of Brucker, said that he was quite amazed because fairly recently in a conversation with Brucker, he was assured that Brucker would support presidential decisions, and Persons said that OSD has in the past liked to knife Brucker and wasn't this another such effort. I assured him that my information came from Brucker's own office; whereupon Persons said he will discuss the matter with Tom Gates. Then I raised my personal problems, and (a) got approval to take a week or so off for medical treatment, if necessary, and (b) got assurance that Persons will sympathetically consider my problem next fall when my stand-in at Harvard leaves and my research students will need more of my personal attention. I assured Persons that I wouldn't simply quit and leave the President in the lurch, and he suggested that maybe some arrangement on a part-time basis could be worked out for the fall, since he conceded that the government wouldn't be very active during the last few months of the administration.

8:30–9:45 AM Meeting with the President, Glennan, and others, at which Glennan made specific proposals boiling down to a change in the law setting up NASA so that the Space Council and the liaison committee between NASA and the Pentagon would be abolished; NASA would be given specific responsibility for space exploration and the President would resolve conflicts between DOD and NASA regarding booster vehicle development and similar issues. The second aspect of the proposal is that the President deliver a 15-minute TV speech on our space program, to emphasize that we are not in a neck-to-neck competition with the Soviets. The President was completely cold on the subject and didn't remember previous conversations with Glennan. So the meeting took a long time, but in the end the President pretty much accepted these recommendations and, what is rather amusing, at the very end was rather favorably considering the idea of adding another $100 million to the Saturn project if that could speed it up. I suppose the Soviet announcement of launching superrockets into the Pacific has got him worried too.

> Until this time Soviet ICBMs had been landing in Kamchatka and the change to a Pacific Ocean target area represented a major increase in the test range.

After this meeting we adjourned into Persons's office and continued discussion of the political aspects of the Glennan proposals, i.e., how to handle the matter in Congress and what should replace the Space Council. I recommended against the statutory requirement that the special assistant for science and technology participate in the resolution of space conflicts between DOD and NASA before the President, since my philosophy is that the special assistant for science and technology would naturally be involved as his staff member, if the President so wished.

> The President later sent a message to Congress recommending changes in the law setting up NASA which were consistent with the recommendations of Glennan. In August he issued a public statement on American achievements in space which rejected the notion of "spectaculars" and stressed solid technical and scientific progress.

10:15 AM   Mr. Williams, the president of Freeport Sulphur Company, introduced by the secretary of the treasury, who tried to enlist my help in saving the big nickel mining venture of his company in Cuba at Moa Bay from seizure by Castro. I was most sympathetic, of course, but suggested that Governor Hoegh or Secretary of Commerce Mueller were more appropriate for the activity than I.

12:00 M–1:00 PM   Lunch with Tom Nolan at the Cosmos

Club, at which Nolan discussed problems of the standing committee of the FCST. Two points worth noting: one is that Nolan is unhappy about the study of hydrographic problems being undertaken by PSAC or the Academy, as suggested by Berkner and me, and would like it to be done by the Federal Council, through a panel made up of government scientists rather than outsiders. He emphasized that Secretary of the Interior Bennett was much embarrassed by the oceanographic report of the Academy, so that there are now strained relations between Bennett and the director of the Bureau of Fisheries. He is anxious that this doesn't happen again (too bad! but perhaps it can't be avoided from time to time). Another point is that in the view of the standing committee and Nolan, it is unfortunate that PSAC doesn't include in its membership some government scientists. I explained the philosophy behind PSAC, noted that the standing committee was made up of government scientists and that the two organizations were rather complementary, but he still insisted that PSAC should include government scientists, although not in the class of science administrators who make up the standing committee. I said it was too late to do anything about it this year, but I will raise the issue a year hence.

2:30–4:30 PM   Principals' meeting. I summarized the findings and recommendations of the seismological group that was meeting yesterday. These recommendations are that: 1, a definition of a threshold in seismic terms is perfectly feasible and technically realistic; 2, a seismic magnitude of 4.75 on Richter's scale would be an appropriate one to choose; 3, that choice would enable us to achieve effective monitoring to obtain a strong deterrent against cheating with about 10 on-site inspections per year. Following my presentation, a long discussion ensued. I am not happy about the role which Dillon takes in these meetings. He clearly doesn't see eye to eye with Herter and is far more concerned with pleasing AEC and DOD than with developing an effective negotiating position. So, for instance, Dillon said that he would accept a higher threshold than 4.75 if AEC and DOD felt that a higher threshold was necessary for a more effective U.S. test program. This is really an appalling statement from an acting secretary of state who is instructed by the President to be seeking arms limitation measures! The pathetic outcome of this meeting is that even though at the meeting with the President in Augusta it was agreed that the threshold concept would be accepted as a part of our policy if it was technically feasible, it has now been agreed that AEC and DOD will further consider the proposal and that another meeting of the Principals will be convened next week to hear their opinions and then try to form a policy. When the meeting ended I felt the way Napoleon's generals probably felt after the Battle of Waterloo. Dillon, while personally pleasant, is clearly unwilling to push energetically towards seeking a relaxation of tensions with the Soviets, and one can discern hardly any difference between what he means and what Gates means, although the words they use are somewhat different. It is unfortunate that Herter is on a two-week vacation when all of this is being "settled."

At the very end of the meeting, I raised the issue of the VELA project, that is, an expanded R&D on seismic and high-altitude detection of nuclear tests, but naturally got no firm financial commitments from anybody and no decisions as to who will manage the program and how much money will be involved.

5:00–5:30 PM   Meeting with Piore, Kolstad, and our staff about reconvening the High-Energy Particle Accelerators Panel to review its findings of a year ago. We agreed that Mannie will chair the meeting, that all the previous members of the panel will be invited, and that we will add two more theoreticians: G. Chew and Tsung-Dao Lee. I emphasized that my desire was not to overthrow the original findings of the panel, because this might embarrass the President, but I would accept whatever findings the panel makes as long as they were objective.

> George A. Kolstad, physicist, AEC staff; Geoffrey Chew, professor of physics, University of California, Berkeley.

6:00–6:30 PM   Attended cocktail party for NBC foreign correspondents at the Statler Hotel but spent the entire time with Bennie Schriever, who is much upset about the GAO report and said that he has gotten past the point where he cares about his military career and when called upon to testify isn't going to mince words. Said that Secretary Douglas and his chief counsel have been making a careful study of this report and should be included in whatever policy meeting might be set up to plan proper response.

> This report, as I recall, was critical of the Air Force's handling of the missile development program, especially the involvement of the Ramo-Wooldridge Corporation.

Monday, January 11, 1960

On Sunday I flew by a Navy Corsair to the aircraft carrier *Independence,* 100 miles off the coast of Florida, near Jacksonville, to join members of a PSAC panel. That evening was spent in briefings and inspection of parts of the ship. Monday the inspection of the ship continued after the mildly harrowing experience of being taken for a flight in an A3D attack bomber. The unpleasant part was the takeoff from a catapult. The sensation of being pushed faster and faster towards the edge of the deck is thoroughly unpleasant. One gets accelerated from a standstill to 160 miles an hour in 2 seconds and this is also not very pleasant. The flight itself was nice and I didn't even mind the landing and the arresting gear. The pilot (without prior warning) also demonstrated to me what would happen if the arresting gear broke, by hitting the deck with the plane hook raised and bouncing off into the air again. The afternoon was devoted to a fleet air exercise, including the use of air-launched missiles, rockets, and bombs. The Navy is obviously sensitive, having heard that the PSAC panel is recommending against continuing the building of large carriers. They resorted to all sorts of tactics—including the use of a loudspeaker during the air exercise, and reminding me how tough are the landings of fast aircraft on the decks of smaller carriers—to lecture us on the need of large carriers. In the afternoon the Navy suddenly decided to fly me back to Washington on an A3D rather than via a commercial flight as planned. Not particularly looking forward to another catapulting, I tried to get from under, but was made to feel that this was hurting their feelings, so I accepted. The second catapulting I took better, but it turned out that the plane was going to ''bomb'' most of the southern cities on the way to Washington and so we spent nearly two hours in the air at 40,000 feet altitude, including several (camera) bombing runs over our designated city targets. My oxygen mask wasn't working very well; my mike was out of order so I couldn't complain; being strapped in the seat with a heavy parachute harness was not very comfortable, so I had a rather miserable time and got my ears horribly blocked coming down. So this kind of trip is really not recommended, particularly since I didn't gain any time in the end. Something went wrong at Andrews AFB and I waited more than half an hour until my bag was delivered from the aircraft.

Naturally, as befits a budding diplomat, I thanked the pilot profusely and assured him I enjoyed the trip.

⌒ Tuesday, January 12, 1960

9:30 AM–12:00 M   Informal Space Council meeting in which we rehearsed the afternoon's show. The council accepted the Bode panel's summary report with only a little bit of nit-picking, but McCone later in the meeting gave a very impassioned speech urging a dramatic increase in our space efforts and condemning just about everything that was being done now. Some of his statements were of quite dubious correctness, and I said so to Gordon Gray in an aside. After the meeting McCone called me and asked me to have lunch with him and Gray, which we did cordially, but I pointed out the aspects in which he, McCone, could find himself out on a limb if he repeated the same statements in the afternoon. Evidently my warnings took effect, because he said little in the afternoon meeting. (I was probably too soft-hearted, because I could really have paid him back by letting him deliver his attack and then pointing out to the President in what ways it was factually incorrect.)

2:30–5:00 PM   Space Council-NSC joint meeting, which started with Don Ling's presenting the Bode panel report. He did a magnificent job and several people after the meeting said to me that it was the best presentation in a high-level government meeting that they ever heard. The President accepted the report, but noted that while it was factually correct, we should all remember that we started only five years ago, and the people in the program should not be criticized, but lauded for their efforts. Unquestionably this is correct, but it is really not relevant to the panel's conclusion.

This was followed by the presentation of the NSC policy paper on outer space. The split was neatly resolved by the President in a way which seemed to be agreeable to all but BOB. Glennan then presented the long-range plan of NASA, which was followed by an emphatic statement by the President that maximum effort should go into superbooster rockets. Stans objected to the expenditure of so much money and was firmly dressed down by the President. There followed a discussion of Saturn, and McCone expressed doubts about it on account of its complexity, the only part of his forenoon speech that he delivered again. I chimed in and also expressed grave doubts about Saturn B, but conceded that its cancellation and replacement by maximum effort on the Nova would be politically impossible.

5:30–6:30 PM Meeting of Principals in Secretary Gates's office. The most dismal meeting, and that is saying something. Defense presented its position paper on the threshold treaty, which was a rather nasty attack on me, so far as I could figure out. Fortunately it was an exceedingly unintelligent document and thus it gave me an opportunity to pay back double, which I proceeded to do, such that finally Gates said to me: "Do you mean, George, you do not agree with this document?" Which I vigorously confirmed. McCone presented an AEC position paper that accepted thresholds and was quite reasonable. Privately General Starbird said to me afterwards he was unhappy about this position paper. State, represented by Merchant and Farley, was clearly trying to avoid any decisions' being made, and in that they fully succeeded. If one were to evaluate the results of this meeting, I would say it put us neatly back two months.

~~~ *Wednesday, January 13, 1960*

7:30–8:30 AM Pre-press conference breakfast. Nothing much happened except that Persons asked me to investigate personally the Titan–Martin Company situation by going to Denver. I asked for the trip to be postponed several weeks to let them get things going if they can.

9:45–10:30 AM With the President in the regular Hagerty briefing. The President was obviously in a very bad mood, but gradually relaxed and began to joke. Nothing eventful happened except that I was asked about Senator Symington's statement in the "Meet the Press" TV show and wasn't too well prepared to analyze it.

∿ *Thursday, January 14, 1960*

9:00–10:45 AM NSC meeting. The most interesting part of it was a discussion of the Cuban situation and U.S. policy with respect to Castro. Because of the sensitivity of the matter, I won't put it down on paper, except to note that the estimate of the situation in Cuba is that Castro and his advisors are moving very skillfully toward the introduction of outright communist government in Cuba and are doing it in such a way as not to create by any rash act a justification for U.S. intervention. Dulles, as usual, presented an intelligence briefing in which he quoted from the long speech that Khrushchev has just delivered to the Supreme Soviet, outlining the policies of his government involving missiles, aircraft, army, and navy. When Dulles finished reading the translation, the President said "Well, this is just what Khrushchev told me he was going to do when he was here, so that isn't much news to me."

After the meeting I waited almost an hour for my appointment with the President as the latter was busy on the telephone and with the vice-president.

> The Senate Select Committee on Intelligence had released its report on CIA assassination plots the day before this note was being written. As the above entry in my journal indicates, President Eisenhower was extremely concerned about the policies of Fidel Castro and occasionally used uncomplimentary language about him. However, I never heard him say in my presence that Castro must be liquidated or some equivalent to this that could have been taken by Dulles as an instruction to kill Castro.

11:30 AM–12:15 PM Briefing the president on various activities in which I am involved, such as the Federal Council, life sciences panel, cranberry panel, Nike-Zeus panel, etc. I spoke about the potential gain in national prestige if we establish the first astro-observatory on a satellite. The President was very much interested and said he would certainly be in favor of proceeding vigorously, although he would not support a crash program because it was not "something like missiles" but was scientific in intent. When I spoke to him about my idea of establishing through PSAC and myself a firm link between the membership of the National Academy of Sciences, as the most representative scientific body in the U.S., and the federal agencies represented by the Federal Council, to achieve better long-range planning of research activities, the President endorsed it in most vigorous terms and said this is just the sort of thing he was hoping for from his National Goals Committee. During the conversation, the President answered a phone call from somebody and urged him strongly to get moving on the National Goals business.

Had lunch at the Mess with Don Paarlberg, who is extremely alarmed about the developments in the "broad cranberries issue." He said the actions by the Pure Food and Drug Administration are going to create frightful trouble with the farmers. This was later confirmed by Peterson, who said that farm organizations are insistent on a public fact-finding commission, and he, Peterson, may not be able to resist the pressure. I told him the President has endorsed my setting up the cranberry panel, that Bronk and Loeb will be the chairman and vice-chairman respectively, and that before he sets up a public body, he had better let me know so I can discuss it with the President, because two parallel studies may be exceedingly awkward for the administration.

In the afternoon had a long visit from Trevor Gardner. After preparing for the Democratic advisory council a paper on the need for a "peace agency" in the government—a plan which eventually was unanimously approved by the council—he sent a copy to many people, including Teller. He said that Teller has gotten Symington completely under control and that Symington at a meeting of the Democratic council delivered a speech obviously inspired by Teller against cessation of nuclear tests and generally against attempts to reach arms-limitation agreements. The speech was effective and for a time it looked as if the council would be swayed by it, but through vigorous rebuttals by Harriman and a few others, the Symington proposals were rejected. Gardner said he recently spent a full day with Teller at the latter's invitation and was simply appalled by his attitude. He asked whether Teller is acting paranoic, and I assented. Teller is almost hysterical on the subject of nuclear tests and is against arms limitations of any kind. He envisages the future as an ever-intensifying arms race, but refuses to consider what its ultimate outcome will be. In Gardner's estimate Teller is the most dangerous scientist in the U.S.; I agreed. He then talked to me about the plans of the Democrats to attack the administration on missiles and outer-space issues. The former charge simply involves a matter of numbers vis-à-vis the USSR. On the latter issue the criticism is that the outer-space program has been entrusted to a group of incompetents, i.e., Dryden, Silverstein, and Horner; that these men simply do not have the imagination and vigor to put through the program and are merely trying to build a great empire around the old NACA labs. This view is completely in line with the article in *Harper's*, "The Missile Mess," January 1960, and so explains the origin of that article. Unfortunately, the criticism of NASA is not without foundation, in my opinion, and some of the things Gardner said were just the things I said in private meetings on the subject of the NASA budget in Jerry Persons's office and to Dryden.

Morgan spent some time with me discussing the weaknesses of our position on missiles and outer-space matters as regards criticism in Congress. The rest of the day spent in usual minutiae.

> Trevor Gardner, industrialist and former assistant secretary of the air force, who set up von Neumann's Ballistic Missiles Advisory Committee in 1953 and was effective in getting the air staff to accept the recommendations of that committee in 1954.

⟿ Friday, January 15, 1960

Phone call to Dick Morse, who is still quite emotional about Nike Zeus and emphasizes that with only $200 million they cannot go ahead with the Kwajalein test installation, which he feels is most important to the program. Complained that no program should be run under a committee, as is being done by Skifter in York's office. I emphasized my unalterable opposition to the project, which he outlined in a memo to me, of shooting down a satellite, because once we had downed our own satellite, and of course made much to-do about it, the Soviets could easily shoot down one of their own over the Soviet territory, accuse us of doing it, and make a big public issue of it, which would then give them an excuse for shooting down our reconnaissance satellites.

> The recurrent proposals to develop and test a missile system for destroying or "inspecting" satellites was an outgrowth of a rather widespread fear that the Soviet Union would launch satellites with nuclear warheads to be released on targets in an all-out war. It took several years' arguments to convince everybody but the ultra–space-cadets that nuclear warheads in satellites were not the way to prepare for a nuclear war. The reasons for this are largely

technical: a satellite passes over or near a specified aiming point
but once daily; it flies with a speed of about 25,000 feet per second
and the orbits change slowly with time, so that the timing and
thrust of the retrorocket attached to the warhead are quite a
guidance problem if even moderate accuracy is desired; the satellite can be relatively easily damaged and made inoperative by a
surface-to-space missile while passing over the target country
without leaving observable evidence of the act, etc., etc.

These considerations became sufficiently generally accepted by
1972 that very little opposition developed to the conclusion of a
treaty forbidding the placement in orbit of weapons of mass
destruction, which went into effect that year.

1:50–2:30 PM   With Gordon Gray, to whom I spoke about the
total lack of policy directives regarding public statements on the
peaceful nature of reconnaissance satellites, and cited this proposal of Morse as an instance of how we could easily damage
what I consider an extremely important measure of national
security. Gordon agreed and we then discussed reconnaissance
satellites in general. He decided to have a special meeting of the
Security Council to hear reports on reconnaissance satellites and
then discuss policy matters such as I called to his attention.

2:30–3:00 PM   Wallace Brode, who came to report on his
activities. I resent the way he refers to the science attachés as
"my men." He obviously and admittedly does not consider his
job as that of science advisor, but as an administrative one running the science attachés and he spoke without any concern
about the fact that he sees the secretary of state hardly at all. He
then announced that he had decided to stay in the department
and implied he would like to stay beyond the present administration. I emphasized that the position should be a rotating one.
Later in the conversation, when he talked about what to him
were objectionable efforts of the National Academy to liberalize
U.S. rules for attendance at foreign scientific meetings, etc., and
stated that he wasn't sure where his loyalty should be, I pointed
out that this was just the reason why there was need for rotating
personnel in his job and that his mission was not merely to
become a member of the State Department, but also to serve as a
bridge to the scientific community. We didn't get to sharp words,
but nearly so. The man really amazes me in his self-estimate as
performing a "good job."

⌒⌒ *Saturday, January 16, 1960*

Seaborg's panel on Research and Graduate Education, where I spent the entire day. I believe my presence was useful because there was too great a tendency among panel members to limit themselves to ringing statements about the need of more and better support of graduate education by the federal government and not to come down to brass tacks. Also I dimmed somewhat their notion that the universities are sacrosanct from "interference" by the federal government.

⌒⌒ *Sunday, January 17, 1960*

Until one o'clock, continued with the same panel. Then lunch with George Beadle, during which I sort of indoctrinated him into PSAC and convinced him that he would have to do some work. The rest of the day spent writing that damn Oak Ridge speech. When I finished, very little was left of Mr. Simons's (my official speechwriter) text.

⌒⌒ *Monday, January 18, 1960*

9:30 AM–6:00 PM   PSAC meeting, which went quite well, but I don't recall any of the specific details any more. Then I met with the biologists to discuss membership of the carcinogenic chemicals [i.e., cranberries] panel and of the general life sciences panel, the former to be chaired by Bronk and Loeb and the latter by Beadle. Had dinner with several PSAC members and came back to continue PSAC discussion of better methods for monitoring nuclear test cessation. Tukey and Panofsky were the advocates and after saying that I would be the devil's advocate, I gave them a rough time with awkward questions, since the schemes they had in mind were rather unrealistic. At the end of the meeting, about 11 PM, everybody agreed there were no concrete proposals to be made at present, but that further study was indicated. Tuesday morning I called Farley and told him that if State took the initiative in raising the issue as to whether technical preparation on the problems related to nuclear test cessation was necessary for the Summit conference, I would respond by indicating specifically the areas requiring additional study. I might note that later that day State didn't come forth with anything.

Tuesday, January 19, 1960

9:00 AM–1:30 PM   Continuation of PSAC meeting. The most interesting aspect was the discussion as to whether we should offer our help to the State Department because of the apparent failure of the Coolidge study and start preparing for the ten-nations disarmament conference. The general conclusion was that this would be very unwise and that we should not undertake it. This was then confirmed on Wednesday by the Arms Limitation Panel, which however offered its help in dealing with specific issues as they arose. Tuesday afternoon I spent hastily visiting panels on arms limitation and on high-energy accelerators. At 4 PM had a visit from General Burns and Mr. McCordick, who represent Canada at the ten-nation disarmament conference. Had a pleasant chat.

> Gen. E. L. M. Burns, Canadian Foreign Office, whom I met in Geneva at the 1958 conference; Mr. Y. A. McCordick was accompanying him.

5:00–6:30 PM   Principals' meeting. I am learning that being on the offense at times pays in Washington. The Defense Department repeated their objections to the threshold proposal but much milder in tone. I vigorously attacked the objections, suggesting they didn't know what they were talking about. The result was that when State, AEC, and I found ourselves in substantial agreement, Defense, which this time was represented by Loper, announced that they didn't disagree, as they wished to avoid taking the issue to the President. I suspect that when at the previous meeting I jumped hard on them in the presence of Gates, the latter must have decided that their position paper was weak. We have arrived, thus, at quite a reasonable threshold proposal. Everybody appreciates, however, that the British are violently opposed and that the Russians probably wouldn't even discuss it, but may break off the conference.

⤳ Wednesday, January 20, 1960

This was a very hectic day. Morning was mostly spent telephoning a variety of people, including prospective members of the "cranberries" panel. Response was good.

Visited the high-energy accelerators panel meeting again and managed to get them to face the issue squarely: namely, how should the government allocate its money between existing projects and the new machines, if the total is below what the panel recommends. We had a lively discussion and I left them still completely divided. A perfect example of a lobbyist bunch: those who have machines want the money to go to the existing machines and those that don't have them want the new ones.

In the afternoon visited them again, and egged them on. Understand from Piore that by the end of the day they reached a sensible set of concrete recommendations in unanimity.

3:00–4:30 PM   Meeting with Secretary Flemming, Assistant Secretary of Agriculture Peterson, and their staffs, in Don Paarlberg's office, to discuss a joint HEW-Agriculture cabinet paper. Atmosphere very tense, and the corrections to the paper prepared by Brad Patterson, all in the nit-picking category. Had to leave long before the meeting was finished, but after their agreement, at Paarlberg's request, to the formation of the PSAC cranberries panel, which was obviously not welcomed by Flemming. He emphasized that even the matter of its existence must be treated as a deep secret and reminded me to clear terms of reference with him and with Benson.

Bradford Patterson, assistant secretary of the cabinet.

Then spent about an hour with the ad hoc panel on missile test cessation, giving my reasons for the need of this panel and trying to define the objectives in a sufficiently limited way that the panel can come up with a sensible and properly qualified report.

Thursday, January 21, 1960

Spent two and a half hours with Fred Eaton, the head of our delegation to the ten-nation disarmament commission. Very cordial meeting, during which I cautiously gave him the facts of life, such as the characteristics of Teller and Urey as extremes in the spectrum of scientists' attitudes on public affairs. Also urged the great importance of having good technical people on his staff as otherwise it will be dominated by Defense. He seemed to agree. Obviously a very cautious individual. He said for instance that he will have State and Defense write for him terms of reference, and I suggested that he may find himself with a totally unworkable document; whereupon he said don't worry he will be around when that is written.

That afternoon I left for Oak Ridge to give a speech.

Friday, January 22, 1960

In the morning visited the Oak Ridge National Laboratory, by far the most interesting part being the biology division with its huge mice colony with which biological and genetic work is being done, mainly on the effects of low-intensity ionizing radiation.

Flew then to Princeton via Newark. Spent the evening with the Oppenheimers and Yangs. Talked a lot about old times with Oppenheimer and it was a pleasant evening.

Chen Ning Yang, theoretical physicist.

Monday, January 25, 1960

10:30 AM Saw Secretary Herter in the White House, who told me, first that the President wants me to go with him to the USSR. Very good news. Second, that he, Herter, decided after much thinking that the continuing study of disarmament will be localized in the State Department because he is afraid that another Harold Stassen in the White House will cause trouble. It will be in effect a joint group since it will include Defense representatives, and he is anxious that PSAC maintain close liaison with it if not actually be represented on it.

Lunch with Glennan to discuss, on instructions of the President, the statement of the scientific advisory group to the Democratic National Committee. We agreed that their statement (on outer space), while very stupid in a technical sense, contains enough truthful analyses so that a substantive reply can't be made in a brief statement and that the document would be best answered by a speech of the President, explaining what our program is all about and why it is as it is.

4:00 PM   Meeting in Secretary Douglas's office regarding the GAO charges of Air Force mismanagement of the missiles program. The Air Force feels that they can take care of the business, but it is essential that the rebuttal preparation be centralized as otherwise some people will get panicky and make undesirable concessions.

6:30–9:30 PM   My dinner for Federal Council members at the Dupont Plaza Hotel. Present were Glennan, York, Waterman, Bennett, Ray, Kreidler, and myself. Ray arrived very late, and just as we sat down to dinner he launched into a most appalling and outrageous attack against science, against government support of scientific research, and against some individuals.

He indulged in a repetitive tirade about the dangers of the growing federal expenditures for science, stated that the American people were at fever pitch in favor of science, that they would support almost anything, but that the ballooning government-supported program of science was going to boomerang. He stated that government science did not contribute anything to the productivity of the country, yet was supported by the businessmen of the free enterprise economy, especially by the little guy who invents something useful and sells it and pays the taxes. Waterman, Glennan, York, Bennett, and I cited various examples of how research contributed to the growth of the country. Ray was in no mood to understand and listen to us. He insistently (and profanely) stated that government support of science was "a leech" on the taxpayer and that someone had to put a stop to "such sucking on the tit of the taxpayer." Moreover, Ray continually interrupted others and was pointedly rude, especially to Glennan and York. Our attempts to soothe him were not successful and Glennan left extremely angry. Toward the end of the evening Ray quieted down but, of course, no substantive discussion of the future of the Federal Council was possible. All he was interested in was to prevent duplication of government projects and insisted that every agency throughout the government must notify everybody else through the Federal Council when it undertakes something new in R&D.

~~~ *Tuesday, January 26, 1960*

7:30–8:30 AM   Pre-press conference breakfast, at which I was charged to prepare a statement for the President on the Democratic group's document. We also discussed the carcinogenic additives, Jerry Persons concluding that this study of ours couldn't be kept secret no matter what Flemming wanted.

9:45–10:30 AM   With the President, who firmly said that if asked a question about carcinogens in cranberries he will announce that his science committee is studying the problem. He accepted my version of reply to the Democrats.

10:30–11:30 AM   Mervin Kelly, whom I sold the job of becoming chairman of a technical manpower misuse panel, providing we will get the Ford Foundation to shell out money for investigations and statistical studies and get it cleared with Secretary Gates, which he feels will not be difficult.

> Mervin Kelly, just retired as president of the Bell Telephone Laboratories.

1:30–4:00 PM   Federal Council. A very poor meeting, Glennan being absent, along with many other Principals. In any case Ray was silent, because, I believe, during the executive session I avoided raising the issue of the future of the Federal Council, which had originally been my intention, but which was obviously impossible because of the events of last night.

5:00–5:30 PM   With Herter, on the President's direct instructions, to discuss our policy regarding the monitoring of very-high-altitude nuclear tests. We agreed on the best approach and Herter, upon my presentation, decided it was urgent to undertake quick technical studies to be ready to come up, perhaps during the Summit meeting, with constructive new suggestions, if for no other purpose than to defend ourselves before the world. I agreed to pull together a group to consider specifically what work should be undertaken and by checking with Good-paster found later that this is just what the President wanted.

6:00–6:30 PM   With Secretary Flemming, discussing the carcinogens in the food panel. He is smooth but evasive and wants to check our terms of reference with his staff, but assures us, of course, of his complete cooperation.

∾ *Friday, January 29, 1960*

Went to see General Persons and give him my memorandum on the Monday night dinner, previously described. He was quite shocked since, he said, Ray had been carefully investigated and thought to be very superior to Secretary Mueller, which was the reason for his selection as undersecretary of commerce. His first reaction was to do some talking to Ray, but when I pointed out that I have to live with him in the Federal Council, he agreed with recommendations in the memorandum which are to leave Ray alone but to insist that Commerce accept the Mervin Kelly committee report, establish the position of assistant secretary in charge of technological activities, appoint a technically competent person to it, and have him serve on the Federal Council.

> Mervin Kelly had chaired a committee advisory to the secretary of commerce which recommended administrative changes in the department.

1:30–3:15 PM   In the New Senate Office Building with the medical consultants [to a Senate appropriations subcommittee chaired by Lester Hill which handled the HEW budget]. They have a chip on their shoulder and began by suggesting that all that PSAC and I were interested in were weapons and military budgets and that the health of the American people was of more importance, but its protection was considered a burden rather than an obligation by the administration. Fortunately, most of them were anxious to have their statements on the record and so instead of asking questions, they delivered speeches full of noble sentiment; after the first few minutes I caught on to this and let them talk themselves out. Nonetheless, the first hour or so I was being pressed hard and not in a friendly fashion. Gradually most of the committee became friendlier and only a Dr. Wilson from

Brown University, who seems to resent Harvard, remained antagonistic. My theme was that expenses for NIH should be compared not with military expenses which have greater urgency (this was challenged time and again) but with expenditures for NSF and such like, and the same standards used. I don't think I made any damaging admissions, but doubt whether I did much good either, as the minds of these people were made up in advance.

Tuesday, February 2, 1960

9:30–11:00 AM With Air Defense Panel. The panel was in doubt as to what its functions were to be in the future, and I tried to define them as the evaluation of specific items to assist York's office in regard to programming, as well as the general evaluation and maintenance of competence to assist with fiscal '62 budget planning.

During the meeting, I was called out to be told that Jim McRae died of a heart attack on his way to the office in New York and the rest of my day was heavily shaded by this news.

Rabi dropped in and talked about the bad effect that our bilateral agreements with the USSR are having on NATO countries, robbing them of initiative and interest in mutual cooperation, although logically just the reverse should be the case. I asked Rabi to hear Spurgeon Keeny's estimate of my situation vis-à-vis Teller in nuclear test cessation issues, which appears to him to be quite critical. Later in the day Rabi dropped in again to say that he didn't think I had committed, as yet, serious tactical errors, but better watch myself very carefully in the future, because Teller is "such a clever SOB."

3:00–4:15 PM Two people from *Fortune* magazine to interview me in connection with an article on chemistry for *Fortune*, to be a part of a general series on science. Partly they wanted biographical data relating to the reasons why I became a scientist and partly general opinions on what I think of American science and ways of improving it.

4:30–5:30 PM Pre-press meeting in General Persons's office, much of which was taken up with the discussion of the speech and testimony by General [Thomas] Power [SAC commander], in which he, in effect, completely repudiated the defense secretary's and the President's position. Everybody felt that the President should fire Power, which would create a tremendous outcry, but would be quickly forgotten by the public. The final conclusion was, however, that it won't be done because it really isn't going to change things in general and it is too late for this administration to start on a general cleanup of the insubordinate behavior of the military before Congress. There was also some discussion of the need for the President to deliver a major speech on missiles and space because political pressures are growing tremendously.

∼ Wednesday, February 3, 1960

7:30–8:30 AM   Pre-press breakfast in which I was personally inactive. The discussion again reverted to General Power, but no momentous recommendations were formulated. It appears that Power's speech was submitted to DOD and cleared by low-level people, whereupon Murray Snyder [assistant secretary of defense] approved it without reading it himself. Rather typical for our overworked "policy level" officials, I regret to say. So now not much can be done about Power.

9:45–10:30 AM   Pre-press briefing of the President, at which he stated that he will dismiss General Power's statement as unimportant. The President also was shown the recommended answer to a question about food additives, which states that the matter is being studied by scientists from Agriculture, HEW, and PSAC. This was the form insisted upon by Flemming. I learned that a question about this has been planted among reporters, but later in the day looked through the press conference record and found that it was never asked. So we are still operating in secrecy.

10:30 AM   With Jim Fisk, talking about his testimony before the Foreign Relations Committee tomorrow and about Jim McRae. I also got Fisk's assurance that he will be my stand-in when I am in the hospital.

Over the phone Pete Scoville told me that evidence of a satellite of unknown, but presumed Soviet origin is now definite. OCB is to discuss the question of our policy toward "enemy" satellites. I was afraid that unwise decisions would be taken and went to see Gordon Gray, who assured me that he will watch the point. He also decided that the special briefing for the President and selected NSC members on our intelligence satellites has become very urgent in view of the Soviet sputnik, the news of which is bound to leak to the press. He therefore arranged for a special meeting this Friday rather than in March as we originally planned. York is to give the general briefing, and I am to follow him, bringing out the issue of the need for a policy toward ''enemy'' satellites.

2:30–4:30 PM Meeting in Phil Strong's office with him, Scoville, and Hi Watters, to hear a detailed account of recent Soviet activities. The sum total of their launchings into the Pacific are several failures and then the sudden appearance of the satellite of unknown origin. This makes a fascinating and puzzling picture. It almost looks as if they tried to fool us with their missile launch activities and got off the satellite on the QT.

> This minor mystery was never resolved, to the best of my recollection.

The nature of the shots into the Pacific remains puzzling, notwithstanding the large concentration of our naval intelligence forces in the impact area. We still don't know exactly what the Soviets launched and what for. Certainly these shots don't look like peaceful outer-space efforts, but seemed advanced missile tests; but why should they have advertised them so much and virtually invited us to observe them is very puzzling. After this session I called Gray and urged him to schedule a briefing this Friday morning for the President on what I heard.

Tried to get Din Land to come here and fill me in on his evaluation of our intelligence satellite program, but rather typically he was too busy to come prior to the Friday meeting and couldn't come then and present his estimates to the President. I first thought of making a quick trip to Cambridge myself, but figured it would take too much time, and so Hi Watters is going in my stead. Poor fellow!

CHAPTER VIII

Advising the President

By THIS TIME—more than six months on the job—I felt reasonably secure and confident of access to the President. When starting on the job I was warned that Eisenhower kept short office hours and that I should not ask to see him privately unless urgently needing his decision. The first few times I entered the Oval Office in considerable awe after devoting much effort to my briefing papers. Gradually I discovered that appointments could be had within a day or shortly thereafter upon asking for them; the President must have issued instructions to that effect with his staff secretary, Goodpaster, and his appointments secretary, Stephens. I also found that private meetings with the President tended to become quite informal so that my briefing papers would remain unused and the President would engage in a friendly conversation lasting longer than my allotted time. Gradually, it ceased to be necessary to clear my requests for an appointment with General Persons, the Assistant to the President and hence my immediate superior. He insisted only on knowing about my appointment and on accompanying me into the Oval Office when the issue involved budgetary commitments by the President.

The private meetings with the President remained relatively infrequent because I saw him regularly on more formal occasions, such as the pre-press briefings, NSC, Principals' or cabinet meetings (I was instructed by the President to attend the NSC meetings on a regular basis), and so on. The President, as time progressed, began to ask me frequently on these occasions to comment on what was being discussed so that my need to communicate with him through private meetings did not expand. A little planning helped being noticed by the President in larger meetings (NSC, cabinet) which took place in the cabinet room. Being of subcabinet rank I sat in the back row and was not to butt into the conversation unless spoken to by the President. The vice-president's chair, frequently empty at these meetings, faced the President's across the oval table. By occupying a chair just about behind that of the vice-president I could easily catch the eye of the President as he frequently looked up from his casual doodling. The assistants who accompanied the secretaries of state and defense to these meetings (and who, like me, were not supposed to speak unless spoken to) sat behind their bosses, thus behind the President who was between them. This gave me a distinct tactical advantage that I quickly learned to use.

Eisenhower, except for one brief flare-up of anger in the summer of 1960 (more about this later), was consistently friendly, attentive, and considerate, but our relations remained largely on the business level and I never felt myself as one of his intimates. Thus the private social occasions to which I was invited were few in number. So were the state events in which I was included. The result was that my purchase of "tails" in July 1959 turned out not to be a good investment, as I could have saved a small margin by renting an outfit for each of the white-tie affairs that I attended. The protocol used to put me in the line waiting to shake hands with the President and the visiting dignitaries in the East Room just behind the chairman of the AEC and in front of the director of the CIA. McCone, Dulles, and I had some spirited conversations while shuffling forward.

Thursday, February 4, 1960

NSC meeting started with Dulles's report, which covered the Soviet shots into the Pacific and our discovery of a satellite of unknown origin. The President asked me to comment on the purpose and meaning of the Russian activities and I did so, suggesting that one possibility may be the testing of an advanced kind of nose cone [ablative: the type that flakes away in re-entry] which resulted in increasing the range of their missiles. I noted, however, that the operation was a very expensive one and would be hard to justify on this basis alone. Dulles then read a summary of the general world estimate for the next few years. The President accepted it, but complained that the estimate did not make allowance for the growing wealth and more comfortable lives of the Soviet population, which he believed would make them less and less willing to risk military operations. He felt this was a terribly important factor that may radically change their policies. McCone gave a rather emotional speech, saying he received many telephone calls from all over the country, complaining about the inadequacy of our deterrent forces. He urged the President to embark on a much larger missile program and also provide money for a complete [SAC] air alert as Power wants. He also wanted vigorous work to increase the explosive yield in Minuteman warheads, although first numbers of Atlases and Titans should be increased. The President spoke firmly, that he would not accept McCone's point of view because: firstly, he was deeply convinced that we have adequate deterrents, and secondly, an increase in military effort would so disrupt the national economy that only a highly regimented society of an armed camp could result, and he was not willing to work for that. The vice-president challenged the air alert and Secretary Douglas explained that the DOD plan would permit 6 B-52s out of each wing to be kept in the air, but not 12 months every year, whereas Congressional pressure is for 12 per wing all the time. There was considerable discussion and it appeared that except for McCone, nobody was in favor of the enlarged program. The President called on me again regarding the capability of the Russians to launch a great salvo-like attack with their missiles, so as to destroy all of our SAC bases within minutes of each other. I suggested that while operational launchings are much simpler than developmental tests, it would be years before anybody could hope to salvo many missiles.

After the meeting, I stayed behind on Herter's request and watched a rather vigorous argument between Herter, the vice-president, and McCone. McCone then left and we continued it and concluded that an air alert of the magnitude requested by McCone would be a very provocative gesture, in addition to its other drawbacks. The vice-president said he would much rather spend the money on additional Atlas bases, even though these would become obsolete a few years after construction, and that this may become necessary because of political pressures. Herter said that he was setting up a meeting to discuss the conclusion of the GAC of AEC that centrifuges could be used for cheap production of weapons-grade U 235 and that this may completely change the whole world power balance. I agreed that this could be so and that this might be a strong argument for a comprehensive treaty to stop nuclear tests. Herter said this had been his immediate reaction which he had conveyed to McCone, but McCone insisted that on the contrary this was justification for further tests. What a mind!

> GAC: General Advisory Committee of the Atomic Energy Commission.

After the meeting I spent much time in my office with Panofsky, who delivered a lecture on the comparative ease of making nuclear weapons tests far out in space and urged me to change my attitude on these. I assured him I would. He then spoke about the high-energy accelerators program and how McCone is wrecking it by his insidious maneuvers with Senator Anderson, whom he privately feeds incorrect data. He asked me if it would be all right for him to talk to Senator Anderson, to which I replied "of course" although this may embarrass McCone.

> Senator Clinton P. Anderson of New Mexico; chairman of the subcommittee on R&D of the Congressional Joint Committee on Atomic Energy.

In the afternoon had a meeting with two men from the Civil Service Commission to justify recommended raises for Beckler, Keeny, and Skolnikoff. I seem to have been convincing. When I finished my story of PSAC and the special assistant's activities, I, myself, believed for a moment that we are the center of the government!

Wakelin and I then rehearsed before Bob Gray our presentation to the cabinet on oceanography. Gray thought it was good and that no more rehearsals would be necessary.

> Robert K. Gray, secretary of the cabinet.

I learned that, thanks to Irene's intervention, the meeting of the Principals has been moved by Herter to Saturday, even though that interferes with McCone's plans for a vacation, so that I can still attend Jim McRae's funeral tomorrow.

Irene Benik, my executive secretary.

Hi Watters returned from Cambridge and discussed the information he obtained from Land and Billings on the intelligence satellite program. Clearly there is too much systems development and not enough component research.

Fisk and Panofsky dropped in to tell me that their testimony before Senator Humphrey went off well, but that Humphrey bore heavily on Farley on two points: One, not enough research to improve detection methods, and two, not enough preparation for disarmament conferences. They, themselves, were not handled roughly, and only one touchy question was asked them: what did they personally think about the desirability of continuing the Geneva negotiations?

∽ Friday, February 5, 1960

8:30–9:00 AM Rehearsal with Gordon Gray of the special NSC meeting that was to start at 10:00. We went over the problems of reconnaissance satellites and I emphasized that the Air Force is trying to develop a much too ambitious system of almost instantaneous transmission of information by means of the so-called read-out system, whereas the technical people feel that this isn't going to be effective for many years to come and that therefore the emphasis should be put on a satellite which drops packages of photographic film. We also discussed the problem of the unknown satellite that has been detected by the so-called satellite fence, and in this connection I raised the question of the dangerous statements about destruction of enemy satellites if they overfly the United States. My point was that later this would prejudice the use of our own reconnaissance satellites.

9:30–9:45 AM Made introductory remarks to the new PSAC Panel on Carcinogenic Chemical Additives, in which I emphasized the conflicting issues the panel will face; namely the protection of health on the one hand and, on the other, the possible damage to agriculture and the halting of research in the field of agricultural chemicals and similar things. I urged the panel to write a report in a scholarly vein, taking into account scientific data or their absence in critical areas and to draw conclusions only on the basis of technical facts, as otherwise the report will be subject to valid criticism.

10:00–11:15 AM Special NSC meeting. Herb York reported on the status of the reconnaissance satellite program. Unfortunately, from my point of view, it was a very superficial presentation, in which he completely concealed the fact that the Air Force is unenthusiastic about the recovery-type satellite and even had it canceled for six months during last spring and summer, with the result that the first recovery-type reconnaissance satellite using the Atlas missile will not be launched until mid-1961. With all of the controversial issues omitted from his presentation, it was very difficult for me to say my piece. I nonetheless emphasized that technically the read-out satellite is quite far in the future and, moreover, it has the inherent weakness of not providing sufficient details of objects on the ground to be a useful instrument for our national security. Joe Charyk then presented the plans for a satellite interceptor, emphasizing the inspection rather than the destruction aspect of this plan. This was followed by a general discussion in which I presented my arguments against any demonstrations of our ability to destroy satellites, etc. The President rather unemphatically agreed with my position, but it didn't sound as if it was a directive. We then came to the question of the unknown satellite, which actually may be the tankage from one of our Discoverer satellites launched some time ago. This point is still uncertain. The question arose as to whether a public announcement of its detection should be made, and the President was emphatically against making such an announcement. After the special NSC meeting, I returned to the chemicals panel and listened for about an hour to the arguments of Secretary Flemming, who is very emphatic in his insistence that no compromises should be made with the potential presence of carcinogens in food and that he could not modify his policy and in fact wouldn't want any changes in law.

At noon left for Anacostia and then flew to Madison, New Jersey, to attend the memorial services for Jim McRae.

⌁ *Saturday, February 6, 1960*

9:00–10:50 AM   Spent with the chemicals panel, listening mostly to presentation of [Dr. T. P.] Carney of the Eli Lilly Co., who made a good impression with his thoughtful argument of the extremes that we can get into if we follow the policies set by Flemming.

11:00 AM–12:00 M   Principals' meeting in Secretary Herter's office, which was devoted to a presentation by AEC on the ease of diversion of fissionable materials by the use of the centrifuge separation method. It appears that the cost of this method is still very much higher than of diffusion, but by vigorous research, they say, it could be reduced to be about the same in the next few years. The danger, according to AEC, is that the centrifuge plants can be built in comparatively small units and hence would be virtually impossible to detect if constructed clandestinely. Moreover, they asserted that the technology involved in such plants is very much simpler than it is in diffusion plants, and hence less-developed and poorer countries could resort to it.

Returned to the chemicals panel a little after noon and listened to the last part of the discussion of the three representatives who talked about agricultural chemicals, rather than pharmaceuticals. They were [Dr. R. E.] Eckhardt, from Esso; [Dr. R. A.] Kehoe, from the University of Cincinnati; and [Dr. R. L.] Metcalf, from the University of California. These people, like Carney, emphasized the extremes to which Flemming is pushing his policy, and that in some instances this could lead to total absurdities.

Ate lunch with the panel and spent the afternoon with them until the meeting was terminated at 4 PM. Most of the afternoon was devoted to executive session, and I discovered that Bob Loeb and the rest of the panel had made up their minds to write a report strongly critical of Flemming and were trying to rush things with a proposition that they write the report without further meetings. I objected to that and emphasized that they should at least meet once more with the HEW people and discuss with them their findings without actually showing them the report, even though it may have been written in the meantime. Anything else would expose the panel to severe criticism and make the report useless. The panel members, without great enthusiasm, agreed with my recommendations.

∽ *Monday, February 8, 1960*

11:30 AM   After a previous telephone call, I was visited by
nine representatives of the meat industry, headed by a Mr. [A.
K.] Mitchell, Republican national committeeman from New Mexico. They are here in Washington to protect the industry against
a catastrophic situation such as developed in the cranberry case,
and had already visited Secretary Benson. Mr. Mitchell began by
noting that he was told I was charged with studying the issue,
and I just looked blank, but gradually the conversation became
friendly and animated, and they gave a good presentation of
their case, asserting that all they wanted was previous warning,
which would enable them to take corrective measures, rather
than a public condemnation. I steered the conversation to the
chemicals involved and from there to agricultural research, with
the result that my visitors agreed that the latter is too much
concentrated on short-range problems and doesn't think of the
really important long-range issues of agriculture, such as finding
brand new chemicals, etc. I urged my visitors to put pressure on
their congressmen to change the congressional attitude on the
USDA research appropriations, and this they eagerly agreed to
do, so the time was not wasted on my part. Just before parting,
as a joke, I said that, of course, if Flemming condemns some of
their steaks as carcinogenic, I would be in the market to buy
them, and one of them, a Mr. Davis, said he would see to it that I
got them.

In the afternoon I had a staff meeting on the seismic and high-
altitude detection R&D and learned that the immediate program
is supposedly well underway, the $7.5 million given to AFTAC
being assigned. However, General Betts has emphasized that
there was not a cent of money for FY 1961, and that the total
program, if properly progressing, would require $60 million,
most of which would be DOD funds (with $4–5 million as the
AEC share). I assured those assembled that I would urge the
Principals to consider the issue soon.

∽ *Tuesday, February 9, 1960*

9:30–10:00 AM   Visit from Wiesner, who is most unhappy
that I have accepted the threshold concept for a nuclear test-ban

treaty, and urged me to review the issue in the next PSAC meeting. He gave me his latest draft of a so-called comprehensive arms control system, which I read and found a little optimistic.

12:30–2:00 PM   Federal Radiation Council luncheon meeting in Flemming's office. Dr. [R. D.] Huntoon presented an outline of a proposed staff report on radiation standards to the council. Discussion was animated, but not sharp, except that Commissioner Floberg repeatedly warned those assembled that AEC had a major interest in the problem and would not accept any compromises with their operational efficiency.

> The Federal Radiation Council, an interagency group, was the arena for airing the persistent conflict between AEC and HEW. To simplify its own operations the AEC years earlier had adopted high radiation standards, i.e., large permissible radiation exposures. Public protests about these and also about the exposures to fallout from weapons tests, backed up by a critical report from the National Academy of Sciences, drove AEC reluctantly to reduce these standards.

> Consumer advocates felt the concessions to be inadequate (the situation, by the way, persists to this day) and the result was the executive and then the statutory creation of the Federal Radiation Council. What Floberg probably objected to was the lowering of exposure limits for uranium miners—a sensitive issue for AEC.

After the meeting, Floberg came to my office, where we were joined by Mr. Schaub of the BOB, to discuss the report of the high-energy accelerators panel which Floberg sent yesterday to the Joint Committee on Atomic Energy in Congress. Schaub was very unhappy about the release, but in the end accepted it as inevitable. We then talked about the report and Floberg insisted that PSAC review the panel report and in particular advise AEC whether the recommended level of support, rising to nearly $200 million by 1965, was necessary. I explained that PSAC was not competent to judge relative priorities and, therefore, relative levels of support in all fields of science, and that it was only in that way possible to reach a balanced program. For instance, PSAC contains nobody in earth sciences, nobody in biochemistry, nobody in oceanography, and nobody in astronomy, etc., etc. We then began to talk about this problem, and to my surprise, Schaub expressed the opinion that too much money is going for developments and not enough for basic research. Wonderful statement from a BOB stalwart, and I talked about my notion of what has to be done to improve our scientific planning and how difficult the problem is.

Late in the afternoon, a package arrived from the Mr. Davis of the meat industry, containing six magnificent steaks, although I had told him on the phone (he called to announce its arrival) that I personally couldn't accept them, so I sent two to Jerry Persons, who had suggested he should be cut in on the graft, and raffled off the remaining four among my staff people.

Was told that the President had decided to visit Cape Canaveral on Wednesday and wanted me to accompany him. Left the office early (6:30 PM) and got a phone call from Herter about 7 to tell me that State is going ahead with the threshold test-ban treaty proposal and that it would be made public very soon. He emphasized that the President wants to include in the treaty all tests that can be monitored and urges that vigorous development of the seismic detection system be pursued until we eliminate or at least reduce the threshold level that the treaty must allow for.

ᵒᔑ Wednesday, February 10, 1960

Got up at 6:15 AM and was driven to Andrews Air Force Base with Gordon Gray. The White House garage is certainly conservative in its time estimates. We arrived twenty minutes ahead of time, and Jerry Persons complained that he got there twenty minutes ahead of us.

Took off at 8:30 AM and almost immediately after the takeoff, Gates, Douglas, Persons, Hagerty, Goodpaster, Harlow, and I were invited into the President's compartment. It is beautifully appointed, and if we get a plane like this for our Key West PSAC meeting, it will be pleasant indeed. We spent in this cabin two hours flying down, and two hours flying back, in a pleasant conversation with the President. The chief subjects discussed at length were the outrageous security breaches by Symington and others in Congress for political and other reasons. The President is exceedingly angry and has talked at length about lack of loyalty to the U.S. of these people. In his estimation Joseph Alsop is about the lowest form of animal life on earth, so I better not have dinner with Joe while I am in office, even though he might invite me. Next, the President got to talking about what he called inflation in the Armed Forces, as regards both decorations and grades, and suggested to Gates that certain decorations, such as the Distinguished Service Medal, should be reserved for wartime only, and we should altogether tighten up on these things. We then talked at length about the unknown satellite that was observed a few days ago. The President was with difficulty persuaded to make an announcement about it, since the news is leaking out to the press anyway (he insisted that the announcement should be made by NASA, so as to make it devoid of all military significance). Conversation came to Dulles's testimony, from which I gathered that something very awkward has happened; namely that somebody in CIA, on his own authority, had changed the figures for USSR missile capabilities from those estimated a year ago and had given the new figures to Dulles. Dulles (who, as I observed many times, knows nothing in detail about what goes on in CIA) gave them to Congress and now he is in trouble and has to appear again. The President talked vehemently about the need for government officials occasionally to flatly refuse answers to pointed questions. Spent a long time reminiscing about his military experiences while young.

We spent nearly four hours at Cape Canaveral, in a whirlwind inspection of most of the facilities. Except for the utterly repulsive crowd of photographers, it was a pleasant occasion, but they followed us from one installation to another and were certainly not a pleasure to watch.

> Joseph Alsop had invited me for a breakfast at his home a while before I took office in 1959, and he spent the time painting a grim picture of a disastrous "missile gap," claiming that the Soviet Union already had 150 operational ICBMs which they were to launch against us in July 1959! Naturally I had to keep mum to avoid leaking real information.

Returned to my office at 4:00 PM. Modern life is really incredible: in one day to Florida and back and still have several hours there and in the office.

Attended a pre-press conference meeting in General Persons's office. Hagerty said the President will make an announcement of the threshold treaty proposal at the conference, and also that the PSAC chemicals panel will be announced one way or another. Hagerty asked me to check up on the nuclear test-ban statement with State, and Farley later told me over the phone that the British have almost accepted our proposal, which is quite a change in attitude from that expressed in Selwyn Lloyd's recent letter to Herter.

⌒⌒⌒ *Thursday, February 11, 1960*

7:30–8:30 AM   Pre-press breakfast, at which nothing exciting happened, but I was charged with the task of editing for the President's use the State Department statement on nuclear test cessation, which reveals the threshold proposal. This I did after breakfast.

9:45–10:30 AM   With the President. Nothing to record, except the explosion of the President when Andy Goodpaster told him of the security leak leading to the story in the *New York Times* on our knowledge of Soviet activities at Tura Tam and also about our inability to launch missiles on time. Later I read the article and it is shocking. About the worst breach of security I ever came across; I talked on the phone with Goodpaster and Gray, urging that a full-scale investigation be undertaken and the guilty person punished.

> The story broadly hinted at U-2 flights over Tura Tam in central Asia, then the secret Soviet launching installation for ICBM and space shots.

11:00 AM–12:15 PM   Ambassador Bohlen, with whom I discussed the issue of nuclear test cessation, the relative importance of continued tests for the U.S.A. and for the USSR, etc. I found him a thoughtful person and very pleasant to deal with, and gave him a lot of information on PSAC, Teller's and other people's attitudes, as well as my own convictions in the matter. It is interesting that Bohlen is seriously thinking of a move in which we would offer the Soviets a threshold treaty as well as a less formal agreement not to resume tests of any yield, since he feels that this is the only kind of proposal the Soviets will accept. I told him that if one were reasonably sure that the Soviets would not cheat, this commitment would be definitely to our advantage. If they cheated, the shoe would be on the other foot, but not catastrophically.

Spent some time with the interagency panel on cessation of missile tests. It certainly is a complicated business, and it is pretty clear that the panel will come out with almost a flat rejection of its feasibility or desirability.

∾ *Friday, February 12, 1960*

9:00–11:00 AM Special meeting with the President to hear the JCS presentation of a study on strategic targeting and the strategic forces required. This special study has now been in preparation for almost a year. It considered three optimized target systems: one consisting of military targets only; another of industrial and government centers only; and a mixed one. The general conclusion of the study is that only an optimized mixed system will achieve the objective of "prevailing" in a general war.

To prevail in the sense defined by this study, it would be necessary to kill over one-third of the population of the USSR (and about 100 million Chinese), in effect totally destroying about 100 cities, since none of them would receive less than a megaton and some several 20-megaton weapons. The overkill proposed is appalling and clearly the objective of the study is simply to prove that building more SAC aircraft and missiles is necessary. On the other hand, they make some phony assumptions about the extent of early warning we could receive and thus the quality of our response to it, which in reality are not likely to be achieved. Hence our response to a surprise attack would be very much weaker than presented.

There was rather little discussion after the presentation. Gordon Gray asked why was it necessary to use at least megaton bombs when a 20-kt bomb pretty much flattened out Nagasaki, but worded his question in such a way that no answer was absolutely required—and none was given. The President spoke with some feeling about the proposed overkill and suggested that as the Soviet Union grows wealthier, a far lesser threat of destruction than is implied in these planning figures would certainly deter it from war, but he nonetheless approved the general concept for planning purposes by JCS. The presentation also pointed out that unless we have very effective defenses, the initial attack by the Soviet Union will so completely destroy this country that there could be no question of prevailing. This was a pitch for Nike-Zeus and bomber defense. The President commented that if such a level of destruction was achieved in both countries, there was hardly any question of who would prevail, and my own thought was that what the result of all this would be is the taking over of the world by the Chinese, since after killing 100 million of them, there would still be 500 million left.

The President asked the individual Chiefs to comment, and Burke fairly strongly objected to overkill. Lemnitzer vaguely joined, but not explicitly. After the meeting, I walked over to Burke and said I was horrified by this plan and offered my help if something could be done. He said he has a long paper presenting arguments against the plan and he would like to show it to me.

After the special meeting I joined the PSAC Foreign Affairs Panel, Bohlen, [J. A.] Lacey, and others of the State Department also present. The panel expressed its doubts concerning two points of State Department policy: insistence on complete reciprocity in the exchange of visits of scientists, and on excluded areas. Bronk suggested that State relax restrictions on travel of visiting Soviet scientists on a temporary basis. The group also objected to the State Department policy of not admitting nationals of unrecognized satellite countries to scientific meetings in this country. State defended its position quite vigorously, but I gathered from subsequent conversation with Bronk that after I left they conceded that some changes in both areas may be possible.

3:00 PM   Visit from Admiral Bennett [ONR], the purpose of which was not completely clear. He spoke on several topics relating to the support of science by ONR, then discussed the joint AEC-ONR program in nuclear physics and expressed concern that AEC is trying to get out. We agreed that he should talk privately to Floberg, whom he knows personally, and find out how adamant the AEC is on the subject and then we will consider what actions should be taken to preserve the program.

Left that evening for Philadelphia.

ᴄᴄ⁓ *Saturday, February 13, to*
Tuesday, February 16, 1960

In Philadelphia, at the University of Pennsylvania, where in
the morning I delivered the graduation address which went, I
believe, quite well. Certainly had warm compliments and even a
request from Bronk for a copy of my speech, so he could, as he
put it, borrow some of the effective paragraphs.

Left Philadelphia in the afternoon after taping an innocuous
broadcast for the local undergraduate station. After considerable
trouble and delay in Washington due to heavy snow and bad
weather, we all (that is, PSAC members and several members of
the staff) except for John Bardeen, who was too late (by 10
minutes!), made Florida about 10 PM in the presidential plane.

The three-day PSAC meeting at the Key West Naval Base was
very good, although we had a dull beginning reviewing the work
of the military panels. I saw that we should change the agenda
and, therefore, moved general discussion to that afternoon. We
put a lot of time into the question of what is to be done about
government support for science. There was certainly no complete
agreement on how to organize this. Some, such as Rabi and
Piore, definitely favored a department of science of unlimited
scope. However, the outline I presented of tying in the National
Academy, strengthening the office of the special assistant, and
possibly the Federal Council, into a chain that links opinion of
active scientists with government officials, seemed to be received
rather favorably, and I committed myself to write it up and to
appoint a special panel to work on the problem further and to
resume the discussion at the next meeting of the PSAC.

Also gave lengthy consideration to the high-energy accelerators program. Over the opposition of Rabi, and with some reservations from Weinberg and Seaborg, we agreed to endorse that
panel's recommendations.

Then we had a long discussion on arms limitations and on
nuclear test cessation. The initial opposition to a threshold proposal tended to die down when I explained that quite likely the
President would not resume nuclear tests and that there would
be a real advantage in having a treaty with effective monitoring,
so as not to surrender this principle for other negotiations with
the Soviets, to win ratification from the Senate, as well as to have
a gentlemen's agreement not to undertake any tests unless the
other side cheats.

Finally, we had a lengthy and active discussion of the carcinogenic additives in food. It was interesting to see that the PSAC members, by and large nonexpert on the problem, took a stand much closer to that of Flemming than either Bronk or Loeb did in presenting the views of the panel. This alerted me again to the danger that a "cold-blooded" scientists' opinion may be very poorly received by the public at large.

Between meetings we spent a pleasant time at the pool and on the beach, and the evenings were enjoyable also, except for my having to stand in a receiving line Sunday night at the commanding admiral's reception. My attempts to learn to water-ski turned out to be successful, but it is not a sport I would prefer over real skiing.

Returned to Washington about 6:00 PM, February 16.

Wednesday, February 17, 1960

I was almost late for the 7:30 AM pre-press breakfast. Got charged with writing the response to a question about the latest Soviet proposal to eliminate criteria from the definition of inspectability of seismic events, but to reduce the annual quota of inspections called for in the treaty. Had the statement approved by Herter.

9:45–10:30 AM   With the President, who accepted the statement. He was also generally in good spirits. As about the last item Jim Hagerty mentioned the report by the AP of my speech in Philadelphia on Saturday, which took some sentences out of context and made me disagree with the President. The President laughed, suggesting that this is what always happens. Forgot to mention that the subject also came up during breakfast and I said to Hagerty that he had approved my speech and made the text into a White House release, and he said he would do it again because it was a perfectly proper speech.

11:30 AM–12:45 PM   At the Pentagon, getting a briefing from [Dr. M. R.] Nagel of the USAF Cambridge Research Center on the Soviet nose cones that fell in the Pacific. Although I had to be polite and compliment him on the tremendous achievement, not much has been learned, except a suggestion that the nose cones are of the ablating type and that no decoys are being experimented with.

2:45–3:15 PM   Keith Glennan asked that one of my staff be appointed acting secretary of the Space Council, a pure formality, since it will have no further meetings, and Dave Beckler agreed to take the job. Then Keith said that he was going to study carefully the contracting operations in NASA, and I encouraged him to do so because of astronomers complaining to me about the way NASA is handling astronomy projects. I told Glennan that I recommended to Staats against a supplementary budget request to accelerate the Rover project because my interpretation of Glennan's letter to Staats was that Glennan himself was not interested in Rover and was passing the buck to the BOB. Keith readily agreed that this was the case, and saw no reason why the project should be accelerated; that the push was part of McCone's attempt to run the big booster space program. McCone is grooming himself to be secretary of defense, it seems.

Had a phone call from Wiesner, who learned through Doty from Farley that Herter is going to see the President this week, before the latter's departure for South America, to get the State Department plan for an internal organization to study disarmament measures approved. On the spur of the moment I went to see Persons and Goodpaster and told them that I was terribly disturbed by this and didn't know what to do, because I hesitated to see the President on the matter without first letting Herter know, and this I couldn't do under the circumstances. They both expressed their strong conviction that the office for arms limitation planning should not be in the State Department which would be incapable of doing a good job. Goodpaster urged me to see the President tomorrow and, taking off from the question of financing research on nuclear test monitoring, about which the President wants to see me, to introduce the subject of the arms limitation study. This I agreed to do, but it may be difficult, since the President decided not to have any further appointments.

Thursday, February 18, 1960

9:00–10:15 AM   Special meeting with the President to consider the State Department proposal for a ten-nation conference to cut off fissionable materials production for weapons purposes.

Defense and AEC turned out in full strength. Herter began by reminding the President that he, Eisenhower, had repeatedly made this proposal in the past. He then emphasized the importance of somehow slowing down the buildup of nuclear weapons. Tom Gates then gave an extraordinarily weak dissent, voicing the usual concern about the inability to satisfy all Service requirements for innumerable warheads, and fearing that the Soviets would manage to turn this proposal into a world-wide demand for a moratorium without inspection. McCone reminded the President of the additional difficulties in monitoring such a cutoff in view of the threatening centrifuge possibilities for enriching U 235. The President then delivered a strong statement, virtually condemning Gates and the Chiefs (who added their own appeals to those of Gates) for their utter inability to see the positive side. He suggested that the Soviets would also stop production and that he saw no alternative but somehow to stop this mad race, etc. After a while, the objections of Defense subsided and became procedural, namely about the specific wording of the proposal, which did not explicitly state that the monitoring system must be not only installed, but also actually operational before production is to stop. The President instructed Herter to coordinate the exact wording with Gates to the latter's satisfaction, but as far as I could figure out, he also gave a go-ahead on the proposal. For a while it looked as if Defense might stall the whole proposal on the matter of tritium, and the fact that its radioactive decay would gradually destroy our H-bomb stockpile. Evidently the President knows nothing about tritium, and so was rather bewildered, but then he pointed out that the Soviets will be in the same boat and moreover, that tritium, not being a fissionable material, could still be produced. I am afraid this last reservation may be a joker that could play the same role in this proposal as the reservations about low-yield weapons tests did in the summer of 1958. I almost got stuck with a job to study the tritium problem, but believe it was too casual a remark by the President and hadn't been taken down by Gordon Gray. Let's hope for the best since this is a stinker.

> Tritium: a heavy radioactive isotope (modification) of hydrogen which can be made in nuclear reactors but cannot be indefinitely stored because of its radioactive disintegration. It was used in H-bombs.

10:15–11:30 AM Regular NSC meeting, but the President left after the first presentation—that on incapacitating but nonlethal CBW agents. York did a good job and General Lemnitzer afterwards gave a dull but plausible description of how the substances might be used in concrete cases of minor conflict. I merely added that our panel endorsed the strengthening of this research.

The President objected to changes in national policy because of the political impact of the use of these weapons. I reminded him that there is a big difference between biological and non-lethal chemical weapons, since, for instance, tear gas, of which the latter are merely an extension, is now an accepted police weapon. The President agreed with this. The rest of the meeting was quite dull.

Had lunch with Ambassador John Davis Lodge and Bryce Harlow. This Lodge appears to be an extreme conservative and discoursed at length about the stupidity and subversive character of the Harvard faculty, which is too liberal. I thought that this was not the place to fight and limited myself to rather mild objections.

In the afternoon had a visit from Dr. Alexander King, who is the scientific director of OEECD. He is obviously a very dynamic person and spoke at length about his organization. Rather than engaging in over-all planning for European scientific activities, it supports special meetings and other efforts aimed at strengthening specific scientific activities by providing funds for secretaries, travel, etc. Rather an impressive summary, and I have a feeling that in the long run this organization will accomplish much more than the NATO Scientific Council. Just before his arrival, I got a letter from Nate Pusey, telling me that Harvard will not agree to putting me on a part-time salary basis next fall, which made me so mad I wasn't paying as much attention to King as I should have.

Friday, February 19, 1960

Spent most of the day with the chemicals panel, listening primarily to Flemming, who held forth very extensively. He is a very elusive character and seems to have a gift for saying things in such a way that one is not quite sure what he means. The general impression is, however, that in the two weeks since the first meeting of the panel, there has been a significant shift in his position. Then he insisted on the extreme application of the Delaney amendment, now he talks about the rule of common sense and reason, and specifically exempts from the Delaney amendment such issues as putting table salt on potato chips, even though table salt probably contains traces of radium, which is a carcinogen. He also conceded that as soon as research can establish thresholds for the safe amounts of carcinogens, he would recommend a change in the amendment. He also is appointing a panel of experts, all from within HEW, to advise him. I took him aside and argued that it would be to his advantage to have this panel include members from Agriculture and from the National Academy, and he conceded that this might be a good idea. He did make a considerable impression on the panel and the panel will soften its recommendations, but my impression of the HEW crowd is not one of complete admiration. Certainly their "scientists" adjust their convictions to the needs of the moment. Thus, two weeks ago [G. B.] Mider [director of the Cancer Institute] insisted that aminotriazole was proscribed by Flemming as a carcinogen, but now that the evidence has been gathered by our panel, he says this is not so, and that the Delaney amendment wasn't involved. Similarly on the diethylstilbestrol, which was deemed a carcinogen two weeks ago, the opinion now is that it was proscribed because of its estrogenic activity. When the HEW people left, I reminded the panel that there is also a Department of Agriculture that they have to think of.

Interspersed with my attendance at the panel meeting, I saw Bill Elliott of Harvard who came with another memo for Secretary Herter, in his capacity as consultant. It is really frightening how ignorant the fellow is about things on which he writes with such authority. I tried to delete a few of the worst proposals, such as his urging an all-out air alert for SAC—but don't think I had much luck.

Wrote a reply to Floberg, informing him of the conclusions of the PSAC on the high-energy accelerators program and took it to Persons before sending because undoubtedly this letter will eventually find its way into the congressional joint committee. Persons took a serious view of the matter and by the end of the day I don't know what he decided about my letter. What is involved, of course, is "executive privilege" regarding White House correspondence.

Since I couldn't see the President I sent Goodpaster my memoranda concerning the organization of continuing arms limitation studies and the fiscal support of technical work on nuclear test-ban monitoring. This afternoon Goodpaster phoned and we seem to have gained something. The President stated, as I requested, that this technical work is one of high priority and should be supported. With this I should be able to get the money from McCone and Gates.

On the matter of an arms control organization, the President, according to Andy, first objected to instituting the agency in the Executive Offices, which we recommended, because of his memories of Stassen, but then reconsidered. He instructed Andy to phone Herter and tell him to explore carefully the possibility of setting one up outside the State Department, which is, of course, not what Herter wants. We may yet be asked to discuss the organization before it is agreed upon. So, it seems, we have won a skirmish and delayed the decision.

The day was ended by a visit from Mr. Cyrus Eaton, whose main purpose was clearly to size me up and to assure me that Khrushchev and every other Soviet official is sincerely desirous of peace, and that we must trust the Soviets more than we do and find fewer nasty things to say about them, because the tone of our press is very offensive to them. I assured him that I was a peace-loving American, but didn't have confidence in the Soviets to the extent that he has.

Saturday, February 20, to
Monday, March 7, 1960

During the past two weeks I was in the Walter Reed Hospital and then at home. By and large, it was a period of inactivity, during most of which the President was traveling in South America.

I had an interesting phone call from McCone. First he talked about the High-Energy Particle Accelerators Program; it was obvious that he was still seeking ways and means of ducking from under. Then we came to nuclear test cessation and McCone surprised me by stating that: firstly, he was convinced that these tests would not be resumed for quite a long time, if at all; and secondly, that he didn't believe current weapons technology made resumption of such tests critically important. He said he

felt the maintenance of the Livermore and Los Alamos laboratories as effective organizations was the more important issue
(with which I agreed) and then proposed that the entire nuclear
test cessation monitoring development, that is Project VELA of
ARPA, be transferred to AEC and given to the two laboratories
to keep them busy. He pointed out that the DOD wasn't enthusiastic about developing monitoring techniques and could not be
made to change its mind about it, so the work would never go
well if left in DOD. I conceded that this may be a very good plan
and expressed desire to think it over and talk again about it.

Herb York came to my apartment and spent about four hours
in very amicable discussion. We talked about our deterrent position in the immediate future, and he said that the studies in his
office indicate that if one is a pessimist and allows the Soviets the
maximum capability that the national intelligence estimates suggest, in mid-1961 we will pass through a critical period when
they may be able to destroy virtually all our retaliatory forces by
a perfectly coordinated surprise attack. He said that he wants to
take some relatively inexpensive steps to avert this danger by
accelerating the construction of a few Polaris subs so that we
could have five on station in 1961, and also by accelerating the
second (Alaska) BMEWS installation to have it operational by
the beginning of 1961, since even 10–15 minutes warning would
enable us to get enough bombers off the ground to provide an
effective deterrent. The whole plan made good sense to me, and I
said I would support it wholeheartedly once I became convinced
of its technical feasibility. He then brought up the subject of the
engineering manpower study of Mervin Kelly and said he was
getting more and more keen on it and hopes that Kelly's panel
would investigate the threat of technological unemployment
among design engineers working now for aircraft and missile
manufacturers. He fears that soon the almost simultaneous termination of so many huge development projects will mean a
rather catastrophic layoff of engineering personnel, especially in
California.

Monday, March 7, 1960

Alan Waterman in the office. We spent a long time selecting
candidates for the National Science Board membership. Then

Alan went into a rather emotional complaint about my rearranging the activities of the PSAC in such a way that he, Alan, is not informed any more about what PSAC does, owing to our executive sessions and how embarrassing it is to him that Bronk and Loeb, who are members of his board, know things that are not known to him. He argued that he was in a different position from the rest of the government employees and should take part in our executive sessions. The whole thing was a little unpleasant but of highest importance to him, so I will have to meet him half way.

He told me that the Achievement of Excellence committee will finally come into being, because Sproul, the former president of the University of California, has tentatively agreed to be the chairman and that Waterman, Flemming, Killian, and I are proposed as members. Flemming, Waterman, and I are to see the President and outline the plan to him. As I am not at all convinced that this will amount to anything, I suggested that Flemming, as the senior in rank of the three, make the presentation to the President.

7:00–10.30 PM   Attended the banquet of the Science Talent Search Awards organization and delivered an after-dinner broadcast address to the winners. A reasonably pleasant affair. My neighbor was Carmichael, and he and I let our hair down a little and discussed the shortcomings of Bronk. Obviously, we see eye to eye on the subject, namely that Bronk is trying to prevent the membership of the Academy as a body doing anything in Washington so that the Academy is identified with himself.

Tuesday, March 8, 1960

12:30–1:45 PM   Federal Radiation Council meeting and luncheon in Flemming's office. Another utterly innocuous meeting. Floberg outlined AEC's plans for transferring federal radiation monitoring activities to the states and as the plan is eminently cautious, it was approved by everybody without discussion. Then Astin reported on the discussions of the working group of the council, limiting himself to the statement that they are not as yet completely agreed on the safety standards, but that they will have agreement before the all-day meeting of the council which is set for April. There followed some unimportant general conversation and the meeting was adjourned.

Allen V. Astin, director of the National Bureau of Standards.

4:00–5:00 PM With PSAC Air Defense Panel, listening to their findings, which, to put it in a nutshell, argue that while a really good air defense is impossible, it is still necessary to work hard to get some air defense, because some air defense is enormously better than no defense at all.

⌒⌒⌒ *Wednesday, March 9, 1960*

9:30–11:00 AM Meeting with General Persons, McCone, Stans, and Goodpaster on the High-Energy Particle Accelerators Program. There was first a long and indecisive discussion as to precisely what happened on April 2, 1959, when McCone, Goodpaster, Persons, Killian, and Staats saw the President regarding the Stanford accelerator. Goodpaster's memo of the meeting does not agree with McCone's and neither agrees with the recollections of the people involved. It seems that the President was apprised of the need of other additions to the existing accelerator building program besides the Stanford accelerator and that no firm cost figures were given to him, although some figures much lower than now considered were mentioned. It seems also that the President acted only to authorize the Stanford accelerator. McCone was quite coy about committing himself to the program. He spent much time explaining to Persons the reasons for the delay on the Stanford accelerator, whereupon Persons read a handwritten note from the President which asked why there was no action on the Stanford accelerator, which he had requested in April 1959. Persons suggested that the people involved better get busy and make sure the Stanford accelerator gets going. We then discussed the relation of that project to other projects, emphasizing that some physicists gave these higher priority. Stans took a firm attitude that it was improper for the scientists to sell a program to the President on assurances that it will cost so many dollars and then to come a year later and say the costs have multiplied several times and he urged that neither the Stanford accelerator nor other projects be recommended to Congress. On that he was very firm. I got a word in edgewise with a remark that this program involved national prestige in competition with the Soviet Union and presented a unique opportunity to lead from strength rather than weakness. These remarks seemed to have an effect on McCone and he came out with pretty much the same remark as I did and then strongly urged the adoption of the program. It was agreed that McCone and I will study the details, so as to minimize the need for a supplemental appropriation request and present the revised program to the President next week.

12:00 M–2:30 PM   Lunch with my staff and discussion of what we should do regarding the NSC request to prepare terms of reference for a planning board study of the means for the maintenance of Western technological leadership. The tentative conclusion was that we should recommend an interdepartmental study of the broad question, in the expectation that no useful recommendations will be generated, and thereupon continue with separate PSAC panel studies of the component issues, such as education, government organization for science, manpower utilization, etc. These studies would be presented to the NSC as PSAC papers rather than as planning board papers, because otherwise they would be emasculated.

4:00–4:30 PM   Bill Patterson of Convair, with the standard spiel on why Atlas is better than Titan and how it can be improved at a much lower cost than Titan and then do exactly everything that Titan II is supposed to do. I was coy and didn't commit myself since the plan involves storing LOX in missiles and I have great doubt about the feasibility of doing so for long periods of time.

ᨠ᷍ᨠ *Thursday, March 10, 1960*

9:00–11:00 AM   NSC meeting. We had more than an hour's presentation by Dulles, some of which created a great deal of interesting discussion, and I will put down a few highlights. The intelligence evaluation is that Khrushchev's Asian trip was definitely not a success, which I hope is not just a report of people trying to please. Supposedly the relations between Khrushchev and President Sukarno in Indonesia were extremely tense, reaching a point where Sukarno told his foreign minister that he couldn't bear it any longer; that throughout the trip Khrushchev was in bad humor and gave out freely on the faults of the capitalist system and how everybody should imitate the Soviet Union.

Dulles then reported on the continuing violent Soviet campaign against the shah of Iran, attacking him personally and stimulating a Communist conspiracy. Whereupon the President commented that all we do is to identify the symptoms but neglect to do anything about correcting the causes. Why should there be a Communist conspiracy against the shah just after he had announced a very liberal land reform? The President said that during his trip, Nehru asked him to tell the shah that in Indian opinion the shah couldn't survive unless he undertook land reform, but that the land reform would straighten out Iranian difficulties. The President said that before he could even deliver this message to the shah, the shah described reform plans that were even more liberal than Nehru recommended. Dillon then said the difficulty was that nobody in Iran believes any more in the shah's promises because the shah is surrounded by the same old gang of grafters that has been around him for years; all the younger people distrust them and yet are unable to displace them.

Most of the time was devoted to the Cuban situation, which is getting worse every day. Dulles said that his organization is now convinced that nothing we can do would improve relations with the Castro regime, which is bent either on forcing us to intervene militarily or upon discrediting us completely in the eyes of South America. On the other hand, no group in Cuba except the Communists has the political strength to displace Castro. Gordon Gray said that the subject had been discussed at length in the planning board and that the latter reached a less pessimistic conclusion—namely that there might be means of easing relations with Castro—but Dillon pretty much agreed with Dulles.

General discussion followed in which even a military blockade came under consideration by the President, and it was agreed that the whole situation will be reviewed on a weekly basis. The President said that during his South American trip he discussed the Cuban situation with all the four presidents; that in each case the president involved expressed violent opposition to Castro, but conceded his impotence in influencing the course of events in Cuba and warned our President about the dangers of intervention because of evil political effects in South America. The sum total of the discussion was the gloomy conclusion that no active policy was possible now and that it was necessary to wait to counteract Castro's initiatives.

The rest of the time was spent discussing the paper on U.S. policy towards Libya, with the characteristic split: the majority opposing Treasury and Budget. The latter, naturally, are pinching pennies, and in the ensuing discussion the President gave a mild but fairly explicit lecture to Anderson and Stans on the need to subordinate penny-pinching to carrying out national objectives.

About noon, I received a phone call from Persons, who said that he had had a visit from Tom Gates, who wants to strengthen our deterrent posture in the immediate future and needs money for it beyond his budget allocation. Jerry added that he and Stans are firmly opposed to asking Congress for a supplemental appropriation for this purpose since the administration has gone so firmly on the record that we had enough deterrent that a request at this time would be a political catastrophe. Therefore, he wants to find out if the money couldn't be obtained by reprogramming, and would I undertake the review of the DOD budget and identify soft items from which money could be taken. I agreed to do so and assured him that I believed the measures recommended by Gates were good ones. Later I checked with York and found that indeed Gates was proposing the measures that York had spoken to me about last Saturday; namely the acceleration of Polaris and of BMEWS, as well as a future increase of the number of missiles in each operational ICBM squadron. We had a rump staff meeting about it and decided to operate on the basis of the November PSAC recommendations for the DOD budget, compare those recommendations with DOD's requests, and where there is excessive programming, recommend reduction. Stans at lunch assured me that BOB will give us full cooperation (not surprising!).

Left for New York, taking with me a staff memo on ways of stretching out the particle accelerator program to avoid a supplemental budget request. Met with Piore, Fisk, Rabi, and Killian at 5:00 PM and stayed with them until 9:00 PM.

It was a rather emotional meeting, largely devoted to the questions of our deterrent posture and of arms limitation. Killian felt strongly that our deterrent posture is inadequate and hence was happy to hear of the events of this morning. I got all those present to approve our plan of action regarding DOD reprogramming. Then we turned to arms limitation. I gave them a brief account of what has happened since the last PSAC meeting, which didn't cheer them, since even if we can prevail on Herter to accept an effective organization for arms limitation planning, its effect will not be felt for a long time to come. In the meantime, the President has only one more Summit meeting at which he can speak with authority. If this meeting fails, several years may pass before another one is arranged and in the meantime the balance of power is likely to shift toward the Soviet Union, so that it will be even more difficult for the U.S. to speak from a position of strength. On the other hand, everybody felt that neither Eaton at the UN nor the President at the forthcoming Summit meeting has much to propose and that not enough was being done to remedy this. Rabi and others urged that the PSAC plunge vigorously into developing proposals for the President and then see him with these proposals. I noted how little time was available for this, whereupon we discussed the idea of a continuous session of PSAC the next few weeks. What fun! In the face of my unenthusiastic acquiescence, we agreed to have a special meeting of PSAC this Sunday evening to consider ways and means of procedure and also to discuss the meeting with Herter which Killian and I will have on Monday regarding an arms control organization.

The sum total of the discussion was great concern and the feeling almost of guilt that we haven't come forth with useful plans, but certainly this session didn't produce any concrete ideas. I hope the Sunday meeting will be more productive, but have my fingers crossed because the time available is becoming very short and it is not clear to me how we can in the meantime turn out really thoroughly documented proposals. To come forth with half-baked ideas of the kind that the State Department handed to Eaton would be disastrous for us.

> The President was scheduled to meet with Khrushchev, Macmillan, and de Gaulle in Paris in May.

ᥱ Friday, March 11, 1960

9:00–10:30 AM   Cabinet meeting which began with presentations by Robert S. McCollum of State and the Right Reverend [Francis B.] Sayre, the dean of the Washington Cathedral, on the refugee program. Both were good, and Sayre especially was very moving. He read a short letter from a young girl in a refugee camp in response to a question whether she wouldn't like to have a home. Her answer was that her parents, brothers, and sisters had a lovely home and were only looking for a place to put it in, which I found rather touching. Bob Merriam then reported on the organization of the National Goals Commission. The President very emphatically requested everybody in the administration to clear all public speeches dealing with U.S. foreign policy with the State Department. The meeting ended by a presentation by Wakelin and myself on the oceanography program for 1961, which went off rather well, although there was no discussion afterwards. I got a big grin from the President when I slightly changed my text and referred to "that painful subject," the budget. Stans looked grim when he saw the figures for fiscal 1962 and Persons whispered to me: "Hey, you pulled a fast one. You shouldn't present budgetary matters to the President before clearing them with me." But he wasn't annoyed.

∽ *Sunday, March 13, to*
Tuesday, March 15, 1960

PSAC Meeting. The meeting Sunday night, as it developed, was devoted to a discussion of the arms limitation issue, not the administrative problem about which we were to see Herter the next day, but the basic issue that an adequate, secure deterrent must be provided and that it is essential to have a better understanding of the purpose and usefulness of monitoring systems in arms control negotiations. A proposal was made to see the President as a body and urge him to agree to a comprehensive nuclear test cessation treaty as a dramatic gesture. There was no complete agreement on this proposal, but we did agree to try to prepare a paper which set out the limitations of monitoring systems and the need to rely on other factors to maintain national security. This paper, to be presented as a briefing to the President, was considered very urgent and was to be prepared before the April meeting of the PSAC and presented to the President prior to that meeting. I asked Killian and Fisk to assume responsibility for coordinating the writing.

Monday morning we had an executive session and went over

various activities of this office. After 11, in the presence of Mervin Kelly, we talked about government organization for science. The concept of a strengthened science agency, with responsibility for planning and defending interagency types of scientific projects, was developed, but there certainly was no clear definition of the interrelated functions of PSAC, the new agency, BOB, etc. However, we agreed to work on the subject, and I asked Al Weinberg to prepare the next draft.

Immediately after lunch, Killian, Fisk, and I went to see Herter, an exceedingly successful visit. I first talked about the need to maintain a secure deterrent to give the President a position of strength in the Summit meeting and expressed the hope that Herter would support the Gates-York plans for bringing up our deterrent. Herter eagerly picked up the theme, endorsed the concept wholeheartedly, and said he has been urging the same thing on Defense. He then launched into a rather frank discussion of his difficulties with Defense, such as the terrible push by McElroy and the JCS to start flying to Berlin above the 10,000-ft. ceiling insisted upon by the Russians. State got certification from the JCS that this was an operational requirement, got the thing approved by the President, and the State Department rammed the concept down the throats of the British and French, who were unenthusiastic, whereupon the paper was accidentally seen by Secretary Douglas, who raised hell about it, called in the JCS, who admitted they had mechanically signed the staff paper. On reconsideration they withdrew the whole plan, about twenty-four hours before a note was to be handed to the Soviets. Then another cute thing: when State was urging Defense to assist the NATO allies in acquiring mobile missiles, DOD would make no recommendation but could only come up with an offer of either Polaris, Pershing, or Minuteman missiles, thus carefully preserving the inter-Service balance and keeping peace in the family. Herter then reported that he had been advised that a land-based Polaris missile is now operational (something which I don't believe, but didn't challenge) and that it could be installed in fair numbers by mid-1961, in Europe. I then introduced the subject of an organization for arms limitation and asked Jim Killian to present the subject. To our surprise, Herter said that he agreed with our thesis to place the planning office in the Executive Office of the President, providing that the negotiating authority remained in the State Department, which we, of course, endorsed. He asked me to send a new memo on the subject, which he will approve after staffing it in State.

As we left, he held me back for a couple of minutes and told me he was seeing Wallace Brode and that he was going to fire him.

When we returned to PSAC, we found that considerable tension between Mervin Kelly and the members had developed during the discussion of Kelly's proposed panel on manpower utilization. He rejected the idea of investigating misuse and stacking of scientific personnel by industry, insisting that that was a congressional function, not his. I soothed the ruffled feelings by properly chosen generalities and everybody parted in good spirits.

> The issue of the waste of technical personnel by industries working on government contracts proved to be unresolvable and Kelly withdrew. The panel never materialized.

⌒ Wednesday, March 16, 1960

The day started at 7:30 AM, with a totally uneventful pre-press breakfast. It almost looked as if everybody had a hangover, the discussion was so slow. The briefing of the President was equally uneventful. He was in a bad mood and was swearing about the reporters who expect him to know an incredible number of trivial details about unimportant things.

After the briefing, Bryce Harlow asked me to come to his office and showed me a document prepared by the Convair people, which purports to prove that the Ballistic Missiles Division of the Air Force has deliberately kept the information on the growth possibilities of the Atlas missile away from the high levels in the Air Force and OSD. He asked me what to do about it. I said I knew nothing about the dates of the particular briefings and how they came about, etc., which was the substance of the document, and so he said that, unfortunately, York and Sharp both confirmed its factual statements. I explained that at least York knew about the technical aspects of the possible growth of Atlas and explained the noncryogenic storable propellants development for Titan, etc. Then Harlow launched into a violent condemnation of CIA, and especially Dulles. He obviously is not well informed because he conceives the CIA as the complete centralization of our intelligence activities, instead of the actual state of affairs, where the three Services do most of the gathering of technical information and the national intelligence estimates are a result of hard bargaining between the Services and CIA.

D. C. Sharp, assistant secretary of the air force.

12:30–2:30 PM   Lunch of the Advertising Council, which was quite enjoyable. Firstly, I got a number of compliments on my speaking performance yesterday, which naturally was soothing, and then the management presented two visiting Russians, with Roscoe Drummond asking them questions regarding U.S.-USSR relations, etc. The Russians did an effective job proving that they were going to win over us economically and technologically and pointing out the weaknesses of our society, etc. This got some of the audience—all of them industrial big shots—very angry, and so the questions from the floor developed into impassioned speeches, delivered in the heat of indignation. Thereupon the chairman of the meeting explained that the Russians were actually working for the USIA and were good Americans. It was very funny to see how the steam went out of the audience, but it gave them a very good hand. I must say I was also completely taken in.

Roscoe Drummond, a syndicated columnist.

3:30–4:30 PM   Visit from nine members of the American National Cattlemen's Association just to get better acquainted and they re-emphasized their concern about HEW activities. I tried to be soothing, but pointed out that the Delaney amendment was here to stay, at least for some time.

⌒⌣ *Thursday, March 17, 1960*

This was one of those days!

Breakfast with Persons, McCone, Goodpaster, and Stans, regarding the High-Energy Particle Accelerators Program. Its beginning was very satisfactory, but after McCone and I had explained that the program aimed at implementing a unanimous recommendation of the physicists, evaluated by PSAC, Stans came up with a bombshell, reading aloud quotations of remarks made by Stan Livingston in which he condemned the Stanford accelerator as premature, unwise, and probably useless to the progress of high-energy physics. This really fixed it and nothing was decided, so we will have more meetings on the subject. After the meeting, I telephoned Piore, who became as annoyed as I was, and later that day I got in touch with Livingston and read him the riot act, pointing out what he had done thoughtlessly and asking him to write me a letter by the weekend which would clearly state his opinion and reasons therefor. I also recommended that he consult with scientists who know more about the subject than he does. The most annoying part of it is that Mc-Cone told me that Livingston is trying to increase the budget of CEA, and so he really got his neck stuck out with these remarks, as he could be accused of trying to kill the Stanford accelerator to get more money for CEA. Hi Watters was also very much upset and, being a good personal friend of Livingston, phoned him and also read him the riot act. Livingston told Hi that by then he had had four calls from Washington on the subject. So that will teach him! The remarks which Stans quoted were made by Livingston to a member of the BOB staff!

> A joint AEC-PSAC panel chaired by E. R. Piore labored before my time in office on a priorities plan for constructing large facilities desired for research in nuclear physics and produced a report which recommended several particle accelerators, the most ambitious being the SLAC (Stanford linear accelerator, the chief advocate of which was W. K. H. Panofsky). Unfortunately the panel members and other nuclear physicists did not reach complete consensus on their plan nor were they keeping the disagreements in the family. This book has already described the opposition of I. I. Rabi; and now the indiscretion of Stanley Livingston, the director of the joint Harvard-M.I.T. Cambridge Electron Accelerator (CEA) then in construction added to my difficulties in trying to get the SLAC project going in compliance with an earlier public commitment of the President.

10:00–11:00 AM   NSC meeting. The high points were the now weekly review of the Cuban situation, which is deteriorating so rapidly there are fears that Castro, in his efforts to embarrass the U.S., may stir up a massacre of Americans; the discussion included military plans of what to do in that case. Then we had a policy paper on the West Indies, with the usual split: Budget and Treasury wanting a language which will reduce our expenditures. The President expressed his displeasure about the insignificance of splits on which he has to make decisions and then spoke feelingly of his concern about the Cuban situation, which he considers the greatest immediate threat to the U.S. because it may be the opening wedge to destroy the Organization of American States and to separate South America from us. He emphasized that he wanted a policy toward the West Indies which would ensure that the same wouldn't happen elsewhere, even though it might cost us money for economic aid, etc. Treasury and Budget retreated in confusion because it was an exceptionally clear and firm statement of his beliefs which I, for one, found very fine.

> The foreign policy papers prepared for NSC by the Policy Planning Board involved usually numerous disagreements of the agencies represented on the board, alternative wordings being printed as splits for the President's choice.

Another Summit
in Camp David

THE GENEVA TEST-BAN CONFERENCE resumed its sessions in January with an exchange of ambassadorial statements not flattering to either the Soviet or the American scientists. Wadsworth emphasized the newer technical evidence of the difficulties in screening underground explosions from earthquakes and suggested that the Soviet experts' position was to subordinate objective scientific analysis to political guidelines. Ambassador Tsarapkin accused American scientists, especially those who participated in Technical Group II, of subservience to political forces in the United States opposed to the treaty. After a few weeks of largely critical exchanges Wadsworth submitted a proposal for a "phased" treaty which incorporated ideas developed by our group in January. The treaty was to include the prohibition of underground tests producing seismic signals above the 4.75 magnitude on Richter's scale (regardless of explosive yield), also an agreed-upon number of annual on-site inspections, which number would be subject to annual revision. The threshold would be progressively lowered as detection techniques improved. Wadsworth invited the Soviet Union to join in the ambitious experimental program to improve seismic detection on which the United States had already (!) embarked. He noted that the proposal did not include a moratorium on tests below the proposed seismic threshold. It did include the prohibition of tests

[275]

in outer space up to the greatest height of effective controls to be agreed upon.

Soviet reaction was negative, Tsarapkin pointing out that no international controls were needed for tests above the threshold, since the national seismic stations would be adequate for monitoring them. Returning to an earlier major issue he offered compromise criteria for approving on-site inspections which came closer to those proposed by the United States. With some amendments the Soviet criteria were accepted by Wadsworth early in March. Thereupon Tsarapkin stated that the on-site inspections should be permitted only in conjunction with a comprehensive ban, not in a threshold treaty.

Some retreat from this position occurred on March 19 when Tsarapkin accepted the concept of a threshold treaty. The on-site inspections issue was to be resolved on the "political level" and a joint R&D effort to improve detection was to be undertaken, providing the parties to the treaty made a commitment not to test weapons below the threshold magnitude. Tsarapkin also stated, however, that this commitment would somehow be included in the language of the treaty. Subsequent questioning revealed that these "moratoria" would really be a part of the treaty. Thus the Soviet proposal went against the basic American tenet of no disarmament agreements without controls. As this chapter indicates, the proposal was not rejected outright. Because of the importance of the issue, Prime Minister Macmillan came to Washington and his visit was largely dedicated to formulating a joint U.K.-U.S.A. position in response to the Soviet Union's initiative.

The meeting of the prime minister with President Eisenhower took place in Camp David and I was an active participant. The subsequent Washington meetings between Eisenhower and President de Gaulle late in April did not involve me but I was invited to the reception for de Gaulle at the French embassy. I shuffled in the line to shake his hand and murmur respectful amenities in my best French. President de Gaulle towered over my 6′3″ there in the garden, virtually immobile and silent, like a statue; even his greeting hand hardly moved. The refreshments indoors were first rate.

Friday, March 18, 1960

12:30–2:00 PM Lunch for Dick Morse, who cried on my shoulder about the difficulties of working under Army Secretary Brucker. All that Brucker is interested in is Michigan politics. All his decisions are influenced by the effect on local politicos, whom he notifies in advance of proposed moves and seeks advice from. The man obviously has very little loyalty to the President because, upon a specific request of the latter, transmitted by Kendall, to stick by the administration position on Nike Zeus, Morse witnessed Brucker suggesting leading questions to Democratic senators so as to bring out in his subsequent testimony his opposition to this position. Kendall, by the way, in his second conversation with Brucker identified me as the source of doubts about Brucker's position, and Brucker in turn accused Morse of being a link in the adversary chain. Then Morse talked about the inefficiency of the present Army R&D organization which is controlled by procurement men, and his plan to separate R&D from the so-called technical services. In his opinion the plan will be rejected by Brucker and hence will require high-level support for effectuation.

2:00–5:00 PM With Bob Loeb and Fred Holtzberg discussing their draft of the carcinogenic chemicals panel report. Bob, while admitting that the present draft was not as good as the first, was initially rather resistant to further changes, but by and by accepted Dave Beckler's version and agreed on a meeting of the panel on Saturday, which would make the whole panel contribute to the report.

Monday, March 21, 1960

I learned from Fred [Holtzberg] that the carcinogens panel had a constructive meeting on Saturday and the report was much better now.

> Dr. Frederic Holtzberg of the PSAC staff, on loan from IBM laboratories.

A long phone call from Secretary Herter, who spoke first of the problem of deterrents and said that Gates does not agree with the information I got from Admiral Raborn's office regarding the availability date of land-based Polaris. In Gates's opinion, Raborn's office is anxious not to interfere with the submarine capability, and therefore deliberately stretches out the availability dates for land-based Polaris. He then spoke of "Skybolt," that is, ALBM [air-launched ballistic missile], as something available in the near future and already offered to the British. With a little perturbation I told him of the sad state of this project, which, on inquiry by York's office, turned out to be largely a figment of an Air Force colonel's imagination. We then came to talk about the latest Russian proposal in Geneva and I expressed some concern about tomorrow's meeting of Principals as being too early for the development of a definite position. I emphasized that I couldn't take any lead at the meeting, but would support a positive response as contrasted with rejection. Herter strongly concurred with my observation that we have to aim at cessation of nuclear tests, as the only alternative to intensification of the cold war and lack of progress in the whole arms limitation field. Finally, Herter spoke about our administrative proposal for the arms limitation office and expressed his concern that the director would interfere with policies of the State Department. He suggested that I discuss these matters with Phil Farley.

> Rear Adm. William F. ("Red") Raborn, Jr., was in charge of the entire Polaris R&D project in the Navy Department.

> The Russian proposal about which Herter phoned me was that of March 19. I was against assuming the initiative because the issues were nontechnical.

2:00–3:00 PM   Navy briefing for the NSC on "Control of the Seas," which turned out to be a rather generalized description of the fleet tactics for defense against manned aircraft and submarine threats and the assertion that the task forces will survive even in the missile age. Not very exciting, and I had a hard time keeping awake most of the time.

4:00–5:00 PM   Rehearsal by Don Ling of his presentation on controls of missile tests for NSC. Again an outstanding performance and the only criticism which those present offered were of detail. Then a long phone conversation with Secretary Ray, trying to pave the way for a smooth Federal Council meeting. The worst of the lot is the squabble between NSF and Commerce, the latter obviously preserving their little empire and fighting tooth and nail against the assumption by NSF of even mild coordinating functions in the science information services organizations as directed by the President.

Tuesday, March 22, 1960

1:30–4:00 PM Meeting of the Federal Council. The morning before I arranged to take off the agenda one item on the management of science information, which is still in dispute between Commerce and NSF, with the understanding that there will be a private meeting between the interested parties to resolve the matter. The rest want off well, although much time was taken up with Secretary Ray's presentation about ways to improve science in government and to strengthen the Federal Council. Unfortunately, what he has really done is only to identify the weaknesses, but did it in such a language as to imply that he had the solutions. So I am probably stuck with trying to do something in a very difficult area.

Talked on the phone with Carmichael, who agreed to be the new chairman of the standing committee if selected, and this was approved by the council.

4:30–6:30 PM Meeting of Principals, at which the latest Soviet proposal in Geneva was indecisively discussed. It was agreed that in effect the proposal still offers a comprehensive treaty without safeguards, but that it is put into such a form as to require positive reaction and Secretary Herter, during the short time he was there, emphasized the great political importance of coming up with a constructive response. After he left, the meeting was taken over by Merchant and later by Dillon, the change of chairmanship not assisting the progress of discussion. The State Department emphasized that the British are putting heavy pressure on us to accept the Soviet proposal as is. It was agreed to meet again next day because of the urgency of the issues involved.

Wednesday, March 23, 1960

This was a day!

9:00–9:30 AM   With Jerry Persons and Bryce Harlow to get instructions for my response to Senator Jackson's inquiry on the administration's position regarding the statutory establishment of PSAC and my office. Instructions were to reply that the administration has no position and that if Senator Jackson moves, it will be necessary for the administration to develop a position. Harlow emphasized strongly his opposition to statutory establishment since it would weaken rather than strengthen the organization and its relation to the President.

9:30–10:30 AM   With the President, Persons, McCone, and Stans, regarding high-energy accelerators program. McCone gave a rambling and poorly organized presentation, much of it being devoted to the justification of his own delays regarding the Stanford accelerator. Result was development of a slightly negative attitude to SLAC on the part of the President, whereupon I spoke in general about the reasons for support of science by the government and how we must, in addition to a moderate support across the board, also select certain areas for vigorous all-out support. I identified high-energy physics as one of them. This seemed to influence the President, and he spoke most emphatically about putting the Stanford project through, even on an accelerated basis. Then he criticized AEC for having started too many accelerators, suggesting that the scientists shouldn't mind having to travel to these big machines. Although he was not very explicit, both McCone and I got the impression that he authorized the rest of the program, but wanted to stretch it out over several budget years, and this was certainly the impression of Stans, because when we left the office the three of us agreed that the Berkeley bevatron should be given higher priority than the Princeton-Penn machine. Next day McCone phoned in high dudgeon, saying that Stans now maintains that the President didn't approve anything but the Stanford accelerator and that Goodpaster seems to agree with him. So I will have to straighten this out; if possible, without going to the President.

11:00 AM–12:30 PM   General Persons, Secretary Douglas, Secretary Sharp, York, and Stans to discuss Defense reprogramming to strengthen our deterrent. The Air Force will cancel SAGE super hardened command centers and reduce Bomarc B, which will make enough money available to accelerate BMEWS and add Atlases to the existing squadrons. Then Secretary

Douglas outlined the Navy plans which involve accelerating seven Polaris subs by about two months, which will not cost much money, and also getting authorization for six additional subs, which involves, of course, almost a billion dollars. Followed a very lengthy and involved discussion about the political hazards of telling Congress about these plans, because they might suggest the weakness of the administration's position. Stans, naturally, was for standing pat. What a man! I finally left before the tactics were agreed upon, although in principle the decision was positive, since it is a good plan.

> Bomarc B: surface-to-air medium-range antiaircraft missile; canceled later.

1:30–2:00 PM Gerard Smith of the State Department to review with me his briefing for Herter on the missile test ban. He proposed to recommend to the secretary to take a positive position and to urge further and more detailed studies.

5:30 PM Another meeting of Principals, which lasted until 7:30 PM. The State Department came up with a three-point proposal for the Geneva conference involving acceptance of a threshold treaty, a joint research program in seismic improvement, and a unilateral, executively declared moratorium on testing of a few years' duration. McCone, and after him Jack Irwin, vigorously objected to the third point. An interesting change in McCone's attitude is that now he doesn't want to test and would accept a complete test ban, but he is afraid that the other side would cheat while we would not and—since there will be no detection below the threshold—this testing by USSR will harm us. Irwin's line is more extreme, insisting on the need for improvement of our weapons. To my surprise and pleasure, Jim Douglas rather sided with the State Department in rejecting rigid adherence to a threshold and agreeing to a moratorium of short duration on all tests. Meeting ended in disagreement on all points except the proposal to meet with the President and present to him the split.

◦◦◦ Thursday, March 24, 1960

9:00–11:00 AM NSC meeting, which started with Don Ling's presentation of the ad hoc study on the missile test suspension which he did extremely well. The discussion was mild and not very long. Burke, speaking for the JCS, took exception to some conclusions and urged further study, which was concurred in by Herter. Jim Douglas merely emphasized that in the analysis of our deterrent posture not enough emphasis was given to the invulnerable Polaris and other weapons besides the ICBMs. The President, to my surprise—since he has spoken in the past about the need of making careful studies of disarmament issues—remarked that it is too bad that these first-class people have to put so much time on a study like this and suggested that further studies be not on a continuous but on an on-and-off basis.

After NSC, the Principals joined the President in his study and McCone stated his objections to the State Department proposal; whereupon the President in a sharp voice rejected McCone's point of view and got obviously angry when McCone suggested that the State proposal was a surrender of our basic policy. The President said that he had decided to offer one or two years' test moratorium on executive basis, that he didn't think that the Soviets would accept it, but felt that this was in the interests of the country, as otherwise all hope of relaxing the cold war would be gone. Douglas agreed with the President's decision but wanted to let the President know that there were strong voices in Defense opposed to it, and Jack Irwin tried to argue the President out of his decision but unsuccessfully.

12:00 m–12:30 pm   Carmichael, with the surprise that he doesn't want to become the chairman of the standing committee of FCST. Ducky! I soothed his ruffled feelings and assured him that he would do a fine job without spending much time on it and assured him that I would try to find him a full-time executive secretary, which became a basis of his acceptance. Maybe I was mistaken in offering him the job. He acted quite irrationally today.

4:00–5:00 pm   Phil Farley here to discuss the PSAC recommendation for a White House office to develop policy proposals in the arms limitation field. He questioned rather sharply how we proposed to avoid this director's becoming a second secretary of state, and I may have answered his doubts only partially but he conceded that a similar organization within the State Department would be ineffectual. All of this was on a friendly level.

Friday, March 25, 1960

This was a quiet day.

The cabinet meeting began with a completely memorized speech by the chairman of the President's Committee Against Discrimination in Federal Employment. Both Saulnier and I thought it was frightfully pompous, but the President said afterwards that it was one of the finest presentations he had heard. OCDM gave a briefing on the need to increase emergency medical supplies, including packaged hospitals, for use after an all-out nuclear attack. The needs would be colossal and what we have amounts to only about 20 percent of the estimated total. Naturally, Stans thought the whole program should be stretched out, but the President said "No, better get it done." An interesting tidbit was provided by General Quesada that the light American plane which was intercepted by Castro's government while landing in Cuba was actually operated by Castro's government, which may backfire on them if we collect enough evidence to make it public. The meeting ended with lengthy economic reports on the state of the country, which were indecisive but not pessimistic. Thereupon, the vice-president pointed out that political opinion is actually set several months before election and therefore all measures should be taken to prevent a temporary recession in the summer, a matter which was weightily discussed.

> Raymond J. Saulnier, chairman of the Council of Economic Advisors.

Sunday, March 27, 1960

9:30 AM–4:00 PM   The Life Sciences Panel meeting devoted to the discussion of the Department of Agriculture program. The panel agreed that what Dr. Shaw and his associates were calling basic research wasn't anything of the kind and that the program was suffering very badly from a lack of imaginative research. There was also much discussion of the Beltsville experimental station, its purposes and functions. The general consensus was that the Agriculture Department must establish a closer contact with the university community, providing research grants, training grants, etc., but that this couldn't be done at the expense of in-house facilities because of the resultant opposition. I kept suggesting that to criticize publicly they would have to come up with good sharp cases as examples of unsatisfactory programs, and suggested that the panel meet with Secretary Peterson first and discuss with him its findings, as he might be much more understanding than Dr. Shaw. The panel's analysis of the reports being distributed by the Agriculture Department is that they are a whitewash of their programs. On my initiative, we discussed the question of radioactive strontium in wheat, and the panel members gave me helpful advice on how to handle this issue with Flemming and others. The comparison with milk is misleading; comparison should be made with milk powder which, of course, would contain much more radioactive strontium per unit weight than whole milk. Another recommendation was that the problem of strontium content of wheat shouldn't be handled on a total food basis, but be considered in relation to a typical American diet and the contribution of individual foodstuffs to this diet.

During the meeting I had a visit from Starbird and English, with papers which they prepared for a meeting with Bill Penney, presenting arguments for a coordinated research program, its character and the inclusion of nuclear explosions in it. The documents were very rough drafts, but I agreed with their general spirit.

> Dr. B. T. Shaw, the administrator of the Agricultural Research Service, Department of Agriculture; Dr. Spofford G. English, chief of the Chemical Branch, Research Division of AEC.

> The presence of radioactive strontium in American wheat— due to the nuclear weapons tests of 1958—had been discovered and because of the flap with the cranberries was causing much concern in the White House.

∞⌒ *Monday, March 28, 1960*

Had a quick glance at revised versions of the Starbird papers of yesterday and was told that Bill Penney does not disagree with the plan of using nuclear explosions in VELA.

12:30–1:45 PM Admiral Hayward for lunch. I gathered from his remarks that he came mainly to convince me that Polaris and additional carrier task forces would take care of all the possible Soviet missile threats and would eliminate the missile gap. We then talked on a variety of subjects, none of which were of far-reaching importance. It might be of some interest to note that he concedes that the JCS position on the need to have "adequate monitoring" in arms limitations agreements is becoming more and more irrational.

3:00–5:15 PM Session with Penney, Farley, McCone, and large retinues, during which we went through the AEC draft of a proposed seismic research program, and agreed on some modifications, since the paper was emphasizing the proof of inadequacies of the existing monitoring techniques rather than an eager search for improvements. We also agreed on the need for nuclear test shots and after the meeting I rewrote the nuclear test-shot argument. An interesting sidelight is Farley's whispering to me that the President is so irritated with McCone about his stand on the whole nuclear test-ban issue that he doesn't want to have him in Camp David and will probably ask me to go there. Toward the end of the day I learned through Goodpaster that I am going to Camp David tomorrow morning, there to present our view on the research program requirements and promises. I understand that this morning McCone spent two hours with Macmillan and Herter in the British Embassy in preliminary conversations. Very curious about what went on. In a phone call McCone didn't say, but said he was thoroughly annoyed with the State Department for abundantly leaking out information on him, McCone, that wasn't favorable. Clearly the makings of a major intra-administration feud. I must stay on the sidelines!

Tuesday, March 29, 1960

After a very hasty effort to refresh my mind on a variety of facts and figures about monitoring of nuclear weapons tests, I left by helicopter at 9:30 AM for Camp David. To my surprise, in view of information I got yesterday from Phil Farley, I saw as the last two arrivals on the White House lawn the vice-president closely followed by McCone, who climbed into our helicopter. It would be interesting to find out who engineered McCone's presence, but it will probably remain a secret.

When we arrived at the Aspen Lodge in Camp David, there was a period of inaction while State and the British Foreign Office people were working on a joint public announcement by the President and the prime minister. I was told that to the Americans' surprise, the prime minister came here with a plan not very dissimilar to what the President had in mind, and hence a press release was readily agreed to.

In an aside Herter said firmly to me that I would be called upon to report on the technical aspects, and so it was fortunate that during the flight I went through the last version of the paper we considered yesterday and even, on the back side in pen, hastily summarized it for presentation.

About 11:00 AM, the President, who was mostly talking to the vice-president about the pleasant aspects of Camp David life, grew impatient and asked when we were going to have that lecture by the scientists, so everybody sat down and I summarized the recommendations of the paper. The President questioned my use of the word "coordinated" research program instead of "joint" and I responded by saying it is sometimes difficult to have joint work even among friends, which seemed to amuse him considerably, and so it was agreed that it will be a coordinated program with some joint exercises, to be determined later in detail. I was very cautious in estimating what would be the accomplishments of one or two years of work, so that the statesmen wouldn't pass the buck to the scientists, but that, nonetheless, was pretty much what followed, everybody talking as if all the technical problems would be solved in two years' time. The vice-president raised the issue as to whether we would go with a flat take-it-or-leave-it duration of moratorium, and the President said he would start with one year, but be willing to compromise on two and the prime minister gently hinted that three would be a reasonable time. It is quite cute how Macmillan, in a most innocent manner, serves notice that the British may have different ideas. For instance, at a later time, when I called attention to another important negotiable item, the number of on-site inspections and the question of inspections of seismic events below the threshold, Macmillan said: "You know, it might happen that you would suggest a hundred, and we ten, and the Soviets might suggest five." It was agreed then that we would start with a one-year proposal, but would be ready to negotiate and that the Russians' performance in trying to improve the system would be a major element in our decision to continue the moratorium or not. During this discussion the text of the press release was brought in, was gone over and changed to emphasize our willingness to start with a coordinated seismic research program on an immediate basis. Followed a long discussion of the need to use nuclear explosions in it and it was agreed that they will be used. The problem of meeting the Russians half way on their refusal to allow "sealed black boxes" to be used for these tests was discussed and the British offered to reveal designs of their obsolete weapons for the purpose. According to McCone, we cannot do so without revealing the design to the whole world by declassifying it or without signing a bilateral military agreement with the USSR, a proposition which made the President quite amused. McCone was rather subdued throughout the discussion, although contributing some lesser items. Obviously he had been told not to raise objections in principle.

The meeting broke up about 12:30 and was followed by casual conversation. The whole affair was really informal and friendly —quite a difference from the session with Khrushchev last September. We each had a drink and then a pleasant and informal luncheon, during which the conversation was dominated by the President and Macmillan, reminiscing about their experiences in World War II.

After lunch, the vice-president, McCone, Dillon, and I went for an almost hour-long walk, during which time I was temporarily paired off with the vice-president and used this opportunity to talk about the complexity of monitoring systems, the fact that they may not be wholly realizable and so will get us into political trouble a year or two hence, and then made the suggestion that we should trade off insistence on monitoring against the "opening" of the USSR. He seemed interested in the idea. He complimented me on the smooth way in which I avoided giving a carefully scientific answer to the President's question in the morning as to the probability of detecting evasion in the monitoring system we discussed. At that time I had answered the President's question by saying I could only have an intuition and that to me the probability wouldn't be great, but would be finite. However, that it was hard to conceive of a great nation planning its actions so that there would be a possibility of proof that it was cheating on its solemn commitments.

Interesting that during the morning's general discussion, the President emphasized heavily the need for this agreement in order to keep the "atomic club" down to four and that only this agreement would make it possible to put pressure on de Gaulle to stop testing. On both points Macmillan firmly agreed.

Before lunch I had a pleasant chat with Macmillan, to whom I was introduced by Bill Penney, although actually Macmillan did all of the talking. Like many other great men, he has clearly reached a point where he really doesn't listen to others.

After the walk there was a brief meeting to look at the draft of instructions to our delegation in Geneva, which is to be telegraphed the same day because the general expectation is that as soon as the press release is made, Tsarapkin will put pressure on our delegation seeking an answer to a number of points. The instructions were noncontroversial and were agreed to in half an hour. The prime minister said they will send essentially the same instructions to their delegation. These instructions, to a large extent, are still delaying in nature.

Another interesting tidbit during the morning arose on the initiative of Herter, in connection with on-site inspections, who said that it may be necessary to leave the number of inspections blank in the treaty and have that number settled at the Summit. The President conceded that this is very likely to be the case. Thus, in a simple quick way, he conceded to the Soviets one of their main contentions, namely, that the number of on-site inspections is a political rather than a technical issue.

Everybody except the principal persons left at 4:00 PM by helicopter, and flying back I meditated on events of the last eight months. I remembered the situation I encountered just after taking over, Killian coming out for only a partial treaty and for resumption of testing; that horrible Bacher panel report saying that the detection of cheating by on-site inspections was virtually impossible, then the July 23 meeting with the President in which he seemingly approved a decision to retreat to a position of an atmospheric treaty only, although fortunately Andy Goodpaster's notes later showed his decision to be more limited. A tremendous change has taken place since then. I think that the McRae panel report had a major influence on the President and started a change in McCone which has now reached a point where he doesn't really insist that we test any more and only objects to the Soviets testing if we don't. Perhaps my conversations with him have helped in addition to the McRae report. Perhaps my conversations with the President have influenced him in accepting imperfect monitoring. These are things I will never know the answer to, but I do have a feeling that I have played some role in this gradual change and hope it was for the good of the country. On the other hand, things might have gone very differently if instead of Herter and Douglas, the Principals' recent meetings had been attended by Dillon and Gates. There is no doubt that Douglas is much closer in his attitude to Herter than is Dillon; Gates and Douglas are much farther apart then Douglas and Herter.

After landing on the White House lawn, McCone, who flew in the other helicopter, caught me and said he was being visited by Berkner and what did I think of getting him involved in the planning of the seismic improvement program, to which I said it might be a good idea. Later, reviewing the events of the day with Spurgeon Keeny, I mentioned this, and he suggested that Berkner would not be a good chairman. I conceded that and called McCone, recommending to him over the phone not to commit himself to Berkner to chair this planning group. McCone asked me to come over, which I did, and spent half an hour with him, Berkner, Starbird, and English. In the course of this conversation, McCone said that it was his conviction that I should organize the group rather than he, and then Berkner asked me who should chair the panel, to which I replied it was either he or Panofsky. Then Berkner started advocating Panofsky and said he would help regardless of being chairman or not. I said I would undertake the organizing job only if desired by State and Defense and approved by the President. With that we parted, McCone escorting me out and privately explaining that his present political position vis-à-vis the President was so difficult that he couldn't assume the management of the research program. He would certainly be accused of parochialism, he said.

Wednesday, March 30, 1960

7:30 AM   Pre-press breakfast, which I could just as well have skipped.

Went to the briefing of the President, in which again I took very little part, as the questions discussed were not relevant to me.

11:30 AM–12:30 PM   Visit from Bill Penney, which was very useful, I thought, as we exchanged observations on a number of issues and found ourselves in complete agreement. Penney reached independently the same definition as I did regarding the one-point safety tests, namely, that so long as the total explosive force (of HE and nuclear charge) is due predominantly to the HE explosion and the nuclear reaction is just observable by instruments, this should be called an experiment and not a weapons test. Penney said that the British intend to use this technique extensively on their proving ground in Australia, because of the hysterical fear of fission products by the British public at home. He thus expects the possibility of major progress in weapons technology, but concedes that nothing can be stockpiled without real weapons tests. Penney then emphasized that the main thing to accomplish in discussions with the Russians about the coordinated program is to ensure a situation where they cannot come back afterwards and insist that we used wrong instruments or wrong procedures and to claim that our results are meaningless. Regarding the centrifuge method of isotope separation, he said he was convinced that this method could be used clandestinely for the production of U 235, but in amounts small enough to be of no interest to the Soviet Union, only to some "nth" nation trying to get into the atomic club, and that it could be done rather cheaply with very low chance of detection. Finally, we talked about support of science and agreed to keep each other informed. We were in agreement that high-energy physics, medical research, and molecular biology were the areas where the West was way ahead of the Soviets and that we would be very foolish not to push them hard and not to use our superiority for propaganda purposes. It was a very pleasant meeting.

HE : chemical high explosives in the implosion atom bombs.

Later I had a phone call from McCone, saying that Herter and Douglas also want me to organize the seismic R&D program, designed eventually to lower the thresholds in the test-ban agreement. Not surprising that they want to toss me the hot potato.

In the afternoon had an interesting meeting with the President, at which Persons and Goodpaster were present. The President approved my assuming the planning role for the seismic R&D, gave that program high priority and assured me that he would have no hesitation in asking Congress for supplementary funds for it for FY 1961 if necessary. We then covered briefly the issue of organization for arms limitation studies and I was pleased to see that he took no exception to my suggestion that the planning office should be in the Executive Office, although he said that he would have preferred to have it under his "first secretary" which however, he said, was impossible to achieve this year.

I then spoke about my concept of trading the "opening" of the USSR against the elaborateness of monitoring organizations because otherwise we may get ourselves into a hopeless political fix by proposing schemes impossible to put into effect because of complexity and cost. Then followed a long discussion about carcinogens in food and radioactive strontium in wheat. Both the President and Persons took the matter very seriously and the President obviously is unhappy about the past assurances by AEC that there was no problem from fallout. Persons asked and got approval for a stern message to Flemming not to do anything rash and to discuss the problem in the Federal Council. I then covered briefly my ideas about laboratory-type nuclear-weapons experiments and the danger of deliberate leaks to the press, to which the President said: "That goddamned joint committee [of Congress, on atomic energy] will certainly do anything in its power to embarrass me." He ended a very pleasant and cordial meeting on a plaintive note, saying nobody comes to see him with good news.

Later in the day, I saw Goodpaster regarding the high-energy accelerators program and was given his record of the earlier conversation, from which it appears that AEC plans for stretching out the building of the machines has to be approved by the PSAC. It is interesting that when I questioned Goodpaster, he assured me that there was no decision against an emergency or new authorization for fiscal '61 in the President's mind, only a preference for avoidance if possible.

I later passed this message by phone to McCone and also got his approval to have Panofsky as chairman of the group to develop plans for seismic R&D. So we are off on another adventure.

ᐁ *Thursday, March 31, 1960*

The entire morning was spent trying to get agreement between McCone and Stans on the modified high-energy accelerators budget. After talking to McDaniel and getting him to accept some reductions, I phoned all members of the Piore panel and got their agreement, and after lunch talked to Stans, who naturally proved to be tough, but finally agreed to the authorizations I proposed (a reduction of $2.6 million from the $23 million proposed by McCone) but wouldn't accept the idea of a supplementary request, since he maintained that McCone could certainly reprogram the required $3 million. Tried this on McCone and to my surprise got acceptance, so my role was not in vain, but it certainly was not a pleasant one. I am getting to be quite skilled in the bedside manner.

Had a visit from Eugene Rabinowitch in the afternoon with a hard sell of the Pugwash conferences and a request that I assist him in raising money for more of them. My response was that I am not at all sure about the value of these conferences when they are on sensitive current topics of political importance, such as the nuclear test cessation. I questioned the wisdom of Wiesner's proposal, telephoned from Moscow, that a conference be held on the concept of ''peace through secure nuclear deterrence.'' I conceded, however, that some less controversial topics, like international cooperation in science and aid to underdeveloped countries, might be desirable for Pugwash, and I would say so if a foundation asked for my opinion.

Then had a visit from Waterman, which lasted an hour and a half. We agreed on the nominees for the National Science Board, on the tentative outline for my speech at the tenth anniversary of the NSF next May, and then talked about the role of NSF in interagency program planning. I found Waterman quite sympathetic and, in fact, he said that my concepts of the future organization resemble very much what he did organize to manage IGY, without, of course, the support of PSAC.

> Dr. Paul McDaniel, chief of the Physics Branch, Research Division of AEC; Dr. Eugene Rabinowitch, professor of biophysics, University of Chicago, editor of the *Bulletin of Atomic Scientists,* and the chairman of the American Committee of Pugwash Conferences.

> As regards the Pugwash conferences my attitude changed with the passage of time and a growing realization of the ineffectiveness of diplomatic efforts in the field of disarmament. I became a fairly active member of Pugwash.

Friday, April 1, 1960

9:00–10:00 AM   Off-the-record NSC meeting to listen to the Navy presentation of "Control of the Seas." The briefing was somewhat better than the one which we heard here a few weeks earlier, but still it was rather thin and qualitative and this was not only my impression, but that of others also.

10:00–11:00 AM   Regular NSC meeting, which was uneventful until the proposal by Secretary Douglas that the President authorize going into production of Minuteman ICBMs so as to achieve an operational force of 150 missiles by mid-1963. The President sharply questioned Douglas about the rate of proposed build-up of the Minuteman force and upon learning that the production capacity will permit nearly 400 missiles a year, remarked in obvious disgust: "Why don't we go completely crazy and plan on a force of 10,000?" The President put me on the spot by asking me point blank what I thought of the proposal and the operational prospects of Minuteman, which I ducked by saying I had no up-to-date information, but what I heard was favorable. I felt very awkward about evading the answer, but I just did not have the information, and so after the meeting I asked the missiles panel, through George Rathjens, to get busy and plan a briefing on Minuteman for the President a month or so from now.

> Speaking of going "completely crazy," in the words of Eisenhower: The United States is now (1976) approaching the deployment of 10,000 strategic nuclear warheads, despite official assurances of the "coming generation of peace," "breakthroughs for peace," etc.

Nothing much happened until I left Washington at 1:20 PM for Kansas City and the annual convention of the National Science Teachers Association. That meeting was dull, so were my dinner companions, and my speech was anything but good. I certainly hope I reached the lowest point in my oratorical efforts with that speech. Professor Samuel Devons, from Manchester, England, spoke after me from rough notes and while his remarks were hesitant at times, they were witty, so he did much better than I. He was horrified when he learned from me before dinner that there would be no drinks and said he had never before delivered an after-dinner speech while sober. That probably explains the hesitancy.

⁓ Saturday, April 2, 1960

Spent an hour with Herter at his invitation in his home. The conversation was pleasant, but Herter served notice that he has decided to put the arms limitation office into the State Department. Perhaps our conversation had a small useful effect in that he agreed at the end that the office director would have the status not of an assistant secretary, but of an undersecretary, and would be quite separate from the regular State Department machinery. His reasons for this decision were not very strong, but I remained conciliatory and assured him of our continuing cooperation. We then talked about a number of subjects casually. I tried on him the idea on trading elaborateness of monitoring systems for the opening of USSR and he is the first one that didn't like it. He thought this was simply playing into the hands of Khrushchev. We discussed also the relations between Defense and State and in his opinion they are improving, but he conceded that they were still very bad. A point on which I was able to set him straight was his notion that when Fisk went to Geneva in the summer of 1958 there was no previous agreement on what the monitoring system should be. He was quite surprised to learn that there was the interagency group which met in February or March of 1958 and recommended a system that is very similar to what Fisk agreed to finally in Geneva.

> The point of this conversation was that it was a defense of Fisk against widespread accusations (in DOD and even State) of having conceded too much to the Soviet Union during the conference of experts in 1958.

The conversation with Herter shed an interesting light on Herter's estimate of McCone. I mentioned that McCone had asked me to undertake the planning of the seismic improvement program and that he said he felt he was now not in a position to do so. Herter said that he was very grateful to me for being willing to do it and that, of course, McCone was not in a position to plan this program. He disqualified himself by attempting to set foreign policy of the U.S. when his business is to run the AEC and he did so in opposition to the President's decision, known to him, to accept a temporary moratorium on nuclear tests. He then commented on the general problem of high government officials trying to set foreign policies instead of running their own agencies, and suggested that that might be an irresistible temptation also for the director of the arms limitations office, if such director were reporting not to the secretary of state, but directly to the President, as PSAC recommended.

> It is an interesting commentary on the persistence and influence of the State Department bureaucracy that the Arms Control and Disarmament Agency (ACDA), established by statute during the Kennedy administration, is in the State Department and the director has the rank of undersecretary, as discussed above.

⮜⮜ Monday, April 4, 1960

Radiation Council luncheon meeting to discuss radioactive strontium in wheat. Flemming's attitude on the subject underwent a substantial change during the meeting. His initial remarks hinted at the need for some regulatory actions, but by the end he was emphasizing that the only problem was to devise a press release that would allay public fears and eliminate the need for any regulatory actions. It was agreed that such a release would be prepared by his office and discussed again by the members of the council.

Toward the end of the day spent some time with Bob Loeb, going over the latest version of the carcinogenic chemicals panel report and pointing out to him a number of places where the report is unclear or evasive. I agreed to prepare a series of questions about what the panel is trying to state with the hope of thus clarifying the fuzzy statements, with the intention of having another meeting of the panel right after the PSAC meeting, to convert this interim report into the final version.

⮜⮜ Tuesday, April 5, 1960

Had a phone call from McCone in which he assured me that he may be able to reprogram so as to get about $3 million additional for the high-energy accelerators program. Had also an interesting visit from Karl Harr, who said, to my utter surprise, that he is working on a policy paper for submission to the President which proposes that the President, at the Summit conference, raise with Khrushchev the issue of opening the USSR as a trade for reduction in our demands on the elaborateness of monitoring systems. I was shook up to discover that our two great minds independently developed the same idea and told him that I had already mentioned it to the President.

~ *Wednesday, April 6, 1960*

Spent the entire day in the office, first working on the Weinberg panel paper on government and science, then talking with Bill Baker about the global communications situation in the Defense Department, and other matters in which he is interested. The rest of the day was spent in conversations with staff and some telephoning, none of which is of such nature as to justify recording it here.

> Alvin M. Weinberg, director of the Oak Ridge National (AEC) Laboratory and a member of PSAC.
>
> The paper referred to dealt with "Science in Government" and was to develop in detail the idea I discussed at the preceding PSAC meeting: that PSAC–FCST–NAS become the conduit for advice from the scientific community on federal science policies and related matters. I appointed Weinberg as chairman of the small panel who were to do this job.

~ *Thursday, April 7, 1960*

Spent the morning with the ad hoc seismic R&D program planning group that I agreed to supervise. Meeting didn't go well, soon developing into an interminable debate between people (mainly Dick Latter and General Starbird) who wanted to push nuclear tests and those who questioned their usefulness.

The morning was interrupted by phone calls from McCone and short conferences regarding the Stanford accelerator, which is being rejected by Senator Anderson's subcommittee of the congressional Joint Committee on Atomic Energy.

3:00–4:30 PM Meeting with Merriam, Paarlberg, Flemming, Benson, and Peterson to discuss the carcinogenic chemicals panel report. Everybody but Flemming thought the report was very good, but Flemming raised a number of objections, some of editorial and some of substantive nature, most of which were well taken. They will require additional thinking and a little rewriting by the panel, at least to justify its conclusions if it doesn't want to change its point of view. In the course of the meeting Flemming was driven into a corner, when he asserted that he had the authority to choose the chemical assay methods to determine the presence or absence of carcinogens in food, even though in principle more sensitive assay methods might be known. He rejected our contention that this was equivalent to setting tolerance levels and hence contrary to the Delaney amendment. He maintained it was simply the "rule of reason" but got quite confused. Then we went into the administrative issues and Flemming first wanted to have the advisory board to be on an ad hoc basis and to report to the director of the Food and Drug Administration. He conceded gradually that the first proposition was not sound and almost agreed to have the board report to him. Then he protested against the publication of the report and said a simple public statement by the President, that the report has been received and that the secretary of HEW will appoint a board in accordance with its recommendations, would be sufficient. After the meeting, Paarlberg and Merriam told me that they would fight like hell to make the report public. After getting back from the meeting, I dictated a long letter to the panel members summarizing the objections of Flemming and urging the panel to think about it. Then had a dinner and spent part of the evening with the VELA seismic program planning panel which by then was writing rough drafts of its report. Went home about 10 PM, but the panel, I learned, worked until about 1 AM.

> The substance of the Delaney amendment and the resulting administrative difficulties have already been reviewed in the introduction to Chapter VII. After much discussion and much rewriting (and some pressure from me, as the reader may have noted), the panel recommended in its draft report discussed at this meeting that the amendment be administered in accord with the "rule of reason," an expression introduced by the Supreme Court when it dealt with some earlier categorically worded acts of Congress.

The panel also urged that the secretary of HEW appoint a technical advisory board that would report to him, to avoid the confusion that happened with cranberries. The board, by the way, was not set up.

Friday, April 8, 1960

Spent morning with AICBM panel, being interrupted by phone calls from McCone and arguments with Panofsky. The gist of the issue is that Senator Anderson will not authorize the Stanford accelerator, but might approve a one-year contract for the engineering design only. Stanford, as represented by Panofsky, is convinced that they couldn't do a good job under these circumstances and do not want a contract of this nature because they couldn't hire good-enough people and would have no facilities for them. AEC is somewhere in the middle and everything is horribly confused. At one time McCone urged me to convince Panofsky of the wisdom of accepting such a contract; whereas simultaneously John Williams was telling Panofsky over another phone to reject the proposal flatly. The absurdity of it all is quite impressive. Finally I washed my hands of it and Panofsky went to AEC to try to get things straightened out. (On Saturday, while in Cambridge, I got a phone call from Paul McDaniel saying that Panofsky wrote a strong statement to AEC rejecting the one-year design contract and that this decision would be acceptable to AEC. He also told me that against AEC's expectations and on its own initiative, the same congressional subcommittee voted to authorize construction of buildings at three universities for the materials research program—that is, the one that McCone doesn't want. McDaniel said that since the initiative was coming from the joint committee, the chairman would certainly accept it, so after all the materials research program may get under way.)

12:00 M–12:30 PM   With Jim Wakelin and Donald C. Snyder, talking about the international oceanographic expedition to the Indian Ocean. We agreed that it is a good idea and that we should support it as a national project, probably through the medium of NSF and IGY. Wakelin wants to determine the extent to which the Navy will be willing to commit ships, which would become the basis for developing over-all plans.

12:30–2:00 PM Lunch with Admiral Burke and his staff, which was noncontroversial and covered a number of harmless topics. Burke, though, asked me pointblank a question, ''What was the objective of the President in agreeing to a nuclear test moratorium?'' and I said it was part of a general objective to feel out the intentions of Khrushchev. After lunch Burke walked downstairs with me and talked privately about thirty minutes trying to determine what were my views on arms limitation, etc. I spoke rather freely regarding my conviction that we are over-elaborating monitoring system concepts and putting too much confidence into them, when in reality we may get into impossible political situations where we ourselves would not be willing to accept them (e.g., lack of ratification by the Senate). I also maintained that security cannot be bought by reliance on machines and that the spirit of the people and unilateral intelligence activities are all-important, and that the opening up of the USSR, which would make evasion more difficult, was more valuable than elaborate monitoring. Burke said that his views were not very different from mine and that he was happy to learn of my beliefs. He thought the Chiefs would be pleased also. I hope what he said is true.

~~~ *Monday, April 11, 1960*

Spent all day in the office doing a lot of telephoning to try to get the Joint Committee on Atomic Energy to revise the decision of its [Clinton P.] Anderson subcommittee; got Ed McCabe active in sounding out and influencing the committee members. However, it probably is all in vain because this morning [A. R.] Luedecke, the general manager of AEC, appeared before the committee and told them that notwithstanding the Stanford attitude, AEC thought that a one-year contract authorization would be okay. What a fantastic double-crossing muddle. So the last I heard from McCone is that the Republicans will put up a fight for SLAC but are resigned to defeat. Several Democratic members were approached over the weekend by various people and responded by saying that they were against Anderson's move, but couldn't take the initiative in opposing him. However, today I learned that they actually were going to vote with Anderson.

Most of the afternoon spent with Bob Loeb going over the chemicals report which was a useful exercise and I think that next meeting of the panel will be productive of the final version.

⌒⌒⌒ *Tuesday, April 12, 1960*

Started the day with a two-hour visit from Bruce Billings, talking about problems he has as the intelligence deputy for Herb York. His main difficulty is that by statute he can challenge a project only on the basis of technical feasibility, but the main source of proliferation and near chaos is that, as he puts it, stupid little people write absurd requirements which then cost the country hundreds of millions of dollars and achieve nothing. For instance, the latest one is for an air-launched ballistic missile to photograph passing Soviet satellites and return the photographs to earth; it might cost a billion dollars and what for? He is very pleased that finally the secretary of the air force has instructed SAC in firmest words to stop developing "Subsystem Eye." Poor Colonel Bob Smith; but certainly this is a very much overdue action. General Power replied with a four-page telegram, screaming like a stuck pig. Bruce tends to be vague, though, and it was difficult to get facts from him. He maintains he needs help from an authoritative body that would evaluate intelligence military requirements; nobody does it now, but it is very questionable whether PSAC could perform this function. I have asked him to present his story to PSAC in May.

> Col. Robert Smith (later brigadier general), deputy for intelligence, Strategic Air Command.

11:45 AM–12:45 PM A very cordial and informative visit from M. Paul-Henri Spaak and a M. Mlieux, introduced by Mr. [I. M.] Tobin of the State Department. I needed only to mention my confidence in Nierenberg and the importance of his office to get Spaak to talk at length. He clearly is skeptical about the importance of civilian science and wants to push practical short-range military effort instead. Complained about the slowness of accomplishing anything through NATO. Toward the end I got him to talk about OEECD, and there he was very frank, maintaining that this is not a useful organization because such neutral nations as Sweden, Austria, and Switzerland would block any projects that had even a tinge of being counter-Communist, e.g., technological help to underdeveloped nations. He maintains that NATO consists of nations with a common purpose and, therefore, is more likely to reach decisions that aren't merely empty words and generalities. I did my talking in English and he in French, and it went pretty well.

> Paul-Henri Spaak, secretary-general of NATO, and his assistant
> M. Mlieux; William Nierenberg, professor of physics at the
> University of California and assistant secretary-general of NATO
> for scientific affairs.

3:00–4:30 PM   Mr. John Finney of the *New York Times*, supposedly to get my personal profile, but actually talking exclusively about the office and its functions. I think on the whole I assuaged some of his suspicions, e.g., my supposed opposition to the space program, and didn't say anything that was really controversial. We stuck pretty closely to problems of science and only glancingly touched on other subjects. I emphasized that while this office and the Federal Council are a step forward and have accomplished useful things, further development of organization of science in government is necessary.

Floberg phoned to say that AEC, on the request of the joint committee, will prepare a letter committing it to undertake work on the Stanford accelerator under the one-year engineering design authorization from the joint committee. I talked with Panofsky and then urged Floberg to be cautious in his commitment, which it may not be possible to carry out. Stanford will consider the restrictions placed on this money in the light of the character of the general report on the Stanford accelerator by the committee. If both are unfavorable, Stanford will reject the design contract, according to Panofsky.

McCabe phoned to say that he had instructed General Luedecke to show to Andy Goodpaster his proposed reply to the joint committee about SLAC, since the President is personally involved, and I enthusiastically endorsed this action. The way things get conveniently distorted in Washington: Luedecke said to McCabe that the reply was being prepared jointly by myself and the AEC!

Finally had a long phone call from Don Paarlberg, who believes Flemming should make no public statements, however reassuring, about the radioactive strontium in wheat until such time as Gamarikian or another feature writer plays up the high strontium content. I tried to argue that this might be very unwise, but neither of us convinced the other. However, I agreed to present Paarlberg's point of view tomorrow at the council meeting.

✺ Wednesday, April 13, 1960

9:30 AM–3:30 PM Meeting of the Federal Radiation Council. About two hours in the morning and an hour in the afternoon were devoted to editing the prepared public release dealing with strontium in wheat. The text was heavy on officialese and clearly would have aroused the fears of the unenlightened rather than reassure them. The discussion was exceedingly uncoordinated and not to the point, but in the end a text was hammered out which appears entirely satisfactory to me, and in effect says that there is no reason for any alarm because not only the average intake of wheat by the U.S. public but also the intake by those choosing very unusual diets, e.g., rich in bran, will still be well within the safe limits as presently defined. The rest of the meeting was devoted to going over part of the report of the working group on radiation exposure safety standards. It is a rather appalling document which takes 140 pages to state the simple fact that since we know virtually nothing about the dangers of low-intensity radiation, we might as well agree that the average population dose from man-made radiation should be no greater than that which the population already receives from natural causes; and that any individual in that population shouldn't be exposed to more than three times that amount, the latter figure being, of course, totally arbitrary. Most of the discussion was in the nature of nit-picking in which I merrily participated. The job was nowhere near finished, so we will have to continue it at the next meeting. At the end, the BOB pointed out that the report as it stood would be rather useless for the President and should be accompanied by evaluations of the interested agencies as to the effect of proposed limitations "which, by the way, are more of the same as presently" upon their activities, so that the basic philosophy of balancing risk with benefits could be presented to the President and used for justification of the safety standards recommended.

Attended the session of the PSAC Panel on Foreign Affairs, which was of no great moment. However, I learned that if I go with the President to the Far East in June, I shall have to deliver several speeches in Japan as otherwise the Japanese will be insulted. There goes my vacation!

ᴓ⌒ᴖ Friday, April 15, 1960

With the Space Sciences Panel, which spent most of the day discussing the Rover project. Bradbury, [D.] Froman, and [R. E.] Schreiber were here from Los Alamos. I had a private conversation with Bradbury who feels that if experiments similar in nature to the one-point safety experiments are permitted, he will be able to make substantial progress in weapons technology and could easily keep his staff busy, so he is not worried about the next two or three years in case of the moratorium; but of course eventually such work will come to an end. He even expressed confidence that he could introduce new designs into the stockpile that way, but, of course, not with revolutionary changes. He does not want Los Alamos to get involved in seismic detection work.

Learned that the NSF budget has been cut by $32 million in a congressional committee and wrote a vigorous memo to Bryce Harlow, who responded that he will move on the subject. Bryce told me over the phone that the letter written by Commissioner Floberg to the joint committee regarding the attitude of AEC on the one-year contract for engineering studies on the Stanford accelerator was so hesitantly worded (and I agreed with him after he read it to me over the phone) that the Republican members of the committee couldn't stage a fight for the full authorization. Bryce explained that Floberg is a staunch Democrat and was helping out Senator Anderson. It was certainly not written in the spirit of my recommendation to Floberg, which was to express grave doubts about the feasibility of achieving anything under the one-year contract and urge full authorization.

∾ *Monday, April 18, 1960*

PSAC meeting all day, which didn't go well from my personal point of view, because I had to spend an inordinate amount of time outside the meeting room, answering phone calls and telephoning others. McCone was crowding me to determine where the money for the VELA seismic research program is coming from, the AEC contribution being some $32 million. I finally established that according to the Goodpaster record, the President definitely assured McCone that he would ask for a supplementary appropriation, but then I ran into the impossibility of getting in touch with Stans or Staats to clear it with them.

The discussion of the Seaborg panel report by PSAC was not very profitable, because the whole text was discussed and too much time was spent nit-picking the language and less time in criticizing the conclusions and recommendations, some of which require major redrafting in my opinion.

> That was done later that summer by a member of the panel, McGeorge Bundy, with my assistance.

∾ *Tuesday, April 19, 1960*

Day started by reaching Stans on the telephone, which developed into quite a heated argument, since he accused me of crowding him, and I replied that I was being squashed into a pancake between him and McCone. We ended on more soothing tones, and he promised to call McCone and tell him how to testify before the Appropriations Committee, which was the reason that McCone gave me yesterday for pressing for an answer.

The meeting of the PSAC continued until 4:00 PM and went much better, possibly because I told the girls not to call out anybody, including me, for phone calls during the session. However, we were exposed to a rather hard sell on an institute for atmospheric sciences by the National Academy committee and spent too much time on the subject. The result was that the rest of the meeting was hurried, and again we didn't get very far on the organization of science in government. This is really rather distressing.

❧ *Wednesday, April 20, 1960*

Spent practically the whole day with the carcinogenic chemicals panel, except for the first half hour, which was with McCabe and Gordon Gray, to discuss the Senator Jackson hearings which are starting next Monday. McCabe maintains that all that Jackson is trying to do is to embarrass the administration, and we discussed the problem of coaching witnesses. I did say that none of the PSAC witnesses, if they are honest, could deny that improvements were possible in the organization of the government for "science policy." That didn't make me more popular.

The chemicals panel meeting was bad, mainly because neither Loeb nor Bronk provided leadership. Bronk was largely passive and I am beginning to doubt that Loeb is qualified for any panel chairmanship, because he doesn't think. There were an awful lot of contradictory statements, and in the end it was Harold Hodge and I who did the pushing and most of the rewriting. I am not at all sure the report is in good shape, but it probably will not be embarrassing now. The shocking thing is that the panel members were all willing to sign their names to the interim report, although some of them emphasized that it was totally erroneous in places. I just don't understand how they can do it.

Harold C. Hodge, professor of pharmacology, University of Rochester.

❧ *Thursday, April 21, 1960*

Spent some time with Andy Goodpaster. He is unhappy about Herter's decision to put the arms limitation organization in the

State Department (at lunch, Gordon Gray said he was also unhappy and was not willing to accept this as a final decision). Goodpaster said that although I am not on the State Department list of people going to the Summit meeting in Paris there is still a possibility that I may be asked to go.

Friday, April 22, 1960

Sent the carcinogenic chemicals panel report to Flemming, Benson, Merriam, and Paarlberg. Later in the day heard through Andy Goodpaster that Paarlberg thinks it is very good and that it should be made public, regardless of what Flemming thinks.

Had a number of phone conversations with Goodpaster, including one on seismic detection. He asked me, after completing the plans for organization of the VELA program and determination of its financing, to prepare a memorandum for the President, after coordinating it with State, AEC, Defense, and Budget, which would request authorization to start the program.

Had a visit from Mr. Joseph Sisco of State, who is replacing Ambassador Lacey at the desk on cultural exchanges with USSR, and discussed with him plans for my personal activities in Russia when accompanying the President. It was agreed, to my complete satisfaction, that the U.S. Embassy in Moscow will make all arrangements.

Monday, April 25, 1960

Discussed with Fisk and Mettler their forthcoming testimony before the Jackson committee. Otherwise the usual chores.

In the middle of the afternoon went to the National Academy, which is holding its annual meeting, and participated in the Chemistry Section meeting to defend the idea of their getting together to identify important problem areas for a future report on proper federal support of chemical research. Unenthusiastic reception to begin with, which softened gradually, and it was agreed to hold such a meeting next fall.

Tuesday, April 26, 1960

Had a visit from Al Hall of the Martin Company, who told me about the wonderful progress that Titan is making and what a fine, simplified modification Titan II is going to be. He questioned me about the political future of Titan and I was sympathetic but noncommittal.

12:30–1:30 PM   Lunch with McCone, at which after hemming and hawing he agreed that the high-energy accelerators program would be better classified as a national program and then come under the jurisdiction of the Federal Council. This was clearly a hard decision for him to make, and I had to paint the role of the Federal Council in very rosy colors indeed. We then talked about the action of the joint committee in authorizing a materials research building at the University of Illinois, and he said he wasn't sure he would go ahead with it as it will require first an engineering study, which is rather silly because the decision certainly doesn't depend on a design, but on whether he wants to go ahead or not. As we were walking to the EOB from the White House Mess, McCone grew emotional about nuclear test cessation and said that we were committing ourselves not to test and not to have controls for at least five years by conservative calculation and that in the meantime the USSR will advance its weapons technology far beyond ours; that it is a national peril; and that he may have to resign his job as he is unwilling to assume that responsibility. He also made nasty remarks about Hans Bethe, maintaining that Bethe made conflicting statements before the joint committee and then last night before the Washington Philosophical Society.

Wednesday, April 27, 1960

7:30 AM   Pre-press breakfast, which was thoroughly dull. It appears that the State Department is completely mixed up as to what is going on in Korea. Whether or not the resignation of Rhee has been accepted by the Assembly seems to be unknown. There was much discussion as to how we can justify the pressure that the U.S.A. put on the Rhee government whereas we don't put similar pressure on Castro, and it was decided to ask State for an explanation. It appears that actually the decision to use pressure was strictly a personal one of the President, who refers to Rhee as "that tough old cookie" and felt it was time to get him under control. Jim Hagerty, who has visited Korea, said he was appalled by the attitude of American officials there. They behave, he said, as if they were in a conquered country.

Syngman Rhee, president of South Korea.

9:00–9:30 AM   Meeting with the President and Secretary Flemming, regarding his press release on radioactive strontium. The President complained that the press release was too long and too technical and asked that it start with a brief statement of the findings, since that was what the reporters need, and also that it emphasize more the routine character of the whole business. Then Flemming and the President got into a vigorous argument about how to proceed with making the stuff public. Flemming obviously wants an imposing press conference, which is precisely what the President doesn't want, and the latter got quite angry near the end. I have also a feeling that the relations of the two are not of the tenderest, because as we walked into the office, the President only nodded to Flemming but said to me, "Hello, George." When we left the office, Flemming was quite upset and asked Andy to interpret for his guidance the final statement of the President which was: "Well, if I were doing this, I would do it the way I told you, but you can do what you please."

4:00–5:30 PM   Meeting with McCone, Stans, York, to discuss the VELA seismic research program. A lot of hemming and hawing, but it was finally agreed that nobody objects to the program which we prepared and that General Betts of ARPA will be either the manager or the coordinator, depending on how the financing is arranged. Both Defense and AEC want a budget split into their two parts, but Stans is firmly opposed. I am to report this disagreement to the President.

After the meeting, Herb York stayed behind and we chatted for about an hour on various uncritical matters. Most interesting is the evolution which has taken place in Herb's thinking since he left Livermore lab, where he was dedicated to the Teller line. He is now strongly opposed to McCone and is strongly for nuclear test cessation, including a moratorium. He is very much afraid that McCone and his friends are maneuvering public opinion, including the Senate, so that the President will have a very difficult time getting a treaty ratified.

Thursday, April 28, 1960

NSC meeting, listening to the Net Evaluation Group [of the Joint Chiefs of Staff], which presented its estimates of a hypothetical war three years hence, initiated by a surprise attack of the USSR 48 hours after we got strategic warning and went on a complete military alert. The conclusion, as was to be expected, was that it wouldn't come out well and we would take a terrible beating but in the end the USSR would be damaged even more. However, the war would not be settled by the first strikes. This of course will now justify the military in setting increased requirements. The whole treatment was very schematized and not very convincing. I had the impression that the President also was rather skeptical since he implied that these presentations may not be continued.

12:15–1:15 PM With the NIE board at CIA. I commented on our misgivings about their making sharp predictions, when the knowledge is vague, and without spelling out in some detail how uncertain the estimates are. I gave several examples for this, but an hour's discussion with the board got nowhere because they maintained that their orders are to provide precise estimates and that when they fail to do so, the consequences are even more disorganized planning than when they give sharp figures. The discussion was definitely a draw, but we parted friends after a pleasant luncheon.

Friday, April 29, 1960

All day in Cambridge at the conference on cultural relations with the USSR, organized by the State Department's [R. A.] Thayer, special assistant to the secretary for cultural relations. Kennan gave a fine analysis of the Soviet society as a starter. This was followed by a speech by Bill Elliott, characterizing the nature of our society and it was appallingly bad. Then followed a number of speeches from the floor, all of which had little point, except for remarks by President Goheen of Princeton and Dean Bundy. A couple of people present, namely Bishop York and a fellow who represented AFL–CIO, condemned cultural relations altogether. Goheen and Bundy, with an assist from some others, made the general point that the State Department made commitments for universities about entertaining visitors, etc., without consulting them in advance, and this not only created serious financial burdens, but also forced universities to operate in ways contrary to their spirit. The main theme of the conference was that the State Department must include universities in their planning process insofar as the universities are to be called upon to do something. Also that the Soviet people, once they reach this country, should be treated as guests no different from other foreigners. The best speeches were those by Ambassador Bohlen, who had just arrived from Moscow and said about the same things that Kennan did, and the summary by Dean Rusk, which was quite masterly. The main thesis of the Kennan and Bohlen talks was that major changes had taken place in the Soviet Union, that the intelligentsia is not as dominated by the Party as it was, and that our contacts with them can lead to further changes in the system.

[Note: Intensive search has failed to identify an American bishop with the family name York. We must therefore assume that the diary is in error.]

In the evening after dinner, President Benezet, Bishop Pike, and I delivered addresses. Afterwards it was mentioned to me that only mine was on the subject of the conference and made some sense. Considering how little I said, it was a terrible reflection on the others. For some strange reason Bishop Pike concentrated mainly on South Africa and apartheid as well as offering numerous clean jokes appropriate to a bishop.

George Kennan, formerly ambassador to the Soviet Union, State Department; Robert Goheen, president of Princeton University; McGeorge Bundy, dean of the faculty, Harvard University; James A. Pike, bishop, Episcopal Church; Louis T. Benezet, president of the Colorado Teachers College.

CHAPTER X

The Lost U-2 and the Summit That Never Was

THE EISENHOWER-MACMILLAN DECLARATION on a threshold treaty agreed to in Camp David was presented to the Geneva conference on March 31 and was questioned extensively by the Soviet delegate. The conference proceeded then to argue inconclusively about several other aspects of the treaty and adjourned on April 14.

It reconvened on April 25 and on May 3 the Soviet Union delivered its formal reply, which was quite conciliatory. The Soviet Union accepted the proposed unilateral moratoria and stipulated a joint (rather than coordinated) seismic research program which might last three years, and could include some nuclear test explosions; the moratoria should last four or even five years. The conference then engaged in a long and sharp discussion of safeguards needed to ensure that nuclear explosions for seismic research purposes would not be useful in weapons development programs.

Macmillan guessed very well while in Camp David: the Soviet Union in June proposed no more than three on-site inspections per annum in USSR and the criteria for such inspections also became the subject of a long dispute.

Starting on May 11, seismic experts from Soviet Russia, the United Kingdom, and the United States met in Geneva to discuss the VELA seismic research program, the preparation of which is referred to in my journal. However, the Soviet position in Geneva began to harden strikingly some time after the U-2 incident, and on May 27 the Soviet delegate stated in the conference that his country never doubted the validity of the conclusions of the original conference of experts in 1958 and that therefore further seismic research was unnecessary. Contradicting his own experts still meeting in Geneva, he stated that no such research would be undertaken on Soviet territory. The conference returned to the discussion of the treaty but the only agreement on specific language was the definition of the 4.75 datum on Richter's seismic scale, reached late in July.

In retrospect, the U-2 crisis signaled a long-term chilling of Russian attitudes. On May 1 a U-2 plane, piloted by Francis Gary Powers, was shot down near Sverdlovsk in the Urals by a novel type of surface-to-air missile. The U-2 project had originated in the period 1954–1955 when an intelligence panel, chaired by Edwin H. Land, of the TCP study, organized by J. R. Killian, Jr., for the President, learned of an earlier proposal by the Lockheed Corporation to the Air Force to build a reconnaissance plane operating at extreme altitudes. The panel recommended to the President that he approve the project and only a year later an unarmed long-range plane capable of flying at over 70,000 feet altitude and carrying sophisticated photographic equipment came into being. The panel pointed out that as of that time no antiaircraft defenses in the USSR could damage such a plane but that eventually the Soviet Union would inevitably deploy effective countermeasures.

The U-2 project was managed by Richard Bissell in CIA. The planes began overflying Soviet territory in 1956. One of the prime objectives of this photographic intelligence gathering was the status of the Soviet ICBM program, about which extreme ignorance prevailed in the United States. For those in the know the information gathered by the U-2 was of unique value in discounting the alarmist stories, spread especially by some circles in the Air Force but also by others such as Joseph Alsop, about the Soviet Union's being years ahead of the United States. Without this information, especially after Sputnik, the President probably could not have resisted political pressures for massive expansion of our already large strategic arms programs, such as took place in 1961 to discharge the pre-election pledges of Kennedy.

The U-2 flights revealed extensive testing of intermediate- and shorter-range ballistic missiles but only a moderate test effort and, to the end of the flights, no deployment of operational ICBMs.

My own involvement in the U-2 operation was a passive one: I was privy to the photographic "take," was shown the coal-black U-2 planes in 1958, and after joining the White House staff was consulted (probably by General Goodpaster) on the importance, for purposes of "comparative evaluation," of learning details of this or that Soviet installation.

I knew that the Soviet Union was making numerous efforts to shoot down the U-2s. The Soviet Union, however, never lodged a diplomatic protest against the overflights, although their journal *Soviet Aviation* once complained about the "black lady of espionage." In the United States there were a few security leaks regarding the clandestine use of the U-2, which infuriated the President. The information on these flights was limited to relatively few individuals, on the need-to-know basis. Officially released information was that the U-2 was a high-altitude weather research plane, of relatively low performance, operated by NACA [later NASA]. The discussions of future flights, in which I participated, did not neglect the consideration of the risk/benefit ratio. Therefore the confusion and lack of preplanned crisis management which followed the loss of the U-2 on May 1 surprised me.

Under what circumstances and when the White House approved the flight plan that was put in execution on May 1, only two weeks before the scheduled Paris Summit meeting, I do not recollect. That Gary Powers was missing was first learned in Washington in the afternoon on Sunday, May 1. The President was in Camp David, Secretary Herter in Istanbul, Dulles in New York, Dillon somewhere else. A hastily organized relatively low-level meeting decided to follow the pre-existing cover story of a NASA weather research plane that might have strayed over the Soviet border and was lost.

Starting then and ending on May 9 the United States government issued in Washington four or five successive mutually contradictory stories, the first being that the plane was flying for weather research and must have strayed over the border, and the last the admission that it was an intelligence-gathering flight, a part of a continuing program organized on instructions of the President. Khrushchev played a skillful game, releasing bit by bit the information on the shooting down of the plane, then the capture of Gary Powers, then his admissions, then the nature of the plane equipment, etc., so timed that all but the last American statements were embarrassingly shown as untruths. The last statement unfortunately used deliberately fuzzy language so that the impression was created (and not denied) that U-2 flights would continue. Only Walter Lippmann publicly raised his voice in protest against Washington's thus creating a situation on the eve of the Summit meeting which must be intolerable to the Soviet Union.

Eisenhower, it appears, decided several days before the Summit to discontinue the flights but chose not to make this public until after his arrival in Paris on May 15.

Added to this confusion was a mixup by Hagerty who announced on May 7 the start of the VELA seismic program in such a way that the reporters assumed that we were resuming nuclear weapons tests and this was not properly denied until the presidential press conference on May 11.

In the meantime, the Soviet government issued threatening warnings about their involvement in spying to Pakistan, Turkey, and Norway, whereas the United States assured them of our support. The atmosphere was becoming ever less conducive to a Summit meeting.

This journal contains enough comments on the Summit that failed before it began, to make another account of the events in Paris unnecessary here. Suffice it to note that Khrushchev insisted on an American commitment to cease U-2 flights, punishment of those immediately involved, and an apology by Eisenhower, whom he attacked personally. The latter stated that he had terminated the flights but refused to order punishment or make apologies. These positions did not change, notwithstanding the efforts of de Gaulle and Macmillan, and so the Summit ended before it began— unless one counts the preliminary meeting on Monday, May 16, when Khrushchev read the angry speech demanding an apology, etc., as preconditions to a Summit.

My own involvement in the U-2 and Summit events recounted in this chapter was that of a partially and belatedly informed observer. But the reader will see how the affair affected the whole tone of the administration.

Tuesday, May 3, 1960

Arrived in Washington 10:00 AM, prepared for trouble because Beckler called me Monday to say that Bob Loeb is in a state of high excitement and I asked him to invite Loeb to come here.

In the morning went to see Persons, who agreed that I may have two weeks' vacation in the summer and still take four days a month in the fall to see my students in Cambridge. I suggested that this was a good opportunity to fire me, but he said he didn't want to and rather liked having me around. I tossed the bouquet back at him, since he really is a nice boss to have.

At 1:30 Loeb arrived, evidently under great emotional stress, with clenched teeth, etc., and we had a thoroughly unpleasant session. All my attempts—and there were many—to soothe his feelings were to no avail. He resigned from the committee and the chemicals panel and maintained that I had committed an unpardonable action—an act of bad faith—in letting the secretaries see the report on chemicals with his name under it without having him first read the report. I maintained that it was unintentional, but he maintained that I did it deliberately and all my arguments that this was not so, that it was a slip in the office, and that the situation in any case was not irreparable because the report was not yet official and had not been submitted to the President, had no effect. Bronk also helped to stir things up by telling him I had said that I would never have another medical man on the committee. This was not quite accurate, as I had spoken in the office against having those damn medical men on the panels, and how this got to Bronk in the first place I don't know. It was all very distressing and as a later checking up with Holtzberg and telephoning to Bronk established, the panel did give me complete authority at the end of the meeting to present the report to the secretaries, providing that only editorial changes would be made. We did not make substantive changes. Went to see Morgan for advice and was told to accept Loeb's resignation and go ahead with the report anyway, but I told Dave [Beckler], we had to have every member on the panel confirm the report before I would present it to the President.

ᐁᐧ Wednesday, May 4, 1960

11:00 AM Presentation to the President about the Minuteman missile system, which went well. Frank Long was relaxed (although he maintains he was scared as hell) and presented the thing in a clear and simple manner, which obviously was much enjoyed by the President. Result was that a 15-minute presentation, what with questions and discursive remarks by the President, took 47 minutes. At the end I was almost caught in a trap because the President thought it was such an important set of findings that we should present it directly to Gates. So I suggested that wouldn't it be all right to present it to York, feeling that otherwise I would get in Dutch with him, and the President said that since Gates thinks so highly of York, he guessed it was all right.

When Long left I presented to the President the memorandum on the VELA seismic improvement program and its financing, to which Spurgeon [Keeny] got the concurrence of the agencies involved, which evolved very nicely because Stans over the weekend drastically changed his firm position and accepted the budgetary plan of AEC and DOD. I also described to the President the Russian statement in Geneva, which obviously goes a long way to meet our stipulations, since it agrees to nuclear test explosions in VELA and to unilateral declarations of moratorium. The President was unhappy about the supplemental appropriation, but Jerry Persons helped him to make up his mind in the affirmative. He then asked me whether I was going to be able to keep a sharp eye on the program, and I said I proposed to have periodic meetings of the key people and to summarize my findings in brief reports to him. Then he approved and signed the memo, so I have helped to spend about 100 million bucks in the next two years. Poor taxpayers! Because of l'affaire Loeb I did not present the chemicals report, as we are trying to get every panel member to endorse explicitly the last version before submitting it.

2:00–4:30 PM   Federal Radiation Council, going over the abbreviated report on radiation exposure standards, which is to be submitted to the President for approval as a memorandum. The council accepted a number of minor changes I suggested and agreed also to reconsider the table of standards in one respect which I thought was very undesirable to include. In general the meeting was relaxed.

5:00–6:00 PM Mr. John Finney from the *New York Times*, who was told by the editors that his article about me was too political and he should write a more personal one, so he needed lurid details about my life, which I gave him with some reservations.

Thursday, May 5, 1960

The day began at 7:15, when I was called to the telephone
while in the shower and soapy, to be told that I had to be at the
takeoff point for a helicopter flight to the Relocation Center in
one hour's time. I asked the girl to call me back in five minutes
so I could dry myself first, but she refused, but let me dry my
hands so I could write down the instructions. As I got into the
White House car, I saw a large Cadillac blocking the exit from
2500 Q Street, and a highly agitated director of central intelligence hovering over the chauffeur and the engine. He was des-
tined for the same point but his car had broken down, so I gave
him a lift after pushing the Cadillac out of the way. Otherwise
nothing untoward happened, except that of all the regular members of the NSC, it was the chairman of the Joint Chiefs who
failed to make the helicopter and was missing at the meeting in
this supposed period of great tension. The Relocation Center
"Crystal" is a really mammoth underground structure and is
most impressive, but notwithstanding air conditioning, the
briefing room in which we had the NSC meeting was stuffy. The
meeting itself started with long presentations by York and Scoville on the history of long-range ballistic missiles here and in
the USSR. The rest of the meeting was not interesting.

> 2500 Q Street was the address of my apartment. Dulles lived in a
> house a couple blocks up Q Street.

> This "evacuation" exercise, a kind of rehearsal of procedures
> in the event of an impending nuclear attack, was planned, I
> believe, by the Civil Defense authorities before the U-2 incident
> and had no connection with it.

12:00 M–12:45 PM   Meeting of the PSAC Space Sciences
Panel with Glennan, Dryden, and Newell of NASA to discuss
mainly the grievances of outside scientists who felt they were
being completely left out of planning of space missions, etc.
While I was there the atmosphere was slightly tense and the
NASA boys flatly denied the criticism of the panel, but I understand that later the meeting got friendly and cordial and that
substantial progress was made towards resolving the difficulties.

1:20–3:00 PM   Meeting of Principals, which started with a
long briefing by ARPA on high-altitude nuclear-weapons-test detection systems. It largely followed the Panofsky panel findings

in January, but additionally estimated the duration of development and costs, which for a complete system would reach nearly $2 billion and would still permit the evader to test if he was willing to go far enough into space. The DOD position is that there should be no cessation of tests outside the atmosphere, but I certainly hope the State Department will not accept this, because we have already conceded that the tests can be monitored to substantial distances in space. The usual desultory discussion followed, Dillon being in the chair, and at the end McCone gave us figures on possible improvements in our own and the Soviet weapons regarding yield-to-weight ratios predicted by Livermore and Los Alamos. Very impressive figures if true, and if the Russians ever reach similar conclusions they would certainly be strongly tempted to cheat on an agreement not to test. No decisions were reached and we will meet again next week. I was mildly pleased to have been able to embarrass the DOD boys. They spoke of the great difficulties in developing detectors for X-rays emitted during nuclear weapons explosions and when the presentation was finished, I pointed out that I had had similar detectors in operation in my small laboratory at Harvard for the last five years and couldn't see why the job was so terribly difficult. The briefer retreated in confusion. My other contribution to the day consisted of a remark this morning to Tom Gates, loud enough so that Stans could overhear it, that I thought at least in an "emergency" such as the Relocation Center meeting the budget director could be left behind and we could start spending money freely. Everybody thought it was rather funny, including Maury [Stans].

Friday, May 6, 1960

Spent entire day in the office with brief visit from Hagerty to discuss the public release about the enlarged VELA program. Short visit from Dick Bissell to exchange commiserations on the U-2 plane mishap. Both Dick and I, and according to him, Allen, feel that the President should take congressional leaders at least partly into his confidence to prevent the building up of indignation in Congress, which would only pour more oil on the fire. I called Andy about it afterwards, and he said that this has been considered and flatly rejected because the Boss thinks that these congressional fellows will inevitably spill the beans. Dick said that the pilot is apparently alive and is being interrogated, which, of course, won't help the situation.

Next a visit from the whole command of the Coast and Geodetic Survey plus Secretary Ray to convince me that they need an extra million dollars. Clearly an empire building scheme by Coast and Geodetic to get into dominant position in seismic research. In fact, they protested vigorously the involvement of the Air Force in this program.

> I was informed of the U-2 loss soon after returning to Washington on May 3, but the matter was so hush-hush that I made no journal entries until May 6, by which time the incident was being publicly discussed.

Monday, May 9, 1960

A call from Dick Morse, who is recommending to the secretary of the army that Atlas missiles launched from Vandenberg AFB in California be used in the Nike-Zeus tests over Kwajalein Island and that Brucker is violently opposed, and wouldn't I do something about it. I got in touch with Andy Goodpaster, who said it was unthinkable that the Air Force would let the Army shoot at one of its biggest weapons systems, and that the Army, of course, wouldn't want to fail destroying an Air Force weapon, and that this plan was less realistic than asking the Russians to install and test Nike Zeus in Kamchatka.

> The target area for Soviet ICBM tests was in Kamchatka.

I tried it also on Morgan and Persons and was told to forget the idea, that we might as well write off as inevitable the additional $70 million cost so that the Army's Johnson Island installation could fire off their own target missiles. Later in the day I called Morse, who told me he had a rough session with Brucker, and was then told to stay in his office when Brucker received the Michigan congressional delegation and that Brucker in the presence of Morse said that it was a unanimous position of the Army that Jupiter missiles fired from Johnson Island

and not Atlases should be used. Brucker also instructed Morse privately to be uncooperative with York on this whole subject. It really is a fantastic situation.

At 11:00 AM went to see the President with Don Paarlberg regarding the chemicals panel report, which by the way includes the name of Bob Loeb who reinstated himself as vice-chairman over the weeeknd.

The President asked my reactions to the U-2 incident and commented on the difficulty of getting in advance an objective evaluation of the consequences of being caught and also on the unfortunate way in which the statements of the pilot changed the political situation.

I presented the report of the chemicals panel on carcinogenic food additives and the President looked over my accompanying memo, the comments of Secretaries Benson and Flemming, and the conclusions and recommendations of the report. The President pointed out that in recommendation no. 2, on page 10, the wording ''It is recommended that appropriate modifications in the law be made,'' should be changed to ''It is recommended that appropriate modifications in the law be sought,'' since the report was for him and not Congress.

I noted that while there was no disagreements with the substantive parts of the report by Secretary Flemming, he was questioning the desirability of its being made public, whereas Secretary Benson was anxious for this to be done. Don Paarlberg noted that the unanimous staff position in the White House was that it should be made public. The President asked Paarlberg to clear the matter with Flemming. The President thought that it ought to be made public after proper staffing. After the meeting, Paarlberg and I agreed that he will contact Flemming.

The President then said that he was anxious for the Science Advisory Committee to explore the problem as to whether we are doing too little, too much, or just enough basic research. He felt there might be some boondoggling and that maybe we have exhausted the supply of talent and are now putting second-raters on difficult problems and are not getting back our money's worth. He said, he was specifically referring to the medical research of NIH as an example. I noted that actually more than 90 percent of the research funds went for applied research and development and conceded that in that area quite a few second-raters were perhaps being given big money and perhaps weren't producing enough in return. I noted that in basic research this was less likely to be the case, and the President agreed with me.

He instructed me to prepare a document which he would sign, directing PSAC to undertake this major study, and so I explained that we were already involved in it and suggested that I report his wishes orally to the committee next week, with which he agreed. He also agreed to meet with the committee when things get quieter this summer and discuss the whole matter informally and at length. His interest in this study is primarily for the guidance of the next administration, since he feels that nothing useful can be done in this session of Congress on a matter of such over-all importance.

2:30–4:00 PM   NSC meeting, which was started by the President's saying that he did not want to discuss the U-2 incident and that he firmly requested everybody present to insist that there was no such discussion and not to give out any details. He also informed the group that Allen Dulles would be briefing congressional leaders and that the secretary of state will make a public statement tonight. He said rather sadly that, of course, one had to expect that the thing [U-2] would fail one time or another, but that it had to be such a boo-boo and that we would be caught with our pants down was rather painful. There was a discussion of a legal problem brought up by the attorney general, which wasn't interesting, and then Bob Amory, speaking for Dulles, gave the intelligence summary. The CIA analysis of Khrushchev's intentions in making such a fuss is that he is building up a position on which to stand in case the Summit meeting fails.

Robert Amory, Jr., deputy director, CIA.

After the meeting there was a lot of private discussion, and it appears that I am not the only one wondering as to whether or not the pilot was in cahoots with the Soviets. I found also that a number of others, including Tom Gates and McCone, thought there might be an advantage in publishing a few sample pictures we have from U-2 expeditions to show that the other side is not a "peace-loving nation" and that we have been successful in our intelligence gathering. Gordon Gray and I talked about this to Goodpaster later, but Andy was unimpressed because he couldn't see a real advantage in doing so. Maybe he is right.

I got into a sharp altercation with McCone, who violently condemned Glennan's proposal to offer the Soviets a cooperative undertaking on the meteorological satellite. McCone insisted that that will end SAMOS and other intelligence satellites. I suggested that with that attitude about all we can do is to stop making any proposals to the Soviets and stick our heads in the sand and wait for the propaganda blow to fall. For a while the discussion was rather heated, Gates suggesting that we would have another "moratorium" if we followed Glennan's proposal. Andy showed us the proposed text of Herter's statement and that looked OK, but it doesn't explain whether this particular U-2 mission was authorized by Washington or by a field commander, and that is certainly going to create a lot of trouble. [It did.]

After NSC, Jerry Wiesner came and we had a long session in which we talked about the U-2, then about Pugwash conferences. His feeling is that I should support them informally because they could contribute substantially to our position on disarmament if the right people attended them instead of wild-eyed radicals. I reserved judgment. We then talked extensively about our ineffective arms limitation panel and agreed to raise the issue of its further existence and purpose in the coming PSAC meeting. We also talked about the resignation of Bill Foster as the chairman of the board of the planned nonprofit organization [Aerospace Corp.], a partial successor to STL, because of an unpleasant about-face by the Air Force. This certainly is a mess, the Air Force first asking Foster to organize the thing and assuring him there could be no possible conflicts of interest, and when it came to final approval telling him that he and two or three other key people of the proposed board, who were acceptable 24 hours earlier, were unacceptable. Jerry said he certainly would not accept the board chairmanship, something that Bill Foster told me yesterday he has recommended. That conversation with Foster took place while I was on Bill's motor yacht, a pleasant but wet trip because of the continuing rain.

> William C. Foster, the head of the American delegation to the Geneva conference on surprise attack, in 1958.

❧ *Tuesday, May 10, 1960*

10:00 AM–12:15 PM Meeting of Principals. Indecisive as usual. I talked about the on-site inspection quotas and then on high-altitude test detection. I defended the concept of the treaty including only "sensible atmosphere" and limiting this to about 100,000 kilometers, since weapons tests beyond that distance are hardly worth the cost. State meekly followed my lead. Everybody seemed agreed that the proposed elaborate high-altitude and solar satellite monitoring system was unacceptable, but nobody had the guts to propose what to do instead, and as I didn't volunteer that PSAC develop a proposal, nothing was decided. We then came to seismic detection and Doyle Northrup[7] (bless his soul!) sprang on the meeting a cost estimate by some outfit in Pasadena which testified before the Holifield atomic energy subcommittee. The work was done at DOD expense, which, of course, is impartial! (Northrup asserted that it was done at my suggestion, which I denied, and he then conceded that it was he who started it.) They estimate it would take $1–5 billion to install the 21 seismic detection stations in the USSR and will take five years. This, of course, horrified Herter and Dillon and they just about threw in the sponge, but it was possible gradually to bring out of Northrup the fact that the estimate was very preliminary and was based on the assumption of exclusive use of U.S. personnel, the building of large airfields in the USSR to land them and supplies for the construction activities, hiring icebreakers to bring more supplies by the northern sea route, etc., etc. At the end of the meeting Al Latter gave a good presentation of a proposed revision of the original [1958] Geneva seismic system and the major gains in monitoring reliability it would achieve. Gates first acted dumb, pretending not to understand, and Irwin didn't have to pretend: he simply could not understand. It took some effort to explain to him the significance of the proposed changes, which was not lost on Herter and Dillon. The whole meeting was a concentrated attack by McCone and Defense on the test-ban treaty, so after the meeting I said privately to Dillon that while it was perhaps none of my business, I was worried that Herter will be very much isolated in Paris at the Summit with all these people present and me not along, although I certainly wasn't begging to be taken on the ride. Dillon seemed to agree and said he would speak to Herter.

Albert Latter, physicist with the Rand Corporation.

After lunch had a phone call from Gates, who asked us to try to resolve the conflicts on the Kwajalein Nike-Zeus tests, a subject on which he cannot reach an agreement with Brucker. He sounded so pathetic that I agreed to do it, providing Jerry Persons gives his okay, although it is obviously a very hot potato.

Shortly after the Gates call, Herb York came to enlarge on his difficulties with Brucker and to urge us to undertake the Nike-Zeus job. This government is really in a sad state and this is only one small bit of evidence for it. The bigger piece, of course, is the handling of the U-2 incident. The first two public releases, as well as their timing, are a pretty severe indictment.

3:00–4:00 PM   A man from CIA with maps of the flight of the ill-fated U-2 and latest information on what happened. It is pretty clear that it couldn't have been a flameout of the engine, but what did happen is difficult to say. The picture of the crashed plane which Khrushchev passed around is not that of a U-2, so it begins to look as if the pilot might have landed it undamaged. On the other hand, our information about his maneuvers is strongly against the idea that he was a double agent. At most, he decided "to turn state's evidence" when under pressure; and at least, he just chickened out. Very depressing.

4:30–4:50 PM   Pre-press conference, at which we worked out the President's statement to straighten out the unfortunate interpretation of the Project VELA public release on Saturday. Hagerty passed the word that the President didn't need any advice on how to answer questions about the U-2. He wanted to think it through himself tonight and will cut the questions short after the first one. Jerry Persons okayed our study of the Nike-Zeus tests.

Wednesday, May 11, 1960

9:35–10:30 AM   Briefed the President, but I am afraid that the information we gave him on seismic detection research in Project VELA was not effective, since he did get mixed up a little during the press conference. However, he did emphasize that the nuclear tests were only for Project VELA and would be done under joint supervision, so the gaffe isn't too bad.

11:00 AM–12:00 M   Secretary Ray, and, unfortunately, his special assistant, so I wasn't able to be as tough as I intended. I suppose that is why he brought the assistant along. We talked about the conflicts between NSF and Commerce in the field of collecting and distributing scientific information, and I emphasized that the executive order gave the Foundation real leadership, and that Commerce should be more cooperative. I offered to seek further clarification of the meaning of the executive order if Ray was doubtful and hope this remark sank in. He assured me that he will try to be cooperative and suggested that maybe he is being pushed around by his staff people.

2:00–2:30 PM   A Mr. Guy Waterman, to enlist my participation in the preparation of Republican campaign speeches and publications. He received conditional acceptance subject to availability of time, which I told him there wasn't any of.

5:00–5:15 PM   Gordon Gray to discuss the nature of the papers to be prepared by the Federal Council of Science and Technology in response to the NSC request regarding maintenance of Western strength in science and technology. We agreed that identification of problems without their recommended solutions would be sufficient. Talked also about the paper being prepared by Gene Skolnikoff and coming up before the planning board tomorrow re: Science and Foreign Affairs, and the violent opposition of Brode and his gang in the State Department to this activity. Gordon said that he expects fireworks.

Eugene B. Skolnikoff, PSAC staff, concerned with foreign affairs.

ᴓ Thursday, May 12, 1960

9:00–11:00 AM   Cabinet meeting, at which, following Flemming's request of yesterday, I presented a report to the President from the Federal Radiation Council. I made my presentation brief and then asked McCone to explain in more detail the meaning of the specific radiation-exposure safety standards. The discussion was at first quite relaxed, but Attorney General Rogers suddenly said that the third paragraph, which was a summary of underlying biological assumptions, would be very embarrassing if made public. This was backed by Gates, who said we couldn't live with such a statement, and the vice-president agreed. The discussion became heated, a number of people proposing changes in the text and for a while the President lost control of the meeting, several people talking at once. It was obvious that the cabinet was not the place to rewrite a text, so I kept emphasizing that what was written had already been published by the NCRS and that a statement very different in tone would be regarded with suspicion. So I finally accepted McCone's proposal that we eliminate the paragraph altogether and this the President agreed to. After the meeting I spent a lot of time telephoning various people trying to get agreement on this matter.

∽ Friday, May 13, 1960

Went to the White House at 8:30 to attend the staff meeting, and on the steps was caught by Flemming, who said that his agency violently objected to the elimination of the third paragraph of the Radiation Council report (referred to in my account of yesterday) and urged that a modified text is absolutely necessary. We were joined by Don Chadwick who is obviously behind these efforts to include the paragraph, and spent half an hour rewriting Chadwick's text, which was very bad. The discussion was rather heated and I said some uncomplimentary things about making math into a fetish without understanding it. At issue was the reference to a linear relation between dose and effect, which as I still believe is entirely unnecessary for the definition of the current radiation guidelines, since they are pulled out of thin air without any knowledge on which to base them. We finally agreed to leave the linear relation in, but in a very mild form. It was this text with which we went to see the President, who cut out a few more words, making the text even milder and then approved it.

I then hung around for a while, while Flemming was talking to the President about other things with Persons present. Finally Jerry came out to tell me that Flemming had agreed to have the chemical additives report made public, so I didn't have to go back to see the President. Flemming came out and said that there was, however, one sentence which was present in earlier drafts, but which was omitted in the final draft, and to the omission of which HEW objected. Got back into the office about 10 o'clock and did a frantic amount of telephoning. First to McCone and Gates in Paris to get their approval of the text of the Radiation Council report, which I received without any complications, and then to Pond and others at HEW about the missing sentence. It appears that Secretary Benson pulled a fast one on me when he requested the omission of this sentence in his letter. The sentence, which in effect gave more authority to HEW, had been included in the cabinet paper that was supposedly approved by everybody present! Followed long negotiations ending in my pulling Peterson out of a meeting and winning his somewhat unhappy consent to put the sentence back as the price of having the report made public. What a life!

Went to see York at noon and he spent a long time explaining his position on the Pacific test program of Nike Zeus. From the way he presented this, it was obvious that here was a battle to the finish between him and Brucker, that York was wrought up and extremely anxious not to lose. It was also obvious that he was not terribly happy about having Skifter as chairman of our panel since in his opinion, Skifter is unduly favorable to Nike Zeus.

In the afternoon visited Don Paarlberg. He is coordinating the White House press release on the chemical additives report, which is going to be released tomorrow even though it is understood that PSAC hasn't formally approved it.

Saturday, May 14, 1960

Morning spent with the Limited War Panel's executive session. It is clear that continuing sessions have brought a substantial degree of clarity into the views of the panel and that the major unresolved issue is the role of nuclear weapons because of basic conflicts of opinion, for instance between Lauritsen and John Foster. The former contends that any limited war will go into a general war if nuclear weapons are used, and the latter rejects this contention. We agreed that the panel will prepare a summary of its long report, which will be treated as an internal document, and I will arrange for briefings of the secretary of defense and then the President on the findings of the panel. These transcend the purely technical problems and go into such matters as roles and missions of the Services and are therefore quite touchy, but they are little different from the findings of earlier civilian studies on the subject.

Sunday, May 15, 1960

Was pulled out of bed by Allen Dulles, with the news of the latest Russian satellite. He was much alarmed and wondering what we should do, so I dressed in a hurry, bought the *Washington Post*, read the article, and talked a couple of times with him about it. My conclusion was that there wasn't anything we could do about it; that the Soviets would probably try to recover the satellite and might fail, and therefore, made the announcement that they wouldn't try to recover.

Monday, May 16, 1960

All-day PSAC meeting.

About the middle of the morning news from Paris began arriving: first the cancellation of the USSR trip and then ever grimmer news of the violent attack by Khrushchev, thus confirming my pessimistic anticipations and actually exceeding them by far. As I look at what has happened, I am rather convinced that the real reason for the break is a decision in the Kremlin, possibly against Khrushchev's wishes, to radically change the USSR policy toward isolation and intensification of the cold war, one of the reasons for this being preservation of military secrecy. I think that the U-2 incident may have been deliberately planned before the Paris meeting because the April 9 U-2 flight was carefully tracked and subjected to many attempted intercepts, but no public protest was made, thereby virtually inviting us to repeat the sortie, whereas a public protest certainly would have prevented our undertaking the May 1 flight. I see bad times coming, with this crisis sharpening to God knows what extent, and very likely the Soviets testing our readiness to use a nuclear deterrent in the conviction that since we didn't do it in the Berlin crisis, probably we would not now. On receiving the bad news, Jerry Wiesner made an impassioned plea that PSAC appeal to the President to set up a strong organization on arms limitation outside the State Department. I demurred and suggested that if individuals of the PSAC felt like that, they could see the President on their own, but I didn't think this was the time to do it.

By Tuesday afternoon things got so bad in Paris that Jerry didn't raise the issue again.

> Khrushchev's memoirs, *The Last Testament,* note that the April 9 flight of the U-2 just happened to miss all the installations of their high-altitude antiaircraft missiles.

⌁ *Tuesday, May 17, 1960*

The morning political news from Paris very bad, the Summit meeting being almost certainly over before it began. PSAC reconvened in a gloomy mood. We had a two-and-a-half-hour discussion with Bruce Billings, Bill Baker, and Pete Scoville about the problems in the intelligence area, among them the SAMOS satellite projects. Unfortunately, the discussion led to no clearly defined tasks for PSAC. We then had a two-hour session on science in government, which was started with a brief summary by Mannie Piore about what should be done. His proposals are very near to what I am now thinking is right. Then Eugene Wigner gave his speech, which could be summarized, I jokingly suggested, in the following statement: "The more we become like the Soviets, the more likely we are to succeed." Wigner was not happy at that remark, but wasn't able to refute it. The trouble with him, as with many others, is that he can identify weaknesses and areas of trouble, but he doesn't have concrete proposals for their elimination. I asked PSAC to reconvene after lunch and we discussed the participation of PSAC in the intelligence field, with no explicit conclusions; I emphasized to PSAC again as I did in the morning, that I will have to rely on them for greater participation. No objections raised, but let's see how they respond. We agreed that a small group will again try to write the "science in government" paper, and for this purpose I have asked Weinberg, Panofsky, and Piore to join me for a three-day session.

> Eugene P. Wigner, professor of physics, Princeton University, a PSAC panel member.

⌒ *Wednesday, May 18, 1960*

Spent about an hour with Jerry Persons, who agreed, on my insistence, that I will not take part in the Republican campaign, but asked me if I would be willing to give nonpartisan talks in the fall on the role of science in government, etc. I agreed to do so, provided they were with largely technical audiences and not party meetings. I told him about the planned presentation of the Limited War Panel report to the President. He said he himself was much concerned with the situation and was unhappy about it, but that the President was impatient with the subject and we would get a fairly rough reception.

I then talked about the intelligence problems, the administrative chaos, and the technical troubles, particularly those of SAMOS. Jerry said that he was out of that field and knew little, but advised me to first discuss the matter with the President before talking to General Hull, as the President is sensitive. He said that after the U-2 incident he said to the President that things might have gone not quite so badly if some people like himself, who are experienced in the political and public relations areas, could have had knowledge of what was happening. In response the President grunted and agreed that he was probably right. I then mentioned that since I got into the job, I was not consulted on appointments of technical people to top government jobs. Jerry was apologetic and said this was strictly unintentional and will be changed in the future.

> Gen. (ret.) John Edwin Hull, chairman of the President's Board of Consultants on Foreign Intelligence.

Attended a luncheon at the Atoms for Peace Awards affair at the Academy and sat in exceedingly boring company. Had the mild satisfaction of embarrassing Bob Wilson, the new AEC commissioner, for whom I have no special love because of his views on science, nuclear test ban, etc. (somewhat less enlightened than McCone's). During lunch today he said he understood that I had done research in explosives and asked whether I had been in any way involved in the development of the implosion atom bomb during the war, so I announced that yes, I was in charge of it, which so confused Mr. Wilson we had very little to say to each other during the rest of the lunch.

> Robert E. Wilson, ex–oil-industry executive.

ᴑᴄ⌒ᴖ *Thursday, May 19, 1960*

On my initiative spent an hour and a half with Gordon Gray. We talked about the U-2 incident and the current criticisms in the press (e.g., Arthur Krock's article) that NSC machinery has failed again. Gordon feels himself being made into a scapegoat, which is rather unfortunate because he had very little to do with the affair. He explained to me in detail what his role is and how the U-2 flights are decided upon. I said, that if there is to be a scapegoat, Dulles should be it, and that the best solution would be to move him into the position of special assistant for intelligence, where his broad interests and thinking could be utilized to advantage in advising the President and to put a tough executive in the spot where Dulles is now. I emphasized that the whole intelligence business is in chaos and then discussed the things I know about, namely various satellite development projects, the lack of control of the military requirements, poor program management and their wild proliferation. He said that there would probably be a Monday morning special NSC meeting to talk about U-2, Paris, etc., and also suggested that he and I go to the President on this whole intelligence business, but that there was no point talking to the President unless we had specific recommendations. I said that I could make recommendations only in the area of development projects, their technical objectives, technical management, and the like. I felt that a strong committee of experts, both technical and management, should be set up to analyze the projects now under way, whose purpose is to replace U-2. We agreed that such a committee should be established on a high level to report either directly to the President or to him through Gates who, of course, would have to concur completely in its being got up. Its responsibility should transcend technical matters and, therefore, it could not be a function of PSAC or York.

I then emphasized my fear of saber-rattling moves that we might make under the influence of McCone and his ilk, such as a resumption of weapons testing. I urged caution and that no initiatives be taken in that direction because they would merely be used to establish, with the kind help of Khrushchev, an image of us in the world as deliberately seeking war. I suggested that a new session of the McRae panel should be set up to analyze the urgency of resuming nuclear weapons tests before any decision is made.

We also talked about arms limitation. I told Gordon about the phone call from Doug Dillon yesterday, in which he suggested that PSAC set up a continuing study of the possibility of missile controls, to which I responded that this was clearly a job for the arms limitations organization that the State Department is setting up. He conceded, but remarked that creating such an organization now didn't make much sense. I responded that precisely the opposite was the case. That we now had a breathing spell for maybe a year during which we should make the maximum effort to come up with thoughtful positions on arms limitation for the resumption of negotiations later. When I told that to Gordon, he said that he spoke to the President against setting up the arms limitations office in the State Department, but the President felt it should be under State's aegis. Gordon and I agreed that such an organization should report to something like the Committee of Principals, which, naturally, Gordon felt should be a subcommittee of the NSC. I raised no objections to that.

Monday, May 23, 1960

A navy captain from the Joint Chiefs of Staff office came to show me a paper he wrote on camouflage and hiding of our missile sites. This evidently grew out of my telling the President that my inquiries within the Air Force a year ago elicited their statement that nothing could be done in the way of concealment. I argued then that we need only make the sites indeterminate by a mile or so to neutralize the effect of future advanced guidance systems in Soviet missiles, and hence force them to assume that they need many missiles to destroy each one of ours, a point that could be important in their missile force planning. To my amazement, the captain said that this had never occurred to him and so his staff had prepared the negative report. I promised to study the report and comment on it.

Lunch and all of the afternoon spent at Sheraton Park with the National Defense Executive Reserve, i.e., Governor Hoegh and his chums. Thoroughly futile and even unpleasant occasion. My luncheon companions were dull and undoubtedly felt the same way about me. My ten-minute speech before the panel discussion was clearly not to the point, as I didn't tie it in with the

Executive Reserve and the need for more money for OCDM. Followed other speeches which were all of the saber-rattling kind. Particularly obnoxious was that of Leo Cherne, who felt it was wonderful that Khrushchev behaved as he did in Paris, as now we wouldn't fall for the Communist trickery, etc. He, incidentally, soundly condemned Eleanor Roosevelt in a lengthy flowing sentence, which he must have used many times before. [W. F.] Schnitzler of AFL–CIO was milder and not as oratorical, but in spirit not very different. The panel discussion flopped horribly, since no panel member had any questions to ask of anybody else.

Leo Cherne, publicist, economist.

Had a long phone call from Allen Dulles, who said that Pete Scoville told him about the briefing and discussion we had with Bruce Billings about the chaos in the intelligence R&D. Allen said he was most disturbed about the situation and felt that something should be done, but of course he had no authority over the military and could only use persuasion, which didn't always work and what would I suggest. He was fishing to find out what I intend to do, and I was not eager to show my hand, so I merely accepted his suggestion that we have a lunch with Tom Gates and vaguely proposed that a high-powered technical committee ought to look into the development projects because they are not moving ahead as well as they should. Very clearly he is worried and feels himself not in a position to effect any significant changes. He pointed out that five out of ten members of USIB are military and that they generally stick together.

ᕍ *Tuesday, May 24, 1960*

NSC meeting, most of which was devoted to analysis of the Summit by Dulles and comments on same. There is strong evidence that the Soviets made a decision for a big blowup quite early after the U-2 incident. For instance, the pages of the *USSR* journal, published by the Soviet Embassy, which were devoted to the forthcoming visit of President Eisenhower to the USSR, were withdrawn from the printer on May 5, and a new text that had no reference to his trip was substituted. On the other hand, on May 9, the Soviet ambassador in Paris called on de Gaulle to discuss detailed procedures at the Summit meeting, and would not concede to de Gaulle that the Soviets would not take part in it. The President decided on Thursday, May 12, to stop U-2 flights. On Sunday, May 15, Khrushchev showed de Gaulle the main portion of his speech, to be delivered on Monday, already in French translation. Evidently it was prepared in Moscow since it did not refer to the cessation of flights. He did not show the last five pages, containing the personal attack on Eisenhower. Later I heard from Jim Hagerty that Khrushchev was extremely nervous on Monday morning, his hands shaking so that he could hardly hold the manuscript, and that other members of the USSR delegation were nervous, except for Marshal [R. V.] Malinovsky. The latter stuck to Khrushchev like a leech, accompanying him even on the formal farewell call on de Gaulle, which, according to protocol, is always made alone by the head of state. Dulles mentioned that Mikoyan has been absent from public view the last couple of weeks, the suggestion being that he is for a mild policy and is now out of favor.

When Khrushchev complained about the U-2, de Gaulle suggested that the Soviets are sending a satellite over France daily, which Khrushchev dismissed as irrelevant and said he didn't care whether there were photo cameras in satellites. The omission of the usual speech by Khrushchev on his return to Moscow, his mild speech in Berlin, etc., strongly suggest that he feels he overplayed the issue and is now trying to restore the role of USSR as a seeker of peace. Accordingly, attacks in the Soviet press are concentrated on the President and the top members of the administration only.

> According to Khrushchev's memoirs, *The Last Testament,* the decision to take an aggressively uncompromising position in Paris was reached by him while en route to the Summit.
>
> The remark of Khrushchev that he did not object to intelligence satellites became the foundation of a consistent policy of both superpowers, which was eventually formalized in the SALT agreements of 1972. During the period covered in this diary there was considerable occasional pressure to develop antisatellite missiles. I opposed these proposals, successfully using the arguments that for us the satellites were far more important than for the Soviet Union which could get most of the information from the open press, such as *Aviation Week.*

After this briefing, Herter was asked by the President to out-line U.S. policies. It was agreed that there will be no increase in the military budget, especially for long-range items, since the budget was prepared on the assumption that there will be no relaxation of tensions; nevertheless some steps will be taken to improve immediate readiness. Gates defended the military alert he ordered after U-2, saying it was misunderstood, because it was only communications testing and command alertness. It was agreed that such alerts will be continued on a periodic basis. Herter's basic proposal is that we should proceed as if nothing had happened. On that basis, he proposes to continue nuclear-weapons tests-suspension treaty negotiations and other disarma-ment sessions. McCone objected to the former, since it binds us not to resume nuclear weapons testing, whereas we are free to do what we wish on the other armaments. The President rather pushed him aside, but agreed that the self-imposed moratorium cannot be continued indefinitely and that the Soviets must be told that the treaty has to be agreed to soon. Everybody felt that general disarmament negotiations are a farce, used by the So-viets for propaganda purposes, but will have to be continued at least until the next General Assembly meeting in September. On the exchange of visits, some discussion took place as to whether high-level official visits, e.g., Stans's and Kozygin's, should be canceled, while continuing with private people. The vice-presi-dent urged not canceling anything and letting the Soviets take the initative if they wish. There was a long discussion of forth-coming congressional investigations. The President urged coop-eration but firmness in not revealing intelligence activities, except U-2, and in not conceding that administration state-ments about the U-2 loss were errors. He pointed out that all failures happen at the wrong time, and the failure of U-2 on May 1 was no exception. Herter urged vigorous development of intel-ligence satellites, which was followed by Gates's statement that they could use another $50 million to speed it up. The President asked me to discuss that matter with York and report to him.

The deputy director of USIA reported on press reaction abroad to the U-2 incident and the Summit, his conclusion being that the Soviets have lost more than we did. He then commented on the desirability of following Herter's plan of action, i.e., policy as before the Summit, to ''regain our leadership.'' This made the President angry. He lost his temper and said we did not lose the leadership and therefore we didn't have to regain it, and he would appreciate it if that expression were never used again, especially before congressional committees.

∾ Wednesday, May 25, 1960

8:30 AM   White House staff meeting, which heard accounts by Jim Hagerty and Andy Goodpaster of their impressions at the Summit meeting. Generally, they reinforced the impression that Khrushchev was very much under the control of Moscow, was almost afraid of Marshal Malinovsky, and was very nervous.

∾ Thursday, May 26, 1960

10:30–11:30 AM   With the President, Gordon Gray, and Goodpaster. My account of the unsatisfactory progress of the Discoverer and SAMOS satellite projects was cut short by the President, who didn't require more than a small fraction of the ammunition I had to decide in the firmest possible terms that a corrective action was necessary. He instructed Goodpaster to prepare for his signature a directive for a special ad hoc group to study the issues involved, including "military requirements," which would advise him rather than report to him, in order not to get into the political trouble he got into with the Gaither report. He told Goodpaster to check it with Gates and was unresponsive when I suggested that perhaps Gates should set up such a group. He said the group should be under me, sort of side by side with the PSAC, and he said to Andy: "Tell the Defense Department that I won't approve any money for the projects until I have the information." Gordon Gray pointed out that he studied the executive orders and came to the conclusion that the director of central intelligence [Dulles] has no authority to control the issuance of "military requirements" in the intelligence area. The President was chipper and said he had a pleasant meeting with the bipartisan group on legislative matters.

> The highly classified Gaither report on matters relating to national security was prepared by a large summer study chaired by

H. Rowan Gaither in 1957. It made some recommendations which
were rejected by the President (e.g., a country-wide program
of building A-bomb shelters). This leaked out and resulted in a
painful controversy when Congress began demanding to see the
Gaither report and the President refused to release it because
it was a privileged document.

Spent the rest of the morning with the Nike-Zeus panel, except
for a brief visit from Alan Waterman, who wanted to discuss his
forthcoming lunch with Philip Ray. I urged Waterman to be
firm regarding the primary role of the NSF in the activities for
distribution of technical and scientific information but to toss a
bone to Commerce by promising additional work to its Technical
Services Bureau. Waterman pointed out that Green, who heads
the bureau, while smooth, is not a good manager and the bureau
has no technical competence to do the information handling job.

2:30–4:00 PM   Cabinet meeting, devoted to the report on the
Summit. The President gave a longish assessment, emphasizing
that, probably contrary to Khrushchev's expectations, de Gaulle
and Macmillan both stuck firmly by the side of the U.S., so that
the result was not a split among the Allies, but their pulling
together. Even Macmillan made no effort whatever to soften the
U.S. position and only asked for some extension of time so that
an attempt could be made to persuade Khrushchev to withdraw
his demands before the formal announcement of the Summit
failure Tuesday night. The President added the new bit of information that when Khrushchev made his final call on Macmillan,
he said as they were walking up the stairs: "You may wonder
why I have Marshal Malinovsky always with me. You see, my
people back home think that you are such a smart diplomat that
you could wrap me around your little finger, so they sent some
help." The President noted that during the meetings in Paris
Khrushchev did not speak directly to him or he to Khrushchev,
and that no messages of any kind were exchanged between them.
Merchant then gave a factual account of what went on. He
emphasized, as had Herter at the NSC meeting, how thorough
were the preparations of the Allies for the meeting. He said it
was the best-prepared meeting ever and also that the Allies had
reached complete agreement on tactics and objectives, a rather
unusual feat. Then Chip Bohlen gave his analysis, which was
that Khrushchev lost interest in the Summit when he realized
from various speeches in the West that the Allies would stand
firm on the Berlin issue and were in accord with each other. The

second reason may have been objections of his colleagues to the personalized foreign policy he has, a feeling which Bohlen got when he was in Moscow. The third reason was the attempt to use the U-2 incident to split the Allies.

Gates then spent nearly an hour with the Nike-Zeus panel, discussing its findings. The meeting was, I believe, very helpful, because Gates clearly had not seen the importance of the panel's conclusion about the need of Army management of Atlas launchings and this was made clear to him in the discussion.

Discussion with Bethe. Some time ago I told Hans that Mc-Cone, in a meeting of Principals, asserted that both Livermore and Los Alamos laboratories were now almost unanimous in evaluating the probable degree of improvement of weapons that could be achieved by testing, and that they estimated nearly a factor of ten in the explosive yield-to-weight ratio. Hans told me today that he checked with Carson Mark at Los Alamos, and that that assertion was definitely a distortion. Both laboratories agreed that without further testing, it should be possible to improve the ratio by about a factor of two. Livermore believes that, with testing, it will be possible to improve the weapons by another factor of about two by 1965, whereas Los Alamos thinks that such improvement would be difficult to obtain and would take many years. Teller then maintains that beyond the Livermore prediction, a further improvement by a factor of two or three is possible. Thus, Los Alamos thinks that resumption of testing can only gain us a factor of two; whereas Teller wants it to be something like a factor of five, on top of what can be achieved with the moratorium in force.

Tuesday, May 31, 1960

2:30–4:30 PM NSC meeting. Rather lengthy discussion of "splits," i.e., different policy statements, mainly between State and Defense, in the NSC paper on policy toward Japan. The President consistently took the side of the State Department over rather violent objections from Twining. The essence of the splits was that Defense wants to maintain indefinitely our military posture and controls in the Far East, whereas the State Department and the President want to influence Japan into developing its own military strength, and pull back our forces. On the economic issues the President was not sympathetic to the secretary of the treasury who wanted, naturally, to save money. Toward the end of the meeting, the President spoke about his concern regarding the intelligence satellite programs and announced that "his scientists" will make a study and that he wants to hear them first before hearing the Defense Department's presentation on the subject. This clearly didn't set well with Jim Douglas.

Met with Solly Zuckerman at the Cosmos Club. Most of the discussion was about the change in the military posture of the U.K. which Solly would like to introduce. It would lead away from reliance on nuclear weapons, especially for limited wars, and strengthen other limited warfare capability. He suggested that I and some of my associates meet with his study group that is advising him on the problem, and we tentatively agreed that such a meeting, if it does take place, would be just before the Tercentenary Celebration of the Royal Society in July to enable me to attend the latter as well.

> Sir Solly Zuckerman, chief scientist of the British Defense Department.

Wednesday, June 1, 1960

Navy briefing for Antisubmarine Warfare Panel on defense of carrier strike forces against air and submarine attacks. A far superior briefing to what the Navy gave the President under the title "Control of the Seas," although even this one was held at a general level. Subsequent questioning by the panel brought out a number of details.

Saw Gordon Gray about NSC actions regarding intelligence satellites, since I was unsure about what I was supposed to do. It finally came out that I am to consult with York about the desirability of putting additional money into SAMOS to accelerate it and to report on this to the President. Separately I will probably be asked by the President to undertake a broad study of intelligence satellites.

Took to Andy Goodpaster my memo regarding the Indian Ocean Oceanographic Expedition, addressed to the President and endorsed by the Federal Council at its last meeting. Andy thought that approval by the President would be a routine matter and no need of my making a special appointment. I raised the question with Andy about the study of intelligence satellites and he explained that the President is changing his initial intentions because Defense and Budget and CIA are undertaking a broad study of military requirements in the intelligence area, to which Gordon referred also, and which presumably will define the administrative procedures for preventing project proliferation. I explained my predicament: that I couldn't undertake a technical study without a specific directive—with which he agreed and said that one will be certainly forthcoming, given the way the President feels. I then asked him whether I am going with the President to the Far East and he said he still didn't know, but would let me know tomorrow.

CHAPTER XI

The Difficulties
Accumulate

THE SUMMER OF 1960 was a painful period for Eisenhower. Although presenting publicly a calm and confident stance he was deeply troubled about the U-2 incident and the Paris fiasco, as some entries in this diary and an unrecorded conversation I had with him indicate. The foreign policy had to be reoriented to a substantial degree to counter the expected intensification of the cold war by Khrushchev. Perhaps a part of this intensification was the exacerbation of relations with Castro, which were rapidly approaching the breaking point. The President was being attacked domestically for having allowed a (mythical) missile gap to arise and felt himself helpless to squelch these criticisms without disclosing intelligence information which he regarded as highly sensitive.

The dis-invitation of the President by Khrushchev in Paris forced a hasty change in travel plans. The original schedule called for a flight over Siberia to Japan after spending several days in the Soviet Union. The revised plan included state visits to the Philippines, Taiwan, and South Korea, before the pre-arranged visit to Japan.

Unfortunately the political situation in Japan was becoming quite unstable. The leftist elements were opposed to the signing of the new Japan-U.S.A. security treaty that would supersede the treaty of 1952 and would provide a greater role to the Japanese defense establishment.

[341]

With an active participation of the Japanese Communist Party, mass demonstrations, largely of students, were organized, quite disorderly in character and seemingly beyond the powers of the police to control.

The President's advance party headed by James Hagerty and Thomas Stephens, which flew to Japan in June to work out the details of the visit, was greeted at the Tokyo airport by unruly mobs. Their limousine was blocked by the crowds of demonstrators armed with sticks and stones and they had to be rescued finally by a helicopter. Although the President was advised against going to Japan, both the American ambassador there and Premier Kishi assured him of adequate protection and the visit to Japan was still on our schedule when we left Washington on June 12.

However, the violent demonstrations continued in Tokyo, especially outside the Diet, and in the afternoon of June 16 Premier Nobusuke Kishi announced that the Eisenhower visit was being "postponed," without giving prior warning to the President. Our other stops were largely ceremonial. Accordingly, private meetings with heads of state and the caretaker prime minister in Korea were limited and the President's speeches dealt with noncontroversial topics—amity between the United States and the host country, the importance of a democratic form of society based on sound nationalism, the need of efforts to improve local economies, the dangers of communism, etc. Throughout the travels Eisenhower was greeted by large friendly crowds and only on Okinawa did he encounter a moderate-sized unfriendly "snake-dance" demonstration in the Japanese student style.

I was greatly pleased to be included in the President's party, having been curious to learn what these travels involve, but my duties on the trip were minimal; I had lots of free time and spent some of it making long entries in the diary. These, in a somewhat condensed form, are reproduced in this chapter.

Khrushchev followed the breakup of the Paris Summit by a violent press campaign against the United States and the President personally. The tactics of Soviet negotiators in Geneva grew more intransigent. The gravest act was the shooting down of our RB-47, which is a B-47 bomber equipped with electronic and other sensors for reconnaissance instead of bomb racks, which was flying in the Arctic north of the Soviet Union well outside its territorial waters. The Soviet responded to our inquiries about the fate of the RB-47 with much delay and then coupled the announcement that two crew members had been captured (and several others perished) with a violent verbal attack on United States policies.

The toughening of the Soviet position in Geneva paralleled similar changes in Washington. AEC and Defense went on a counteroffensive against our efforts to achieve the test ban. The main issue here was the procedure for making nuclear explosions in the joint or coordinated seismic-research program on which the United States insisted and which were being challenged as unnecessary by the Soviet Union. The first proposal by our delegate involved the "black box" concept, that is, using nuclear devices sealed by the host nation for seismic experiments. The Soviet Union attacked this as a subterfuge to continue weapons development. On June 2, the United States then proposed that each country establish soonest a special depository of nuclear warheads from which the appropriate devices would be later withdrawn as needed but would not be inspected. The reason advanced was that the opening for inspection of American warheads could not be done lawfully unless their design was first declassified and so revealed to the entire world. On June 15, the USSR responded that complete blueprints of the devices must be made available and all participants in the treaty must witness the assembly of the devices, but that no USSR devices would be offered for tests. On July 12, the United States proposed "pooling" in advance of the research program a number of older nuclear devices which later could be inspected by the parties. On August 2, the USSR complimented the United States on its pooling proposal of July 12 as more realistic but declined to offer Soviet weapons for the pool, because all that research was unnecessary.

Pressure was also being exerted by the Pentagon to go ahead with a nuclear explosion as a part of our VELA program, regardless of what was being done in Geneva. The first such test, code-named Lollipop, was to involve a "small" nuclear explosion —only 5 kilotons—and was scheduled for later that summer. We learned, however, that very little seismic instrumentation would be used but that some special tunnels near the explosion chamber would be used for the study of weapons effects on military structures or equipment. Fending off these offensives, which were clearly against the President's wishes, took considerable time on my part.

This chapter contains also a greatly condensed account of our trip to England, undertaken at the invitation of Sir Solly

Zuckerman. It may be worth noting that in Great Britain at that time science and technology had three (not always agreeing) foci of power in the government: Lord Hailsham (Quentin Hogg), the minister for science and technology; Sir Alexander Todd, professor of organic chemistry at Cambridge University and chairman of the science advisory committee; and Zuckerman, the science advisor to the secretary of defense. While our scheduled meetings were only with the last, I took the opportunity to call on all three, as well as to sign the membership book of the Royal Society of London, to which I had the honor to be elected in the spring of 1959.

Thursday, June 2, and
Friday, June 3, 1960

Spent all Thursday with Piore, Fisk, and Weinberg, discussing the concepts of the paper on science in government, and had dinner with Fisk, Piore, and General Schriever, at which Schriever talked at length about his troubles with Air Force's Air Materiel Command, which tries to duplicate ARDC. I mentioned to Bennie that I may get a directive from the President to study the intelligence satellites and he seemed quite agreeable to the idea. Was told in the morning by Goodpaster that the President invited me to go to the Far East if I could find time to do so. Talked to a number of people, every one of whom encouraged me to go.

Friday morning informed Goodpaster that I will go to the Far East; had appropriate shots and saw the itinerary. Then had a talk with Harry Kelly of NSF on whom to see, etc. Spent rest of the day having visitors, phoning, talking, but it was a terribly cut-up day with nothing useful accomplished.

Saturday, June 4, 1960

Had a long session with Panofsky, Press, Bacher, Killian, Tukey, and Keeny, regarding the nuclear weapons test suspension problems, specifically the high-altitude tests and the VELA program for seismic improvement. The meeting not only provided me with facts, but also clarified my thinking, except for one fundamental issue, on which I explicitly asked those present not to give me any advice. It is whether it is worthwhile, in the Principals' meetings, to continue to push for test-ban agreement in Geneva or to take a neutral attitude. It is certainly a tough one, considering the general lack of favorable prospects after the collapse of the Summit, and I will have to muddle my way through alone. After the meeting worked for several hours on a letter to Herter to alert him about possible consequences of nuclear tests shots in the VELA program, and also dashed downtown to buy a white tuxedo jacket.

Spent the morning reading background papers prepared for the Sprague committee on "The Image of U.S. Science and Technology Abroad." Pretty lousy documents. Rather critical of science administrators and scientists and generally noncreative. Contain also some frightful bloops like discussion of antigravity devices. After that went to the office, picked up draft of my letter to Herter and worked all afternoon to improve it.

Monday, June 6, 1960

Showed my draft to Keeny and Beckler, neither of whom was enthusiastic and who were worried that Herter may embarrass me by referring to this letter in a Principals' meeting. Conceived the plan of making it into an internal memorandum and providing Herter with a copy for his own use. Phoned him and got assurances that he will not reveal to anybody having received this paper. He told me that he is staffing a letter to Brode requesting his resignation and that I should get on the ball and find a good successor for Brode. He also said that four people in a row had refused his offer to become the head of the Arms Limitation Office in the State Department and he guessed I was right in warning him that that would happen. I let Keeny reorganize my letter into an internal memorandum, which he did by early afternoon. I signed this version and a carbon copy was taken to Herter. The memorandum analyzes the current VELA program and notes that proceeding with it unilaterally against

Soviet objections may create serious world-wide criticism of us as engaging in a subterfuge to resume weapons tests. The most awful aspect of it is the first scheduled nuclear test in September. Although VELA is supposedly designed to improve seismic monitoring, the explosion tunnel is surrounded by ''weapons effects'' tunnels built by DOD so that a lot of information of military value will be derived from the test.

> The problem of informing Herter was causing me so much concern because I feared that my own off-the-record informer might suffer, if I was identified as the source of the information. What I learned was that the AEC, while preparing the first underground nuclear test Lollipop in Nevada for project VELA, allowed the Defense Department to dig "effects" tunnels near the explosion center. In these were being built portions of military structures such as missile silos. The test would reveal their resistance to seismic shock, clearly information of some military value. The President, however, in his press conference on May 11 had said referring to VELA tests: "I had a discussion with Dr. Kistiakowsky only this morning. These things are not nuclear weapons testing. . . . They are not expected to have anything to do either with weapons development or the Plowshare project."

> The juxtaposition of these assurances with an inevitable leak after the Lollipop shot would clearly have been highly embarrassing to the President.

ᖰᘜ᳐ᒧ *Tuesday, June 7, 1960*

Spent all day working on the science in government paper, then on a list of my successors to send to Nixon, and talking to staff people.

Went over to see Persons about my trip to London and got approval except probably for travel funds. While I was there, Dillon came in and at an opportune moment I brought up the issue of the draft agreement between Ambassador Douglas Mac-Arthur II and the Pentagon to allow the military to set up research offices in Japan and peddle research contracts to Japanese universities. Both agreed that it would be a grave mistake to let this happen, so maybe some good will come of my bringing it up.

Wednesday, June 8, 1960

Spent morning in office doing chores and had Rabi to lunch, discussing nuclear weapons tests, about which he is intensely negative, and our coming meeting with the President.

The meeting with the President started on a very friendly tone with Rabi relating that he just had an operation for hemorrhoids by an outstanding specialist, whom he described as "the rear admiral." Thereupon the President said that when he had his ileitis operation, he was being fed rectally and became exceedingly uncomfortable thereabouts, and when he began to feel better, he completely embarrassed a nurse when he said to her that all his troubles were behind him. Rabi then summarized for the President the Armand report on NATO science, which makes good recommendations and represents a major achievement in view of the need to get unanimous consent. The President was complimentary and authorized me to tell the secretary of state that he supports the recommendations of the report.

2:30–4:00 PM NSC meeting. The highlight was a strong stand by the President when the policy on Hong Kong was discussed. Everybody but the Budget wanted to approve the use of U.S. military forces on the invitation of the British if there are severe disorders in Hong Kong instigated by the Communists, but the President took a firm stand against it and said he would never approve sending U.S. forces to Hong Kong under these conditions, but only if the colony was invaded by Chicoms. He won. There was an estimate on what reception the President will meet on his trip and assurances that the disturbances will be negligible. Premier Kishi sent word through Ambassador MacArthur that he will resign if there are major disturbances and in the worst case even commit hara-kiri. The President remarked that he was probably headed for trouble, except in Formosa, but that he will take it.

Afterwards Goodpaster asked me to look over a draft of a "directive to undertake a study of the SAMOS." As I sensed in previous conversations with Goodpaster, he was against giving the job to us even though that was clearly the desire of the President, and, sure enough, the directive gives us only a small part of the job, with no influence over real issues of management structure and military requirements.

The Principals' meeting from 10 to 12. It was a pretty awful mess, at which McCone and Doyle Northrup discounted the Rand Corporation (Al Latter's) report on improvements of the Geneva system for seismic installations in the Soviet Union. Then I forced Northrup to admit that his so-called cost estimate of $1–5 billion for installing the system had no foundation in fact, whereupon finally Secretary Herter lost his temper and complained about listening in good faith to technical presentations which were then almost immediately declared invalid; and how could he conduct foreign policy on this basis? The Rand report has been given to the British and was to be the basis of negotiations at the Summit. I then pointed out that the Rand report was not wrong except for the uncertainty about the number of earthquakes in USSR, and Farley joined me. I don't know where we stand. After some argument, I managed to win agreement from Gates to request from Rand Corporation an official statement as to the significance of their report.

We came then to the VELA program and Herter was rather insistent that McCone develop safeguards to be proposed in Geneva other than the black box concept. Naturally dear John was most reluctant, but had to back down somewhat when I brought up the idea of the original gun-type bomb design, since declassification of it could not help any nation join the "club." Surprisingly, McCone and others conceded that in the present VELA program the number of nuclear shots could be reduced and the small-yield shots eliminated altogether. It was agreed that an estimate of the minimum required number and of yields will be made by the Panofsky panel or its equivalent.

It was agreed that during the next month our delegation in Geneva will try to force the Soviets to respond to the proposals we have presented.

Finally, we came to the high-altitude testing problem, and I urged that consideration be postponed until more information is available on the simplified monitoring system proposed by Lockheed. Not much discussion followed, except that Northrup was clearly trying to get a cost estimate from a stacked group, mostly from Livermore, and I told him privately afterward that I would not accept the judgment of such a panel. Northrup is becoming hopeless. He is obviously dominated by the Air Force officials and exercises no judgment of his own whatever. After the meeting Romney, for instance, said that he agreed with my evaluation of the Rand report as essentially correct, but of course couldn't say so in the meeting.

After the meeting, Herter told me privately that (*a*) he has asked Brode to resign not later than September 1, and (*b*) that, of course, he will flatly object to firing the Lollipop 5-kt shot next September because DOD has coupled this supposedly seismic experiment with a lot of weapons-effects installations. So I was glad I didn't raise this subject. During the meeting, Gordon Gray, who sat next to me, had passed me a note asking when I was going to bring up the issue of the Lollipop and then said that State would manage it so that I would have to raise it and that my blood would flow just as his had at yesterday's NSC meeting. At that session Dillon managed it so that Gray had to bring up the issue of a horrible blooper by CINCCARIB, who sent messages in the clear through Western Union channels to all [our] military attachés in Central and South America instructing them to use classified State Department documents in their possession to induce the participation of all members of OAS in a move against Castro. The President at the time got extremely mad and Jim Douglas looked like thunder and said they were already investigating to determine who is responsible.

> The black box concept refers to the first American proposal in Geneva about the nuclear shots in the VELA program: the warheads were to be used without inspection.
>
> CINCCARIB: American military commander-in-chief of the Caribbean zone.

⌖ Sunday, June 12, 1960

Left Washington at 8:50 AM. After seven hours arrived in Anchorage, Alaska, at 10:30 AM local time. Trip uninteresting, with little conversation. At Anchorage went to BOQ ("Chateau Alaska") while the President was taken into town and then to SAC headquarters, spending there a total of three and a half hours. I went sight-seeing with McCann and Montgomery, but just nothing to see. When we returned, the President was finishing lunch. Stopped me to say, "All they [i.e., Goodpaster and Hagerty] talk about is Japan—it is a difficult decision to make."

He was deeply troubled. Hagerty is worried too, but thinks we must go through with the visit. Reception for the President at the base by 300 military and civilian "leaders"; very dull. A female National Republican Committee member from Alaska complimented me on my appearance before Republican women in Washington. Had another King Crab (French fried, not so good as Crab Louis, which I had at lunch) and to bed at 9:00 PM. Light all night. Up at 4:45 and takeoff at 6:50. After seven hours landed on Wake Island. Wake would be a fine place to live, sunbathe and swim, if the problem of boredom could be solved. Only spectacular sights were the snow-covered volcanoes on the Kurile Islands. So far a total waste of time as regards my job. Tried on Goodpaster whether the President would talk to me about arms limitations. Reply: "No, in all probability." Of the various people on board, Mrs. Whitman, Goodpaster, and the young Eisenhowers are the only ones to spend time in the President's stateroom. Another seven hours of uneventful flight landed us at Clark Field, Philippine Republic. Small reception ceremony by local military for the President and a small civilian crowd beyond the fence.

> Kevin McCann, presidential speech writer; Robert Montgomery, movie actor, advisor on speechmaking; Ann Whitman, the President's personal secretary.

> The Republican Women's Convention in Washington that spring had invited Ambassador Henry Cabot Lodge and me to address one of the sessions. Lodge hogged my allotted time, displaying himself as a vice-presidential candidate so I cut my speech greatly. Afterwards there was a question-and-answer period and a (written) question came: "Why does the White House have somebody with such a strong foreign accent as Kistiakowsky on its staff?" I read it aloud and said that after coming from Europe to Princeton in 1925 I started taking English lessons but various charming local ladies pleaded with me not to lose my darling Cossack accent—so I had to stop the lessons. This got a bigger laugh than Lodge got.

Tuesday, June 14, 1960

Nearly seventeen hours after leaving Anchorage we landed at Manila. The crowd was not large and some of the signs none too friendly, e.g., "Give us economic freedom," "We want to be equals." Short speeches by the presidents, full military honors, and the motorcade started. My escort was secretary of defense Alfredo M. Santos, with two guards plus driver in front. Santos had friends all along the route yelling "Santos, Santos"; he waved back; he shook 10,000 hands stretched to the window. For the first half hour he did it on both sides, passing his big, beefy arm in front of my face. So then I sat straight up the rest of the way and he missed half of all possible shakes. The car was unbelievably hot. I started dripping sweat in the first ten minutes and when, an hour and a half later, Santos led me into the Malacanang Palace to greet the presidents and their families, even my suit, not to mention underwear, was wet all through.

The President's reception was impressive. All the way from the airport—some ten miles—people stood five to ten deep. The official motorcade was soon surrounded with hundreds of motor bikes and scooters (male and female on each; some cute), weaving in and out, also a few police cars attempting to shoo the scooters away—utter madhouse. Normal clearance between passing cars or motorcycles here is one-half inch it seems. As we were well back in the motorcade, we were not moving steadily; brakes were jammed on every few yards. Only self-control saved me from having the heebie-jeebies before we even got downtown. There the crowds were tremendous; people at all windows too. The people pressed toward the cars with hands outstretched. I got my share of greetings: "Hello Joe," Joe being any male U.S. citizen. I waved and grinned madly in return, and stopped being jealous of Santos. The motorcade was moving through a sea of people, tons of confetti coming from above. Crowds picked the stuff up and tossed it into the cars. Ours had a foot-deep layer even though we tossed the stuff back as fast as we could. More enthusiasm and now cartons (probably the ones that had contained the confetti) were being tossed onto the streets and into the cars, but all with shouts of greeting and laughter. These Philippine girls do know how to make you feel their smile is just for you alone! But the cartons got too much for us and we closed the windows for about the last ten blocks. Temperature in that car must have gone to 200°. Finally we did get to the palace.

The next day learned that over a million people were on the streets; many came from far away, and pretty spontaneously, although schools were let off for the afternoon and some factories closed. Goodpaster and Hagerty say that this reception matched anything they saw before, even in New Delhi. I was so bushed that after my introduction to the Philippine president and family by President Eisenhower, I beat it immediately to the hotel and with the aid of pills, slept twelve hours (after shower).

✑ *Wednesday, June 15, 1960*

Spent morning with Dr. Paulino Garcia and about twenty institute directors, members of his National Science Development Board, and a few college presidents. Meeting went well. Nice welcome from Garcia; then two-minute speeches by each one present explaining what they do and then my "remarks." Having been well-primed by the stuff collected for me by Gene Skolnikoff I said the right things and in the right way—the college presidents were obviously happy, so were the institute directors and still Garcia was pleased too. After my remarks there was questioning on all subjects, including radioactive fallout and on who (U.S.A. or USSR) is going to be tops technologically. We got to be thick friends with Dr. Garcia; he insisted on taking me to the informal lunch at the palace and introduced me there to all the Philippine bigwigs as his American counterpart. I smiled sweetly, naturally. He also presented me with a most handsome formal Philippine shirt—Barong Tagalog—which I shall wear instead of a tux tonight to state dinner, as even President Eisenhower will have one on, his being a present from President Garcia.

After lunch, and I perspiring freely all the time, the President went to a joint session of Congress. Mob of politicians jumped into cars assigned to the President's party, and I barely managed to get into the last one in line. Crowds around the legislative building and a mob trying to get in. Some local guys formed a flying wedge and got Goodpaster and me inside and into our seats. Jam-packed! The press photographers en masse. They are the worst feature of all these functions, fighting each other, disregarding orders, etc. The President gave a fine address, much of it ad-libbing, or so it seemed. Lots of applause when P.R. and U.S. friendship and equality were mentioned.

After Congress, the President went to the embassy and spoke to the staff (1000!) pleading that each one think of himself or herself as the ambassador of the U.S.A.—the country of freedom and equality. Hope it penetrated a few heads, but doubt it, as there was much gum chewing and snapshot taking.

At 8:00 PM at the palace again for the state dinner. Dinner downstairs in open roofed-over place after a thronged and disorderly reception upstairs. My table companions were a red-headed British wife of a West European minister and a fat wife of a Philippine cabinet member. The British lady commented that her husband complains about her not speaking clearly—the understatement of the year! She mumbles the first half of each word and swallows the rest. Pure anguish to converse. The other's English was about what mine was after six months in the U.S.A., so we three had ''animated'' conversation. Luckily much and very loud music. Formal toasts by Presidents Garcia and Eisenhower.

The President looked dead tired sitting with President Garcia. Understand that after the party (11:00 PM) which followed the dinner he had to work for two hours on speeches for tomorrow.

◌◠◡◠◡◠ Thursday, June 16, 1960

9:30 AM Joined the President's party and helicoptered to the University of the Philippines. Big crowd. The theater jammed. President was awarded LL.D. degree and gave fine speech on same theme as before—freedom, enlightenment, friendship with U.S.A. As he was leaving, a girl student next to me got his autograph and then kissed him. He beamed. When the President left, I stayed behind and the vice-president of the university took me to the chemistry labs. Most of them were locked as the university was not in session. What I saw was pretty poor—even though these are postwar buildings; the labs are shabby, space very limited and little equipment. Saw a few students in the biochemical and pharmaceutical chemistry labs— all girls! Beastly hot day. Dripping all over. Was driven to hotel and at 1:00 PM attended huge luncheon of chambers of commerce, American and Philippine, naturally with speeches. First time that my name got badly mispronounced here on public introduction, but the same guy (president of the Philippine Chamber of Commerce) got Goodpaster as ''Gotpaste'' so my feelings were soothed. Electric power went off and on, which produced queer effects on speeches. After lunch was driven by two young fellows from Dr. Garcia's office on a sightseeing tour while the President was at an open-air rally. Saw rather elegant new residential section—much like new ones in the U.S.A. in the $50,000 to $100,000 class—and also slums the like of which I never saw before. People here squat on public land on the edge of highways, so one sees first a row of horrible shacks built of left-over iron sheeting, etc., and back of them the fine houses. The old walled section of Manila is a blend of acres and acres of incredibly densely packed shacks and fine modern multistory buildings among them.

Got on board the aircraft carrier *Yorktown* about 10:00 PM. Talked to Captain Gibson (whose quarters I occupy) and Admiral Black and learned that Premier Kishi had canceled our invitation—and in a rather rude manner too. Apparently it became known first from Japanese newspapers and caused quite a flap by Hagerty. Big press conference about 5:00 PM, according to Kevin McCann.

> Charles Edward Gibson, captain in command of the *Yorktown*; J. D. Black, an admiral commanding our task force, aboard the *Yorktown*.

Friday, June 17, 1960

Aboard *Yorktown*—lazy day. Went on flag bridge, which gave me a fine view in calm, pleasant weather. We are steaming in elegant formation: two destroyers a couple of miles ahead and a couple of miles apart. Ahead of them four helicopters hovering with dipping Sonar gear; behind them the cruiser *St. Paul* and a thousand yards astern the *Yorktown*. Some forty miles away an attack carrier with fighter A/C aloft.

Helicoptered with McCann and Admirals Black and [Charles Donald] Griffin to the *St. Paul* for lunch with the President. Most pleasant and the President quite relaxed about Kishi. Said to Goodpaster that we must not do anything to cause Kishi more trouble; the main thing is to get the Japanese-U.S. treaty ratified. Otherwise conversation either on military-technical matters (the President seemed deliberately to emphasize my advice in his decisions) or on golf and the like. After such presidential remarks no wonder the Pentagon thinks I am a sinister character. The President said he had to change clothes completely after every function, i.e., four times a day, so I guess I was not the only one having trouble with perspiration in Manila.

The treaty was signed within the next few weeks.

Saturday, June 18, 1960

In the morning flew by helicopter to the *St. Paul* and then flew to Sungshan Airport near Taipei. Reception ceremony just like others, except that red carpet was rolled out all the way to the plane. Was told to get off behind the President through the front entrance, but it was rather pointless, as none of us except the Eisenhowers were introduced. Chiang Kai-shek looks trim and military. The usual troop inspection and speeches. Chiang talked in Chinese and it did not sound graceful or musical—very sharp sounds. Smooth organization of motorcade and fairly fast driving through orderly crowds. Most of the young people were in school uniforms, boys in khaki shorts, girls in white blouses and blue or brown skirts. Sexes strictly separated and the whole well organized. Here and there music and groups of trained dancers, with masks, dragons, stilts, etc. Rather colorful, but numbers of people much smaller than in Manila and enthusiasm lacking.

Was contacted at the airport by our cultural attaché, who arranged lunch with six Chinese professors and a vice-minister of education. Conversation of no great moment. After lunch, inspection of Taipei University, which is quite impressive—new large chemistry building with adequate labs and quite a bit of equipment. Research obviously in progress. The main complaint is that Chinese students studying abroad usually don't return. The university has large number of overseas Chinese students and ICA built dorms for them. The whole looks like it could be developed into a class A outfit. One of the Americans in our group said he quit the Philippines after 22 months because they "only talked" about nuclear energy (his specialty) but here in Taiwan things are moving forward.

Later was driven with Tom Stephens by a Chinese official to the rally. Strict organization of the affair: groups such as schools or offices or factories met far away from rally and then marched in military formations to assigned places on the great square. The view from the terrace of the administration building impressive—a sea of placards and banners—all in Chinese red; rockets exploding high up and releasing American and Chinese flags under parachutes. Thousands of toy balloons floating up. Loudspeaker system not working well and the translations of speeches dragged out the affair. Crowd not very responsive, but certainly large.

The cost of the rally in banners, etc., must have been high—especially for this poor country. Must say though that the children looked healthy and not underfed—better than in the Philippines. Army formations, of which I saw many, had young soldiers. Wonder where the veterans of the continental campaigns were.

Back to hotel in President's motorcade, moving fast, as crowds were orderly and mostly on sidewalks.

In hotel dressed for state banquet. This one very formal. The President, the young Eisenhowers, and Goodpaster with the Chiang Kai-sheks in a separate room. The rest of us in a small reception room. Sat between vice-president of executive yuan (council of ministers) and the Chinese representative at the UN. Polite conversation. Toasts long and with translation, so I dozed off a couple times. The food was European, the wines French; both good, but not exceptional. After dinner had a pleasant conversation (mostly about Conant) with Hu Shih—a charming fellow.

Hu Shih, a distinguished Chinese philosopher and writer.

ᦒᦒᦒ *Sunday, June 19, 1960*

To the airport, and after a very long farewell ceremony because of the enormous number of officials, to each of whom the President had to say goodbye, took off for Okinawa.

Reception at Okinawa airport all American. Driven to Officer's Club for reception and lunch. The President went on a motorcade and encountered the only hostile demonstration of the trip. According to Hagerty, the hostile crowd was small and unimportant, but others in the motorcade evaluated it more seriously. All agreed that it was well organized and led.

Arrived at Kimpo Airport in Korea at 3:45 PM. The President and some of the staff went by helicopter into town and then motorcaded to the embassy compound. Later I heard that the crowds were beyond belief. The motorcade stalled for nearly a quarter of an hour in the dense crowd and finally took a different route to the embassy compound. A few injured but no dead and no "anti" demonstrations—just curiosity and the magic effect of the Eisenhower name.

I went into town by bus and saw on the highway many people —not very demonstrative, but friendly. Stayed in BOQ in the U.S. military reservation, a city within a city. So is the embassy compound and I wonder how the Koreans, who are so poor, react to the comparatively luxurious life of U.S. personnel. Korean servant girls have no sensitivity about U.S. male modesty. I recovered after initial surprise, but Colonel Schulz suffered repeated anguish in undershorts while we all had drinks in Tom Stephens's room.

Col. Robert L. Schulz, military aide to the President.

In the morning was driven to the embassy compound where the President gave a good off-the-cuff speech to employees (about 1000 present). Then Goodpaster and I talked to him about the cable from Tom Gates which I got yesterday about troubles with the Nike-Zeus test program. The President was emphatic: no launch installation on Johnson Island; launch Air Force ICBM missiles from Vandenberg AF Base in California; make sure Army and Air Force cooperate. Sent off cable embodying this to Gates.

Met academic Koreans for lunch in compound. In Korea an enormous demand for education but then no jobs. Korean scientists very anxious to get U.S. money for research—must do something about this.

Was driven to the university through the "Forbidden Gardens" back of the old king's palace. With care it could become a beautiful park. In a clearing encountered a large group of older school girls in fancy court dress with fans and all. They were to perform for the President yesterday en route of motorcade, but crowds made this impossible. They were waiting here to perform for a distinguished visitor (I believe Mrs. John Eisenhower, who however did not stop). So I consented to act the role, was seated on a high throne, and saw a lovely oriental dance.

At the university, after meeting more professors and officials, addressed a small rally of students gathered on my behalf. The gist of my very much unprepared remarks: congratulations to "young tigers" on their political victory. Now they must become young scholars to serve their country best. Talked about my own life as a mixture of scholarship and of national service when called to oppose tyranny.

Then was driven to Simsoltang racecourse with many detours because of crowds waiting for Eisenhower. Got into back streets of Seoul; very depressing; streets unpaved, narrow, dirty; shops all junky; poverty striking.

> The reference to "young tigers" was intended to recognize the role of students in the overthrow of Syngman Rhee the preceding April. Students' questions to me were quite sharp, almost openly antagonistic.

Took off by helicopter for Sixth Corps headquarters and troop review. Fine show with massed flags, large band, and six battalions of troops; U.S. and ROK—the latter better drilled. By this time, suffering badly from dysentery which started this morning or last night. Again by helicopter to Kimpo Airport and, after ceremony, takeoff for Hawaii. With aspirins and sleeping pills survived the night.

⌀⌁ Thursday, June 23, and
Friday, June 24, 1960

Returned from Far East trip and started preparing for the PSAC and FCST meetings next week.

⌀⌁ Saturday, June 25, 1960

Did more work on the meetings and went to see Secretary Herter. First talked about the science advisor candidates. Then I mentioned my concern about the Plowshare program as not being of sufficient importance to influence major national policy decisions; to my surprise Herter agreed that it was probably exaggerated by partisans. He told me then that in his opinion the Russians would almost certainly break off both Geneva conferences. He asked me to make a quick evaluation of the Rand Corporation proposal to install Polaris missiles on submersible barges as this may have a major effect on our proposals to NATO Allies, and I agreed to do so. He told me that he was definitely leaving January 20, but did not see himself between now and then as a "caretaker" and he certainly expected to continue developing policy. Pleasant, cordial meeting.

> The Soviet Union did not break off either of the Geneva conferences—the ten-nation conference on general disarmament or that on a nuclear test ban; the latter dragged on without much progress until December, with a month-long recess in September.

ᐁᐣᐤ Sunday, June 26, 1960

Spent about two hours talking with eight PSAC members about subjects to be discussed with the President. Three were identified: problems relating to nuclear weapons test cessation; organization of arms limitation studies; the need to support basic science on an expanding scale. On the whole it was not an effective meeting, and I am not sure that the first two subjects are the best.

ᐁᐣᐤ Monday, June 27, 1960

All-day PSAC meeting, which went well.

ᐁᐣᐤ Tuesday, June 28, 1960

More of the same until 1:30 PM. I am getting more and more troubled about Weinberg. He was sold to me by Rabi and others as a rather extraordinarily wise and brilliant fellow, but I find his attitudes provincial. The world seems to start and end for him in Oak Ridge, and he gets into the discussion largely when he feels that PSAC is advocating something that may have an effect on his Oak Ridge National Laboratory.

Thursday, June 30, 1960

9:00–10:15 AM   NSC meeting which was uneventful, except for remarks of the President on the Far East trip. His concern was about the guidance which young people get in schools there and his conviction that the teachers should be paid decent salaries, as otherwise they will preach discontent. Somebody else on the council made a remark which implied criticism of the State Department for general lack of action and that led Herter to a long explanation of what the State Department is trying to do.

10:15–11:30 AM   Principals' meeting to discuss possible actions in view of the (hidden) AEC effort to prevent declassification of gun-type weapons that would be used in VELA tests. Also the negative action of the congressional joint committee on the resolution submitted by the administration to permit the President to reveal details of these weapons to the U.K. and USSR without declassification. I suspect that McCone engineered the joint committee's veto of weapons disclosure by the clever trick of discussing the sophisticated Mark VII implosion weapon instead of the old gun-type device when he supposedly defended the administration resolution. The Principals decided to propose to U.K. and USSR that weapons of all three countries be used in the VELA program and their designs revealed, thus making the operation reciprocal. There was general agreement that Congress would approve this, while it would not approve unilateral revelation of our own designs. At the end Gates urged that if, as expected, the Soviets reject this proposal, we should unilaterally proceed with nuclear tests for the VELA program and carry out the first at the earliest possible time, i.e., in August. I objected, and later in the day called Herter, urging him to resist this, since the Lollipop shot would not be instrumented for seismic measurements, but would involve weapons effects installations, and thus would embarrass the President.

4:00–4:30 PM   Went to see Mr. Stans and offered the services of the PSAC in R&D budget evaluations. The offer was firmly rejected because, as Stans put it, it would merely add another pressure group to the BOB problems. Thereupon I served notice that I would not accept his philosophy of an essentially fixed R&D budget and would fight for necessary changes and increases on my own.

In the afternoon I was in the FCST meeting, which, against my expectations, went well.

After this meeting George Beadle came in and started by saying he wanted to resign from the committee because he had too much to do, but I persuaded him to stay at least as long as I am on the job, in return for a promise not to put him on any new panels. Bad omen. I can also see reluctance of other members of PSAC to get involved in additional activities, which fits poorly with their statements that PSAC and the special assistant should have more influence in government.

ℴ⌣ *Wednesday, July 6, 1960*

Pre-press breakfast, which was uninformative. The discussion centered on the exact timing of the proclamation which the President is to issue in connection with Cuba's sugar quota. Since I didn't know quite what the proclamation contained and this was not disclosed, I was out of the discussion. I gather that it will rescind our agreement to purchase three million tons of Cuban sugar at a price higher than the world market level. The same subject was being discussed between the President and Dillon as I went into the President's office at 9:45. The President emphasized that we mustn't appear to be simply punishing the Cuban people and must give better justification for our action, although an intelligent person could conclude that this is a measure against the Castro government. The latter withdrew the permit for our nickel-ore-loaded ship to leave Cuba, so they evidently will counter our proclamation with the ship seizure. When the question was raised as to what we do if the Cubans seize the water pumping station three miles outside Guantanamo Marine Base, the President said that he felt we should counter that vigorously regardless of consequences. Obviously the Cuban situation is reaching the explosion point.

Between the breakfast and the presidential briefing, General

Strong came to say that the Russian missile fired into the Pacific was fairly similar to previous shots. Tremendous amount of telemetry from the nose cone. We have a large concentration of aircraft, etc., to observe the Soviet tests and will have a lot of information shortly.

The rest of the day had several visits and worked on the "science in government" paper, since Piore, notwithstanding his solemn promise repeated to me on July 4, when I phoned him from home, did not revise the draft of his half of the paper the way I did mine, but merely added a couple of hastily dictated paragraphs. I talked to him on the phone and was as unpleasant as I could possibly be without directly calling him an SOB. I believe he got the point. Clearly IBM is the one corporation for which I will never be invited to consult!

5 :00–6 :30 PM   Principals' meeting to discuss the safeguards of nuclear explosions for the VELA program. The British responded to our proposal by rejecting it, because by announcing that we would proceed with our own test program if the Soviets do not accept a joint one, in effect our invitation would represent an ultimatum following a rather unreasonable demand that the Soviets provide their weapons for tests they themselves consider unnecessary. Wadsworth, too, had cabled that this proposal would certainly mean the breakup of the negotiations.

Over opposition from Gates the long meeting came to the following conclusions: we will instruct our delegation to make a proposal for the use of weapons provided by all three parties, but will not stipulate that we will go ahead on our own if this is rejected. When it is rejected, we will make an offer of permitting visual and manual inspection of our own nuclear devices, providing that Congress approves. McCone and Dillon both feel that by moving in this way, Congress will be more ready to accept this proposal than if we make it first without trying a reciprocal proposition. If the Soviets reject this proposal, too, then we shall move on our own unilaterally with VELA.

Toward the end, Dillon and I raised the issue of the timing of the first shot and the fact that this blast is now planned so as to include galleries for weapons effects tests next to the explosion chamber. Gates said Defense was willing to declassify these installations and permit their inspection before and after the explosion. I insisted that the shot must be fired only after it is well-instrumented seismically, and Gates maintained that some instrumentation would suffice and that the important thing was to fire it soon. We are to see the President tomorrow on this whole mess. During the meeting, McCone gave a long explanation of what went on in the Congressional Joint Committee on Atomic Energy. According to him, just about everybody was completely agreeable to the resolution which would permit showing our devices (visual and manual inspection but no blueprints) to the USSR and the U.K., but overnight the sentiment changed radically for purely political reasons. McCone maintains that he was told that senators and congressmen feel that they would be in trouble in November if they passed such a resolution, although they concede that it is absolutely okay. McCone said also that AEC has again considered the proposal of declassifying gun-type weapons and decided it was contrary to national interests. The whole tenor of McCone's remarks was to establish that he was working hard in every direction to get favorable action.

Thursday, July 7, 1960

9:00–10:45 AM NSC meeting. After Dulles's briefing, more than an hour was spent on Cuba and the Dominican Republic, the discussion being led by Dillon. Although the reaction of Castro to the withdrawal of the sugar purchase quota by the presidential proclamation was milder than expected, the general anticipation is that this is the beginning of at least an all-out economic war, and there was much discussion about a proclamation declaring a national emergency, since this would permit many steps to be taken, such as prohibition of exports to Cuba, the blocking of Cuban funds in the U.S.A., etc. Gates spoke also of the military being ready to move with all sorts of emergency actions; the least being evacuation of U.S. citizens, the largest being a full-scale occupation of Cuba. Treasury Secretary Anderson gave a fairly bloodthirsty long speech about the need to declare a national emergency with all the actions I mentioned above and to take military steps to protect U.S. citizens in Cuba before they are molested. He argued that what is happening in Cuba represents an aggressive action by the USSR. Hoegh went even further and suggested we occupy Cuba, since the Monroe Doctrine was being infringed. The President objected and gave quite a lecture to Hoegh, noting that Bohlen is convinced that the USSR will not establish military bases in Cuba. Randall, who sat next to me, whispered that Hoegh is the stupidest man in Washington. Dillon stated that the Latin Americans feel convinced that Trujillo must first be gotten rid of, as steps against Castro while Trujillo is in power would be very dangerous politically. Our State Department wants both out at the same time. President Betancourt of Venezuela is working to get Trujillo out and will propose drastic steps at the OAS meeting of foreign ministers a few days hence. He will have strong support. The vice-president said that Betancourt is an SOB and as soon as Trujillo is out he will start supporting Castro. Dillon did not agree. He also said that now there is solid evidence that the attempt to kill Betancourt was organized by Trujillo. Towards the end of the long discussion, the President spoke of his desire to change our support of Latin America so that we wouldn't be just supporting governments and the wealthy classes and perpetuating the feudal order of things. We must take new steps and advertise them widely. These would be aimed specifically at land reform and raising the standards of living. The vice-president picked up the subject and urged that the initial announcement be made this weekend before the Democratic platform is revealed on Monday. It was agreed to do so and the mechanism is to give more money to the Latin American Bank in "soft" loans. Needless to say, Anderson and Stans winced and suggested a more cautious approach. My impression from the whole meeting was that everybody expects grave trouble, and the President was even urged to cancel his vacation sojourn in Newport but decided to go but to keep the jet there so as to be only an hour away from Washington.

Clarence B. Randall, special assistant to the President.

10:45–11:30 AM   Principals' meeting with President, at which Dillon outlined the recommendations agreed to yesterday. The President accepted a two-step approach: first a proposal to share weapons for VELA tests and, when that is rejected, an offer to allow visual inspection of our own devices. But he firmly rejected the next step agreed upon yesterday, namely to inform the Soviets that we will start our own tests if they don't accept that last offer. The President said that this clause would amount to an ultimatum and that if the Soviets rejected an offer of inspection we would merely start the program and then announce it. The vice-president was asked when the first shot should be made, and he said it should be done in September or early October. There was some discussion also of the Plowshare program, McCone emphasizing that this will involve highly refined weapons technology, and, therefore, the development would have to be done in secrecy. God only knows how he separates it from outright weapons development. From a remark of the President, I gather that he intends to address the UN general Assembly and at that time (September) announce that we are going ahead with the Plowshare program for digging a second Atlantic-Pacific canal through Mexico, which has agreed to the project. It was obvious that both the President and the vice-president think Plowshare is terribly important. The vice-president also remarked that in his estimate Symington and also Johnson, if elected president, will be for resumption of weapons tests and only Kennedy's attitude is uncertain. After listening to this, I had an unhappy feeling that we in the PSAC had failed to reach the really important people—the candidates for the presidency—and let Teller sell them a bill of goods. The undoing will be difficult if possible at all.

> The President was premature in his reference to the canal through Mexico, which had not formally agreed to this project.

11:30 AM–12 M   With Glennan, Dryden, and people from the British Embassy introducing Professor Bernard Lovell (the Jodrell Bank radio-telescope man) to the President. For the first few minutes the President was obviously troubled about all that had gone on in the morning, but then switched and was charming and interested in Lovell's work, which included the communications with Pioneer V to a distance of 23 million miles. The President created a very good impression, as I learned later from Lovell, while we were having lunch at Lord Hood's residence. The lunch included too many drinks and too much food, but was very interesting because I got to talk to a fellow from the British Embassy who handles disarmament.

Friday, July 8, 1960

9:00–10:30 AM   A retired colonel, who is a planner in the military applications division of Chrysler Corporation, came on a fishing expedition to learn if it was true that the administration had decided to get Chrysler out of military work and concentrate the latter in the aircraft companies. I solemnly assured him that neither in the White House nor at the secretary's level was any such thing even discussed. Followed a long talk with his bemoaning the problems of Chrysler, because it isn't getting contracts although it looks to be the best bidder, etc. I said that the aircraft industry was overexpanded and that the missile business would probably contract and urged Chrysler to look for work in other military fields. Doubt whether I was successful.

11:00–11:30 AM   Elmer Staats, with comments on our paper on science in government, which were mild, and with some proposals. He wants to abolish the standing committee of the FCST, replacing it by an advisory group with himself as chairman, to deal with administrative problems. I fear this might be dangerous and suggested that ad hoc groups to deal with specific problems would be more effective because different agencies would be involved in different issues. Don't know whether the idea took or not. Elmer was not impressed with our emphasis on assistant secretaries for R&D, but not strongly opposed either. A great surprise was his invitation to PSAC to participate in the '62 civilian R&D budget review. I suspect that my serving notice on Maury Stans a week ago that he can't tell me how to advise the President and that I would maintain freedom of action, made him have second thoughts about his rejection of our offer of cooperation. We agreed now to assist on a selective basis, not limiting our involvement to basic research, but largely concentrating on interagency areas like oceanography, medical research, etc. Dave Beckler is to work out the details, and I later emphasized to Dave that he mustn't bite off too much because of the reluctance of my friends in PSAC to do much work.

2:30–3:45 PM   With the National Intelligence Estimates Board, discussing the newest NIE on missiles. I like the new approach and it is the result of my criticism of previous versions. The present one is frank about our lack of firm knowledge and gives a fairly wide range of values, explaining reasons for and against each presumed USSR program. They told me they also liked the text but that the Services are now making so many reservations and objections that the whole thing is becoming almost a farce. I urged nonetheless to go ahead with this format, let all the reservations be printed, and to hope that at least an intelligent planner will see what is right and what is silly. We then had a discussion of substantive issues relating to the USSR missile programs. It is really too bad we don't fly the U-2 any more. The future NIEs will be even more trouble to prepare, because of the total absence of observed operational sites, as contrasted with Mr. K's and other Soviet assertions of operational ICBM forces. The character of their test program of ICBMs is baffling. It is the most leisurely and relaxed operation I have heard of! During the meeting I came back to my assertion of a year and a half ago, that caused such a stir, that we must have missed another test range.

4:00–4:30 PM Generals Starbird and Loper, largely to impress upon me the gravity of unauthorized disclosure of AEC restricted data to the British on our visit to London, and also with assurances that they will give me sharply defined authority to discuss those matters which they think are pertinent to the meeting, i.e., not including detailed weapons technology.

A little vignette on my activities is that a while back, Bob Merriam called me to tell me that the Republican Convention Planning Committee intends to have a series of nongovernment speakers, "distinguished citizens," to start the convention and that one speech was to be on science. The plan was to invite Edward Teller and Bob wanted to know what I thought of the idea. I limited myself to a polite but very firm objection. He asked for substitute speakers and I mentioned two: Lee DuBridge and Det Bronk. Clarence Randall dropped in today to tell me that Lee DuBridge will be the science speaker. I wonder whether this will have any effect on the vote in November and if it does, whether I will have done some good to the country or not.

Monday, July 11, 1960

2:00–2:30 PM Met with Staats, Bronk, and Waterman on NSF '62 budget. Reasonably friendly attitude on the part of Elmer. Waterman is already compromising on his proposal of $290 million but suggests figures adding up to more than $200 million. A major problem is that Elmer and Stans, even more so, of course, don't see why the government should assume a large share of the support of basic research and why the training of new scientists should also, to a large extent, be a government responsibility.

3:30–5:00 PM Meeting with staffers from ICA, and Gene

Skolnikoff. Clearly they want to have a panel, and while not being very specific emphasize the value of a public statement by some influential group such as PSAC on the need for this country to develop more effective technical assistance to underdeveloped countries. I agreed and committed myself to set up an appropriate panel.

During the day learned that the Soviets had shot down our RB-47 which disappeared over the Barents Sea on July 1. I had heard before from our military that the plane never goes nearer than 30 miles to Soviet territory. If it was shot down even farther out, then the intensification of conflict with USSR is almost unavoidable. Very troublesome.

Tuesday, July 12, 1960

Most of the day PSAC spent in Newport at the naval base, preparing and then having a meeting with the President. We left Washington at 7:30 AM and the beginning of the preparatory meeting was somewhat disorganized, but by the time we went to lunch both spokesmen I selected, Panofsky for the nuclear test cessation problem and Weinberg for support of basic research, had their speeches well in hand, and the irresistible urge of other PSAC members to have themselves heard subsided somewhat. My general appraisal of the meeting is one of a very successful occasion. The President was friendly and animated, converting our briefings into informal conversations. I heard him say, for the first time, that financial responsibility for basic research has to be assumed to a large extent by the federal government.

Our briefing on the nuclear test ban argued that no monitoring system can be perfect; that nonetheless a ban would be advantageous to the United States, and that the Soviet Union would not cheat if there was any risk of being caught. The President was noncommittal but did not reject the idea outright.

I had a private conversation with Jerry Wiesner, which began by my saying he should influence his friends, who in turn influence Kennedy, to have the latter pick out a special assistant for science and technology early enough so that as soon as the election is over, the man can start getting acquainted with our work. Jerry replied that if he raised this subject, he would undoubtedly be told that he, Jerry, is the choice, but that he wanted me to stay for at least six months longer on the job. This idea I demolished firmly and urged him either to accept such an invitation or to have Kennedy people approach me for my list of good candidates. I added that he, Jerry, was on the list, but that I wouldn't tell him the order. We will see what comes out of this. Another private conversation I had was with Jim Killian, who said he was approached to work for Nixon and should he accept. My reaction after a short hesitation was to urge him to do so as a counterweight to the influence of Teller and Bill Elliott. I am going to phone him and emphasize that this is important because according to people who work for Nixon, Teller is very active in his entourage and I see the possibility that he may be selected as the next special assistant. If Killian can prevent this, he will perform a real national service.

ᕰᖇᕰ *Thursday, July 14, to*
Thursday, July 21, 1960

A worthwhile trip to London. We had a half-day meeting on Thursday, a full-day meeting on Friday, and a half-day each on Monday and Tuesday with Solly Zuckerman and his group who formulated fourteen questions to consider. Basically what concerned Zuckerman was whether small nuclear arms, such as Davy Crockett, could be employed by NATO forces as localized weapons that did not necessarily lead to escalation and to all-out nuclear war. Meetings on this subject occupied Thursday and Friday. It developed that only Harold Brown, out of a dozen present, maintained that small nuclear weapons could be used in a land campaign without escalation. One of the useful results of the meeting seems to be that Harold had begun to doubt his own viewpoint.

The question then raised was what purpose, if any, small nuclear weapons served in a NATO arsenal. We concluded that their chief function was to make the nuclear deterrent more plausible than if we had only the multimegaton weapons of the SAC. For the Soviets know as well as we do that escalation is inevitable, and thus if we announced that small nuclear weapons would be used from the outset, the possibility of resorting to our major nuclear deterrent is strengthened. The British also consider that NATO has no lasting capability without nuclear weapons, for if we plan only on using conventional weapons when the Soviets start a limited aggression, then NATO at best can only force a stalemate that really involves considerable loss. Accepting this, however, means the breakup of the alliance, since clearly the most exposed members will not remain in NATO. On the other hand, a decision to use small tactical nuclear weapons, which could force the Soviets back to their starting line, and thus would preserve the alliance incentive, makes escalation inevitable.

We then had a less conclusive discussion of tactical nuclear warfare in Southeast Asia and there Harold Brown was on solider ground. On Tuesday we discussed the use of small nuclear weapons in naval warfare and air defense. In naval warfare in international waters, we concluded, the danger of escalation was low; and so far as air defense was concerned it was clear that nuclear weapons would not be employed unless the other side was staging its own all-out nuclear attack. Finally we discussed advanced radiation-type nuclear weapons and whether the British should develop their own. Zuckerman and his people took a negative attitude.

I understand now why the meeting was set up by Solly: there is a struggle in England as to where to put their defense money —into small nuclear weapons or conventional forces. The minutes of the meeting will be used heavily, I suspect. When our JCS see it, they will hit the roof, because we definitely poached on their territory, and I understand that the British chiefs of staff made a determined effort to kill the meeting.

Went to see Lord Hailsham on Monday. Hailsham is a smart fellow, but his mind is completely focused on space. The hot political issue in England is cancellation of the Blue Streak missile system (two-engine, single-stage, double-sized Thor). Tremendous lobby trying to restore it. Pressure groups argue that the British, for prestige purposes, must have their own long-range vehicle. I told Hailsham that they should concentrate on upper rocket stages or on guidance and design of satellites and buy our Atlases and Titans because soon there would be very little prestige connected with them. Don't know how much influence I had. Talked also about this to Alex Todd (chairman of their science advisory committee) and found enthusiastic acceptance by Alex of my argument. Also talked to Hailsham about the possibility of a NATO community meeting of the science officials and he seemed favorable to the idea. We mentioned areas such as space, oceanography, atmospheric sciences, and felt that topics like these should be chosen rather than delving into the internal problems of the participants. Left it that we will explore the reactions of our respective governments to such a meeting and then communicate, to arrange something.

Weekend spent in social entertainment and ended with a colossal hangover. Monday afternoon I signed the Members' Book of the Royal Society—the Big Mace was brought out with the book, which starts with the signature of Charles II. Rumor has it that the society always pastes in more pages, so they never quite reach the end. Monday night was the reception for the Royal Society by Lord Hailsham. Tremendous party—champagne flowed like a river. Another very nice hangover on Tuesday, but mind was clear as crystal. Tuesday afternoon was the official opening of the Royal Society Tercentenary Celebration, a magnificent sight to behold. Beautiful floral decorations. Five hundred dignitaries in academic gowns. Very impressive. Finally the Queen and the Duke of Edinburgh, the Swedish royal couple, and president Sir Cyril Hinshelwood entered. He looked terrific in his vestments. The Queen read a speech. It took the Swedish king, who was elected this year, five minutes to sign the book and his signature covered a quarter of a page. He gave a short speech and then the royal guests went into the royal loge, except the Queen, who left to entertain the king of Siam. President Hinshelwood delivered his address, which unfortunately nobody heard because something went wrong with the PA system. Followed a tea at the Imperial College, with the Duke of Edinburgh, to which all members of the Royal Society were invited. Tremendous difference in the duke's and the Queen's behavior: during the ceremonies the Queen sat like a statue with a totally immobile face; the duke was very relaxed and natural. At the tea he was just one of the chums.

On Tuesday made a call on [Con Douglas Walter] O'Neal of the Foreign Office who is in charge of British disarmament negotiations. He believes that the Soviets are anxious to reach agreement on cessation of nuclear tests because of their fear of Communist China, and that we should make every effort to reach a mutually satisfactory compromise but he admits that the prospects are not good. On the other hand, he considers the ten-nation disarmament conference a fraud with neither side really trying to reach useful results. I then mentioned my feeling that we went to these disarmament conferences poorly prepared. His reaction was that this was not so and that the only trouble was that the positions of the Western Allies were not adequately coordinated. I asked him whether the Foreign Office had any technical competence in nuclear weapons and military security problems. He said no, none at all; for this they relied completely on their atomic energy authority and the Ministry of Defense. I asked whether he could rely on their judgment in matters involving disarmament and he laughed and said not quite, but what was he to do.

Back to Washington on Wednesday.

Friday, July 22, 1960

9:30 AM   McCone came and started by inquiring about what we had discussed in London. I told him briefly, but he was obviously not interested and immediately went over to the briefing paper on the test ban the PSAC gave to the President in Newport and expressed violent disagreement with it. He insisted that if I subscribed to that paper, it meant a major change in my views since I was now in favor of an agreement on nuclear test cessation without adequate safeguards. I said that I always considered the insistence on so-called foolproof inspection safeguards not a realistic policy and that if we wanted to enter into arms limitation agreements, we should think of monitoring as only one factor, and that unilateral intelligence gathering and our own technological developments to counteract the possible effects of evasion were very important factors also. McCone was then pressing me to state whether I advocated surrendering the insistence on nuclear explosions in the VELA program if that was necessary to prevent the breakup of the Geneva conference. I said that our plan to use the black box scheme was politically unwise and especially if it is connected with the use of the Lollipop site. McCone dismissed Lollipop by saying he would be willing to fill in the excavations for the weapons effects experiments and that he said so at the last Principals' meeting, but I don't remember his saying so. McCone insisted that the whole issue has to be flushed out and presented to the President so that he would be forced to make a decision. I agreed in principle, but said it was desirable to postpone the breakup of the conference until after the election, so that the next President would have a free hand. McCone vehemently said that on the contrary Eisenhower must break up the conference so as to give the new president a free hand in deciding whether to continue negotiations or resume testing. Throughout he argued that the Soviets must be making clandestine weapons tests and perfecting their weapons, and that this represents a major threat to our security. He was opposed to permitting Wadsworth to state in an open session in Geneva that the U.S. has no plans for resumption of weapons tests and solemnly undertakes to use nuclear explosions only for the VELA and Plowshare programs. It appears that unbeknown to our office there has been some discussion between Herter and McCone, about Herter authorizing Wadsworth to make this public statement. McCone is violently opposed to it and said he will insist that all of this be put on the agenda of the NSC, which is to meet in Newport next Monday. (When I later mentioned this to Herter, he said: ''Oh, no!'')

Immediately after McCone left I had a phone call from Chris Herter, in which—after telling me that he had an extremely favorable impression of Whitman and hoped very much we could persuade him to take the science advisor's job—he began to talk about his difficulties with McCone and said that McCone is emotional in his insistence that we reverse our tactics and strategy in Geneva and confront the Soviets with a hard choice of either accepting our proposals or breaking up the conference. Herter is getting to a point where he finds it difficult to talk to McCone. On the other hand McCone said to me that he has the statutory responsibility for developing better nuclear weapons and cannot continue in his job under the present circumstances.

The meeting with McCone was generally difficult because he was pinning me down to a flat statement and I just didn't want to make any flat statements which would leave me no room for maneuver. We did agree that the Principals' meetings are bad, the chief reason being that, instead of going to the President with sharply delineated opposing views to let him make a choice, we hammer out compromise papers, shoving a lot of unresolved issues under the rug. According to McCone he has spoken to ''hundreds'' of Republicans in California and without exception they are critical of the administration policies and feel it is high time our entire foreign policy be changed and made tougher. In our discussion of the VELA program I emphasized that nuclear shots could be delayed as much as twelve months without seriously affecting the success of the program if it is a two- to three-year program; McCone felt the tests must be started this fall. He asked me a blunt question: Did I consider that we should give up Plowshare and VELA shots in order to preserve negotiations at Geneva? I answered him evasively.

∽ *Saturday, July 23, 1960*

I went to Gordon Gray's office because he phoned me and said he wanted me to go to Newport on Monday to attend the NSC meeting. I gave him a cautious account of what went on between me and McCone and also told him the essence of Herter's attitude regarding McCone's insistence that the nuclear test cessation issue be brought before the NSC. Gordon said he knew about it and that McCone and Herter will have another session on Monday to thrash it out, but that he, Gordon, thought McCone would prevail and have the issue brought before the NSC. He, Gordon, wanted me to come to Newport for several other reasons. One is that there will be a discussion of how to counter Khrushchev's offensive, i.e., should there be increased military expenditures and if so, on what, etc. He thought that the President is likely to ask my opinion on some of these matters and pointed out how difficult the decision will be because of the consistent position of the administration that our military expenditures are adequate. He then said it was important for me to be in Newport to add to the impression that the President is working. He said he didn't mind the President playing golf and understood he needed it for his health, but he found it unfortunate that the President never has enough time even when in the city to discuss matters thoroughly. He said Tom Stephens would always allot him too little time because the President had such short office hours. Of course, Eisenhower was a far better man than Truman and might have been recognized as a great president if it wasn't for the events of the last eight months, but he, Gordon, felt that actually Truman did a better job of the presidency because he really applied himself, attempting to understand the problems which he had to resolve. I expressed my sympathetic understanding of his concern and said that what upset me since getting into the inner circles of government was the effort of the top people to make meaningless compromises before going to the President, rather than presenting him with sharp alternative choices. Gordon spoke about his efforts to prevent what he called paralysis of government when the President goes on vacation and complained that after the President went to Newport he tried to hold two NSC meetings, but the vice-president refused to attend, claiming that he was too busy, and so both NSC meetings had to be off the record, in other words not contributing to the record of the administration.

> Some time after the Paris Summit I had a conversation with the President—not recorded in the diary—which is relevant to the above. I was alone in his office on some unrelated matter when he said, referring to the U-2, that we scientists had failed him. I responded that the scientists had consistently warned about the U-2 eventually being shot down and that it was the management of the project that failed. The President flared up, evidently thinking I accused him, and used some strong uncomplimentary language. I assured him that my reference was to the bureaucrats that ran the show. Cooling off, the President began to talk with much feeling about how he had concentrated his efforts the last few years on ending the cold war, how he felt that he was making big progress, and how the stupid U-2 mess had ruined all his efforts. He ended very sadly that he saw nothing worthwhile left for him to do now until the end of his presidency.

∾ Sunday, July 24, Monday, July 25, and Tuesday, July 26, 1960

Left Sunday for Camp David and pleasantly loafed with our staff at the pool. After dinner, by which time the other members of PSAC arrived, had discussion of the impending NSC meeting, namely the item on nuclear test cessation. I succeeded in presenting myself as having been put into grave jeopardy by that briefing paper on the test ban, given to the President at the Newport PSAC meeting. Could see that Panofsky and Keeny were thoroughly uncomfortable and thoroughly enjoyed it.

Monday was spoiled by having to fly to Washington and then on to Newport. Contrary to expectations I didn't get into heated arguments, as McCone and Gates were discussing other subjects. McCone talked to Herter about the test-ban topic, but referred to our briefing paper only glancingly. He said, however, that Mr. Shepley, now working for Nixon, showed him a draft of a reference to Project Plowshare in the proposed acceptance speech.

According to McCone the claims were so extravagant that he had to tone them down drastically (wonder what the original contained?) as otherwise Nixon would have gotten himself into trouble. Especially as he would have announced the building of the canal through Mexico without any prior discussions with Mexico.

J. R. Shepley, journalist.

The NSC meeting never came to the nuclear test cessation. Dillon spoke of his talks with the foreign minister of Austria, who thinks of himself as a go-between between the West and Khrushchev. According to the Austrian, Khrushchev asserts that the U.S.A. will not be able to move until after the change of administration. In the meantime he can make moves on Berlin so long as they are not too provocative because ''the Allies won't fight over who is inspecting their passports.'' The President got quite annoyed and asked Herter to deny this in a public speech. He then spoke about his being still the President and being ready and willing to make critical decisions up to January 20. He said he will give the necessary intelligence information to the two candidates, but he would not ask for their advice on any decisions, because they are only ''engaged in cutting each other's throats.'' Herter raised the tragicomic problem of [Patrice] Lumumba, the premier of Congo [now Zaire], who arrived here to visit the UN but is penniless and has nobody to support him. The President said we should entertain him and give him a good time on a modest scale. The general estimate was that he is quite unbalanced. There was a long discussion of a Mr. Detweiler, who signed a $2 million contract with Lumumba, but who is a promoter without money and backers.

Gates said that after extensive discussions with the JCS, he wants to recommend strengthening our military readiness. He admitted at least six months' delay of the Atlas base activation date and proposed a series of interim measures to compensate for this. The President was unhappy about the situation and wanted to hear about ''feasible'' remedies. I suggested declaring Atlas bases operational even though they were not in the condition of ultimate reliability. This started quite a discussion between the President and Gates and ended with the President firmly saying that he wanted done what I suggested. The concluding words, by Gates, were: ''Yes, sir.'' So I wonder where I stand now in the Pentagon. In the discussion the President emphasized the need to strengthen our deterrent rather than other things in the military establishment and Jim Douglas said that in his opinion this was the only thing that was of interest to the American people. Herter talked about the OAS organization's impending action on Trujillo and the President stated that in his opinion Trujillo must be eliminated first before tackling Castro, because otherwise we will enrage all the Latin Americans. Herter said that Representative Cooley allegedly gets money regularly from Trujillo, and so do others in Congress. The result is that it is very difficult to get any action there. The President got very angry and said that he won't let even his agency heads take action in the matter of Trujillo and Castro and that all plans should be brought to him for decision. The flight back was uneventful and I had a drink and dinner in Camp David before giving the PSAC an account of what happened on the nuclear test cessation problem, which of course, was that nothing happened.

On Tuesday we had one of the best PSAC meetings I remember.

> My recollection about the somewhat cryptic entry on Castro and Trujillo is that there was some discussion about the use of force and that this was what made the President angry.

∾ Thursday, July 28, 1960

12:00 M–2:00 P.M.   Luncheon with the Kimpton committee, which is to report on NASA research to Glennan. Cordial discussion of purposes and weaknesses of NASA, in the light of the basic issue of its relation to university research people.

L. A. Kimpton, industrialist.

4:30–5:00 PM   Visit from Staats and Bill Carey, mainly relating to the reorganization of the standing committee of FCST. Evidently BOB wants to get into the saddle as far as administrative aspects of R&D are concerned and wants to use parts of the standing committee for this purpose. This may be a good move, but I emphasized that whatever is proposed should be tied to the Federal Council (because then it wouldn't be in conflict with our paper on R&D organization).

William D. Carey, assistant director, BOB.

CHAPTER XII

The Last Tasks

WITH THE COLLAPSE of the Summit conference, the test-ban negotiations grounded, and the approaching end of the administration, our efforts focused on a variety of tasks, which looked
toward more effective administrative controls and the establishment of greater coherence in our military and civilian research
programs.

One of the major unfinished tasks at this point was our effort
to put order into the chaotic proliferation of projects for the use
of outer space for military intelligence purposes.

The Discoverer project, involving the recovery of photographic
film retroejected from the satellite, was suffering from mysteri-
ous failures in orbit and from losses of the photographic packages over the Pacific. The more advanced SAMOS project was
actually a large collection of efforts, each responding to some
particular military requirements, over which no effective controls
had been exercised so that some plans were totally unrealistic.
The multiple layers of military hierarchy involved caused delays
by sudden changes in specifications, and other difficulties. It was
this situation that our panel was trying to change; the Air
Force, aware of our intentions, was coming up with its own plans
to clean up some aspects of the situation.

Our efforts coalesced, and we could recommend a clear administrative arrangement, bypassing various military echelons
and enabling the individual in charge of development projects to

report directly to a new National Reconnaissance Office within the Office of the Secretary of the Air Force. The cleanup of the technical proliferation was somewhat less complete, as the reader will learn later.

The President asked us to report to him on the probable costs of the various man-in-space projects or proposals of NASA, and this we did.

The PSAC report on "Scientific Progress, the Universities and the Federal Government" (the Seaborg panel report), in the development of which I took great personal interest, went through the clearing process by the many interested parties in the government, not the least by the President, who took it very seriously and finally warmly endorsed it for publication.

Unbeknown to me, the Navy, seeing itself as an underdog, stimulated the President into sending me to the headquarters of the Strategic Air Command where a large joint SAC-Navy-Army team was developing the first Single Integrated Operational Plan (SIOP) for waging strategic warfare. The President gave me broad authority to inquire into what was being done and with my chosen associates, H. E. Scoville, Jr., and George W. Rathjens, I spent three days at the Offutt Air Force Base, following on a reconnoitering visit by Rathjens, and then prepared a detailed report to the President.

PSAC panels, our staff, and I remained busy in the fall of 1960 with those parts of the budget related to the support of science and with more immediate research and development projects. I recall that we provided substantial inputs and that our influence was not negligible but it was all done with a feeling of unreality; we knew that after the election of Kennedy the Eisenhower budget would have a very short life span.

The diary contains very few entries dealing with the 1960 campaign. I was rather walking a tightrope, with Eisenhower's knowledge, remaining nonpartisan so that the president-elect, whoever he happened to be, would not liquidate PSAC and the special assistantship as an unfriendly outfit. After the President excused me at my request from participating in Nixon's campaign, I ceased to be invited, as I recall, to the cabinet meetings where the Republican campaign was being discussed, but otherwise this aloofness had no effect on my relations with the President, my regular attendance at NSC meetings, or other prerogatives. I am sure that my neutral position was frowned upon by quite a few members of the White House staff, but I was not criticized to my face.

Friday, July 29, and
Saturday, July 30, 1960

At Cape Canaveral and at sea. Spent Friday morning and all of Saturday in the SSN *George Washington*. The missile launching had to be postponed Friday because of bad weather, but the crew performance and functioning of the system were both most impressive on Saturday. It is really a great achievement. Got to Washington 3 :00 AM on Sunday.

> This was the first launching of the Polaris missile from a submarine.

Monday, August 1, 1960

Morning meeting in Killian's office in Cambridge on SAMOS. My impression is that the group is moving ahead well and is coming up with good recommendations, but the report they are preparing has to be tightened.

Afternoon in Newport (picked up by helicopter on the M.I.T. athletic field) at NSC meeting. Not terribly interesting. It started by Jack Irwin's briefing on our commitment to NATO, which was concluded by the remark of the President that it was high time the emergency sending of our troops to Europe in 1950 was reversed.

Next a briefing by Dillon on military assistance to five lesser European countries which, however, have been certified by the Treasury as able to pay for weapons themselves. Dillon emphasized that this aid had to be continued for a number of years as the countries were actually not capable of building up their own military forces. Secretary Anderson, starting with the remark that he would be the lone dissenter, launched into a vigorous attack of the proposal and recommended that we cut out, or at least cut down this aid because of the precarious state of our trade balance and the need to increase support in other areas, like South America. Dillon merely grinned his engaging grin, and the affair ended with the President saying that in principle he completely agreed with Anderson, but that as a matter of practice he was going to continue the support.

Then the President launched into a vigorous statement that he wants to give nuclear weapons to our NATO Allies, and over the silence of the members of the council he instructed McCone and Gray to prepare legislation for submission to Congress. I understand that the planning board meeting on Tuesday was the most violent ever, since both State and AEC violently oppose such legislation, although they kept their mouths shut at the NSC meeting. Jerry Smith, who sat next to me, leaned and whispered that this will be the end of NATO.

On the flight back to Washington sat next to [C. H.] Timberlake, our ambassador to the Congo, who made a brief presentation to NSC and who talked to me at length about the situation in the Congo. He is very charming and seems to be very capable. Maybe this opinion is influenced by his telling me he heard so many nice things about me from Ambassador Dowling in Bonn. Anyway, I found his talk fascinating and the problems faced by the Congo challenging. He was rather optimistic about our ability to keep communism away from there.

∽ Tuesday, August 2, 1960

Principals' meeting in the morning, which ran its usual indecisive course, with nothing decided except to insist that the Soviets answer our proposal on safeguards, which they did later in the day anyway, and so will require another Principals' meeting later this week.

Federal Council meeting in the afternoon, at which the Seaborg report ["Scientific Progress, the Universities and the Federal Government"] took up much time. On the whole, surprisingly favorable reception and there is a good chance that this report will be endorsed by the council. The meeting was about the best since I started chairing the council.

Then a long phone call from Joe Charyk, telling me about the plans of the Air Force in connection with the SAMOS satellite project. Clearly, they are trying hard to freeze the management of SAMOS in such a way that our briefing to NSC couldn't change it.

ᴏᴄ∿ *Wednesday, August 3, 1960*

The notable event of the day was a series of phone calls from such as Charyk and Getting, the result of a rumor spreading in the Pentagon concerning the supposed recommendation of our SAMOS panel to transfer its management to CIA. I assured everybody of my innocence, but urged Charyk that the organization should have a clear line of authority and that on the top level the direction be of a national character, including OSD and CIA and not the Air Force alone. Quite obviously, the Air Force is trying to freeze the organization so as to make a change more difficult by the time the NSC is briefed.

Left early with Walt Whitman and was joined in my apartment by Keith Glennan. Those two then got to talking about McCone. I was interested to observe that Glennan's opinion of McCone is even lower than mine. He considers him thoroughly unreliable and unpredictable.

> Walter G. Whitman, professor of chemical engineering at the
> Massachusetts Institute of Technology and former chairman of the
> Defense Research and Development Board; successor to Brode
> in the State Department; Ivan Getting, president of Aerospace
> Corporation.

ᴏᴄ∿ *Thursday, August 4, 1960*

Talked over lunch with Bob Bowie, who is preparing a report for the State Department on the future of the NATO alliance. We discussed scientific and educational aspects of NATO and I argued that a European MIT, starting with the graduate school, is about the most feasible and most promising joint NATO venture, rather than a European university, which has been discussed, that would include humanities, etc. We then came to military topics and found ourselves in complete agreement about the futility—in fact danger—of building up large stocks of nuclear weapons in Europe, but the need to have some of each kind for an effective deterrent against aggression. We also agreed that the nonnuclear capability should be greatly

strengthened (a conclusion which was later in the day rejected
by Secretary Douglas in my meeting with him).

> Robert R. Bowie, director of the Center for International Affairs,
> Harvard University, and formerly in the State Department.

↪ Friday, August 5, 1960

Listened to Air Force briefing on the ARDC space program
and was shocked by the incredible wastage of taxpayers' money.
For instance, $8 million spent in paper studies such as lunar
defense systems.

↪ Monday, August 22, 1960

I returned to Washington Friday night after a two-week vacation, interrupted by a short visit to Washington to attend an
NSC meeting—shifted, however, to another day after I had left
Cambridge. I was also questioned by FBI agents about my possible connection with a leak to the *New York Times* concerning
what transpired at that NSC session, information which I eventually received by phone from Spurgeon Keeny. Also received a
phone call from Tom Gates who proposed that I temporarily
assume Herb York's duties, for York has had a heart attack and
will be out for several months.

On Saturday after a two-hour briefing by Spurgeon and Dave
on everything that went on in my absence, I went to see Gates,
who meanwhile decided that recruiting me would be impracti-
cable and has decided to make John Rubel the acting director on
a temporary basis. I offered him technical help from our panels,
which seemed to be accepted enthusiastically. The most interesting point of Spurgeon's account is that during the NSC meeting,
both the President and the vice-president came out in favor of
not carrying out any nuclear explosions in the VELA project
until after the elections, which made McCone unhappy.

Today had a long visit with Secretary Charyk in company with Land, Overhage, Watters, Rathjens, and Bissell. We went over the briefing about intelligence satellites for the Security Council on Thursday, so it is now in good shape. Last Thursday on my ''vacation'' I attended an all-day meeting in Land's office in Cambridge, where the same people, minus Bissell, were present. At that time the whole briefing paper was still in such a lousy state that I spoke rather harshly. I gather the result of my uncomplimentary remarks was that Land, Overhage, and Watters worked on the paper all through the weekend and with good result. I think that on the whole the study has come out very well, because we will make an unanimous presentation, and the Air Force, i.e., Charyk, have been sufficiently influenced by our findings to develop a plan which both technically and in terms of management will be endorsed by our panel.

> Carl F. G. Overhage, director, Lincoln Laboratory, Massachusetts Institute of Technology.

Had a phone call from Dave Kendall suggesting that in view of Wiesner's involvement in the Kennedy campaign, he should be excluded from PSAC activities until after the election. It took some persuasion to convince him that this would be (a) very awkward, as I would have to explicitly invite Wiesner out of our stated meetings, and (b) would be unnecessary since everything we do is classified and, therefore, Wiesner could not and would not use it in the campaign. I also noted that about half of our committee are Democrats, to which Kendall replied: ''And don't I know it well enough!'' After some discussion, he accepted my recommendation and approved our continuing as normal, providing Wiesner's political activities are nonpublic.

Had a visit from Glennan, who talked about moving ahead on communications satellites. Evidently my prodding had substantial results. He then discussed the awkward situation in which he and Dryden find themselves because they know nothing about Discoverer and SAMOS. I assured them that I would try to make arrangements and let them in on the secrets.

Had Dick Morse for lunch, at which time he told about his continuing troubles with Brucker, whom he rates as acting dishonestly and impossibly, since he can be persuaded to accept good proposals only by threats of public resignation. In this way Morse managed to get Brucker to accept some of his recommendations for reorganization of Army R&D.

Tuesday, August 23, 1960

At lunch talked to Jim Hagerty, who is worried that the Russians will put a man in space just before election and so compromise Nixon's chances. I argued that the sensible course for us is to start de-emphasizing Project Mercury and prepare everybody for this unpleasant event. He agreed.

3:00 PM Went to pre-press meeting in Persons's office, at which the same subject was discussed and everybody agreed I was right, so I was told to prepare a statement for the President's press conference the next day.

4:00 PM Mr. Murray Marder of the *Washington Post* to find out what I am doing for Nixon as a science advisor. I should note that there was a press release to that effect by Nixon's press officer Herb Klein. At my request Dave Beckler called Klein a few days ago to inform him of Marder's forthcoming visit and of my intention to be honest and explain that all I am doing results from our both being members of the same administration. Mr. Klein accepted this, and I told so to Marder. After that I fenced off a lot of questions, including one as to whether I was a Republican or Democrat, although I suppose he can find out I was once a registered Republican.

Then prepared a rather good, if I may say so, statement for the President on our space effort and went to Glennan's office to listen to an analysis of how the Russians achieved what they have achieved. The meeting told me nothing new, but I cleared my statement with Glennan.

Had a phone call from Tom Gates, who started in a highly irritated tone that he heard we are briefing the President on limited warfare, and wasn't that completely improper because that was the province of the Defense Department. I said that I had shown our summary report or briefing paper to Jim Douglas three weeks ago and left it with the request that he, Gates, see it also. Gates denied having heard about it, but finally remembered that Douglas had spoken to him about the paper and softened a bit, although he suggested that General Twining should be invited. I demurred on the ground that we are talking weapons technology rather than strategy and tactics.

Then followed a near comedy, because Tom Stephens phoned me that the President wanted to change the briefing on limited warfare from 11:15 AM to 8:30 AM tomorrow and I couldn't locate Robertson to rehearse the meeting with him in the evening. Did a lot of telephoning and finally reached Bob to have him meet me at 7:30 AM in my office to rehearse.

⌒⌒ Wednesday, August 24, 1960

8:30–9:15 AM   Meeting with the President. Present were also Gates, Douglas, and Gordon Gray, the former two looking most disapproving. In the beginning the President looked rather grim, but Bob Robertson rather skillfully relaxed him by references to bygone days of WW II when he had met General Eisenhower, and the briefing went off exceedingly well, the President being friendly and sympathetic. Afterward Gordon Gray said he thought that the meeting was helpful because it will force the Defense Department to come up with its own promised briefing on limited warfare, on which they have been dragging their feet for more than a year. At that time our report could be used as source material for a directive by the President. As a matter of fact, at the end of the meeting, the President, referring to the promised JCS briefing, said that he wanted Robertson and me to be present. During the briefing, the only interruption from Defense was when Bob spoke of undue secrecy hindering the development of optimum fragmentation weapons, and Gates said that that wasn't so, because they were already at the disposal of the troops, and Bob pointed out that factually this was not yet the case, since the commanders in the Far East don't even know what they are. Thereupon the President commented on the occasional silliness of overclassification and told a story of how a few boxes of proximity fuses were left to the Germans when we made a temporary retreat in Alsace-Lorraine during the German Ardennes offensive. This caused agony in the top echelons of the allied forces, but when we reoccupied the area, we found the boxes still where they were left originally. After the meeting, Robertson and I patted each other on the back. As we were leaving the President's office, he came to me and said he hadn't seen me for quite some time, to which I responded that I had had a vacation, and he said: ''You should have more of it.''

9:45–10:30 AM   Pre-press briefing of the President, who looked over my statement about Project Mercury, etc., and said that he was going to release it to the press.

At 3 :00 PM in Charyk's office, where I met with Killian, Land, and Watters, we went over the details of the briefing on SAMOS with Charyk and I discovered that Killian proposed to make the introductory remarks, which by every tradition I should be doing. I realized that there was some tension between him and Land as to who was to make the presentation and decided to be gracious and not lay any claims of my own.

4 :30–6 :00 PM   Briefing rehearsal on SAMOS before Gates and Douglas, which went off smoothly. The only discussion was of the proposed management, when Gates insisted it be of the simplest possible form, not resulting in policy boards. I am not very happy about that, but at least we are getting our point that there should be little interference from various military echelons.

Thursday, August 25, 1960

8 :15–8 :30 AM   In President's office with Dulles, Killian, Land, and Gray, to show him certain intelligence information and remind him of certain projects related to SAMOS which we propose not to discuss at the following briefing, but which influenced our recommendations.

8 :30–9 :30 AM   Special NSC meeting for the SAMOS presentation, which went exceedingly well. Land brought the house down in laughter when Stans was temporarily called out and Land remarked that Mr. Stans was leaving while it was still cheap. When we recommended that the line of command be directly from the secretary of the air force to the officer in charge of the project, the President remarked that this is the way to do it, but of course, it was an internal matter for Defense, and Gates said flatly that that is how it will be. At the end without further ado the President said he approved all our recommendations. Turning to us he also said that the only unfortunate thing was that we didn't make these recommendations two years ago, so that he would be the guy to see the lovely pictures we will be making. As it is, it will probably be Dick and not him, and he won't even have the clearances necessary to see them. After the meeting, the vice-president was most complimentary on the presentation and on the recommendations.

I discovered later that the recommendations were randomly

scattered throughout the text of the briefing paper, rather than pulled together at the end. After a little difficulty with Killian and Land, who wanted me to rewrite the recommendations even more strongly than Land had stated them, I pulled them together and wrote them in the form in which they were presented. Took a little time to convince Jimmy Lay that all of these recommendations had really been made, but I guess I managed it. A lesson never to trust anybody to write a formal report without checking it myself. I certainly thought that between Killian, Land, and Watters they would know better than to leave the paper in such a casual form.

James L. Lay, Jr., executive secretary of NSC.

Had Si Ramo for lunch, who talked about STL's proposal for AICBM and its importance. It consists of a swarm of tens of thousands of little satellites that would pounce upon any missile as it is being launched. I think it is fantastically expensive, but I may be wrong and if I am, the whole idea may be important.

Am ending the day with a feeling that in the last couple of days we have accomplished quite a bit. If the Defense Department really sticks by its agreement with our recommendations on SAMOS, which will now be reinforced by an NSC directive, this may be a major accomplishment of my eighteen months in office.

Friday, August 26, 1960

9:00–11:00 AM   Cabinet meeting, most of which was taken up by an effective presentation by General Quesada of plans for the Dulles International Airport. Except for complaints from the President about the modernistic building design, everybody seemed to be very pleased. The President began the meeting by a violent condemnation of Congressman [O. E.] Passman, whom he described as the lowest form of SOB, etc., because of Passman's blocking of the foreign aid bill. He spoke bitterly of the ability of a few vicious and incompetent people in Congress to block desperately needed programs. After the meeting, Doug Dillon told me that they are activating the office on arms limitation in State as of September 1, even though they don't have the directive. I said that we shall, of course, give all help possible, but that we don't think it will work. With a grin, he agreed with me.

Had a visit from Dick Leghorn, who covered two subjects. One was a current photographic development of his company, Itek; the other is his proposal for a speech by the President to the UN, offering all the data obtained with the aid of SAMOS, etc., as an aid in peace preservation. I read the proposal and it is a very bold and interesting one. Regretfully, I had to tell him that this is something that only the incoming administration could consider seriously.

Monday, August 29, 1960

This has been a quiet day. Of interest was a meeting in Jerry Morgan's office, with General Quesada, somebody from NASA, a general from the Air Force in charge of the B-70 program, Stans, and Persons. The subject was a request by Quesada that the President ask him in writing to assume leadership in determining the proper national policy with regard to development of a supersonic transport aircraft based on the B-70. Interesting was that Persons, Morgan, and Stans all rejected the proposition, the reason being that they didn't want to do anything that would look like a commitment on behalf of the next administration. Good old caretaker spirit! Quesada as usual was very persuasive, but he didn't win this one.

Had lunch with Don Paarlberg, who has a violently negative attitude toward Department of Agriculture's research. He says that no matter what they say about changing the program, it is always the same old stuff that they are doing, and besides, he doesn't believe that we need much agricultural research because we will have enough foodstuffs in the foreseeable future. Clearly the recommendations of our Life Sciences Panel have fallen on deaf ears here.

Tuesday, August 30, 1960

Had Carmichael for lunch. He talked much, but here and there and edgewise I managed to convey to him the contents of the PSAC paper on government organization for science, as regards the role of the standing committee of the FCST, and to my relief found that he doesn't expect the standing committee to accomplish much and is quite agreeable to downgrading it. Each time Leonard and I are alone, he has some unfriendly things to say about NSF. Not Waterman, who he thinks is wonderful, but the staff, who he says are a bunch of incompetent empire builders. We then talked about the role of the special assistant and he was horrified to learn that we believe it should be strengthened in the next administration. He believes there is a danger that the next president will select a politically desirable nincompoop and that this job is too influential to permit such a man to have even more authority.

Saw Goodpaster, who told me I have to clear it with Herter and Gates before inviting Sir Solly and his gang from England for discussions here, but otherwise the plan was OK.

Ended the day with a nearly-two-hour discussion with Mesthene, Beckler, Westrate of the papers being prepared by the Federal Council at the request of NSC on the subject of strengthening American and Western science and technology. Mesthene is having a rough time, as I expected, because of the tendency of the agencies charged with writing the papers to write nothing but platitudes and to avoid bringing real problems into the open. If Mesthene can succeed in producing a substantive paper, he is really a first-class staff man.

> Emmanuel G. Mesthene and John L. Westrate, members of PSAC–FCST staff.

Wednesday, August 31, 1960

Another slow day. Spent much time on the telephone arguing with Dartmouth people about signing for them an unconditional release for the use of TV rebroadcast of the convocation which is to take place ten days from now. Refused to sign and asked to be excused from participation. Finally acceded after repeated arguments and agreed to go there with the provision that I will be excluded from TV.

Jim Hagerty probably did not trust my Republican sentiments on a TV panel show.

Had a pleasant lunch with Whitman and Farley at which we touched on a number of topics, including the problem of halting nuclear weapons tests. Both Farley and I agreed that the thing is in a hopeless mess and that we have been boxed in by the Soviets pretty badly. Conceded that except for the effect on world opinion it might be best to resume underground testing and then make a second try at a treaty. Whitman says he is enjoying himself hugely in his new job and spends most of his time making personal acquaintances in and out of State and likes State much better than Defense. He says people are friendlier and less formal.

✺ Thursday, September 1, to Monday, September 26, 1960

My notes for this period are very brief and lack substantive information. I accompanied the President on a trip to the former Redstone Arsenal in Huntsville, Ala., which the President dedicated as the George Marshall Space Center of NASA; I attended a convocation at Dartmouth College; made three other out-of-town speeches; and spent several days in Cambridge. I also chaired a two-day PSAC meeting and attended an NSC meeting and sessions of several PSAC panels examining parts of the budget for fiscal 1962, this time dedicating more effort to civilian R&D than a year ago.

✺ Tuesday, September 27, 1960

12:15–1:30 PM   Lunch with Glennan and Dryden. Discussion of the magnitude of commitments we are making in outer space.

I told them of the undertaking of PSAC to make an estimate for the President of the cost of the man-in-space program of NASA during the next ten years. They didn't object, but felt we should include military programs as well, to which I demurred since military programs cannot be predicted in advance. We agreed that even the civilian program represents a very major commitment and that this should be made clear to the incoming administration.

1:30–4:30 PM   FCST meeting. The Seaborg report was discussed at considerable length, but except for one proposal by HEW, which I rejected, the proposed changes were minor and acceptable. The council unanimously voted to recommend to the President a public release of the report, which is a victory. Then Mesthene outlined to the council how he will summarize the nearly 300 pages of papers prepared by the agencies on the general subject of strengthening the Free World position in science and technology (a statement of problems and issues). Discussion was mild and there was general acceptance of his plan. Other items were dealt with quickly and didn't amount to much, except perhaps the notice on the council by Elmer Staats, which he and I prearranged a few days ago in a private meeting, that the BOB wants to strengthen its activity in the area of management of research and its administrative problems and intends to do it in close coordination with the council, using an advisory group made up essentially of the members of the standing committee. An interesting sidelight on the standing committee emerged from the brief report by Carmichael, whom I had warned in private a couple of weeks ago that the standing committee wasn't doing enough to justify its existence. Carmichael promised vigorous action, so the committee is obviously worried. Prior to the meeting, Bob Kreidler warned me that through underground channels he heard that Secretary Ray of Commerce would disapprove the Seaborg report's publication on the grounds that it conflicts with the vice-president's statement on basic research. I instructed Kreidler that when he presented to the council the changes made in the Seaborg paper since the last draft was distributed, he should point out, without waiting for Ray's reaction, that the final version of the vice-president's paper is not in conflict with the Seaborg report; and that he should read excerpts from the vice-president's paper. Bob did it very well and as a result, Ray did not object to our paper as a whole.

Had a phone call from Jerry Wiesner, who said that he was in a quandary because he is getting more and more alarmed about some of the things that Kennedy is saying, obviously under the influence of the advisors around him, like Symington, Murray, etc. That he, Wiesner, is just not in the inner circle and has no way of communicating with Kennedy. That his conclusion is that he should join Kennedy's party for a while. He asked what reaction in the White House would be. I said it all depended on whether his presence would be made a source of publicity or not. I expressed my grave concern about the lack of competent advice that evidently both candidates are suffering from and suggested to Wiesner that he act as his conscience dictates, but try to arrange it so that his presence in the Kennedy party remains in the dark. Finally, I suggested that if, through inadvertence or by design, his presence is made a source of publicity, he forthwith send me a letter describing what happened, noting that it happened against his wishes, but requesting instructions from the White House on whether he should resign from PSAC or not.

> I recall speaking about Wiesner's activities with the President and his agreement, after some hesitation, that Jerry remain on the PSAC so long as he does not publicly criticize the Eisenhower administration.

⌇ *Thursday, September 29, 1960*

8:30–10:00 AM   NSC meeting, started with a Dulles briefing. It appears that in Laos the Pathet Lao movement has scored a substantial victory and the forces of the Western-oriented government have suffered a major defeat. Things certainly look bad. Also from Cuba and the Congo the news is anything but cheerful. Most of the meeting was devoted to discussion of policies toward various countries and was not interesting, but at the end the President emphasized his desire to liquidate gradually our foreign bases for SAC as unnecessary and a source of many difficulties and a great expense.

2:00–3:30 PM   Walsh McDermott, Skolnikoff, and Beckler to discuss the new technical assistance panel. I invited McDermott to become chairman and he agreed. After discussion of the magnitude of the problem involved and difficulties of identifying particular tasks of the panel, we decided to start with a one-shot

meeting of quite a few people, such as I arranged before setting
up the Life Sciences Panel. At this meeting we will hear ideas
various people have, see how much they are interested, etc., and
on that basis will select the members of the panel.

> Walsh McDermott, professor of medical sciences at the Cornell
> University Medical College.

3:30–4:30 PM   Bill Baker about General Greer's proposed
plan for developing SAMOS satellites. In September General
Greer and his staff briefed me on their plan, which shocked me
because it involves development of at least nine different satellite
payloads. It looked like an appalling proliferation and is certainly different from the straightforward plan recommended by
our panel. Greer told me that he had briefed Charyk the day
before on this plan and Charyk approved in general. Unfortunately, Charyk left for the Far East, so I haven't been able to
discuss the plan with him. Last weekend in Cambridge I had a
long session with Din Land, and he also expressed his misgivings
about the plan. Baker's reaction was even stronger and so he and
I went to see Goodpaster. I let Bill do most of the talking, since I
had already talked about the matter with Goodpaster. Goodpaster now decided to telephone Secretary Douglas and warn
him that this is a bad plan, also because it involves some programs "in the black" [i.e., secret]. These are considered by the
President to be improper for the military and he also believes the
military incapable of keeping developments in the black, because
of their irresistible urge for publicity. I asked Goodpaster to be
cautious so it wouldn't look as if we were going over Charyk's
head and he agreed to bear this in mind.

Bill Baker and I talked also about the utilization of the photographic "take" and Goodpaster agreed that this will not be left
in the hands of the Air Force, as is being pushed by General
Walsh. This reminds me of an interesting episode during September. I had a visit from General [Robert] Smith of SAC, who
began by saying that they are aware at SAC of my trying to
reorient the SAMOS program so as to exclude SAC from intelligence activities. Bob thereupon pointed out the great difference
between what he calls development and technical intelligence on
the one hand and operational intelligence on the other. He maintained that only SAC could utilize the "take" for the latter. I
agreed with him and explained the substance of our proposal,
approved by the President, which left him satisfied. I then

learned from him of a violent struggle going on between SAC and General Walsh, the latter being anxious to control all Air Force intelligence, as well as the processing of the "take," whereas SAC is merely anxious that it have the right to study the "take" independently from other centers. Our meeting ended on a friendly tone, Bob urging me repeatedly to visit both SAC and the Vandenberg base. He assured me that General Power does not think for a moment that I am subversive. This refers to an unpleasant remark he made to me in the summer of 1958 when the interagency committee preparing the conference on the prevention of surprise attack visited SAC headquarters. Power at that time suggested to me that this activity was subversive because it could end in nothing but the weakening of SAC, and at a later date I said in Geneva to General Butch that I still resented the remark.

As a result of the President's instructions following our briefing on August 25 the Air Force reorganized the management of the intelligence satellite projects, putting Brig. Gen. Robert E. Greer in charge, who was reporting to a National Reconnaissance office within the Office of the Secretary of the Air Force.

Brig. Gen. Robert Smith was one of the many military who accompanied us to the surprise attack conference in Geneva in 1958, as was Lt. Gen. F. H. ("Butch") Griswold, deputy commander of SAC.

Friday, September 30, 1960

9:00–10:15 AM   Meeting with Jerry Wiesner and Ed Gullion of the newly established disarmament office. Jerry and I made some impression on him with our argument that he not set up a separate military division in addition to political and technical, but should organize the office along functional lines, mixing military with civilians to prevent their simply following JCS orders. On the subject of what kind of scientists he needs, we made less of a dent. He talks about "names" and is rather naive about the chances of getting them, as well as about the costs of modern R&D budgets that his office will need.

Pending the establishment of a statutory disarmament agency, the State Department created an office or desk to deal with disarmament, with Edmund A. Gullion (later our ambassador to the Congo) in charge. The office was in the organizing stage to the end of the Eisenhower administration.

10:30–11:30 AM   Admiral Raborn. He is frightfully exercised because through unofficial channels he has heard that John Rubel is trying to suppress the long-range Polaris missile project and push instead Air Force programs like the Skybolt and the Midgetman. He feels Rubel is influencing the secretary of defense. He is especially indignant because when Secretary Douglas was on board the *Patrick Henry* Polaris submarine and both missile launchings failed, SAC headquarters was informed by direct wire from Cape Canaveral about it and held a big celebration that evening. All this would be very funny if it weren't a little sad. I explained, deliberately minimizing my influence in this matter, that I was on his side.

Midgetman—a sort of miniature Minuteman—died unborn.

Captain Aurand, who was with Raborn, stayed behind and picked up the subject which we had discussed in September when he flew me to the Dartmouth convocation in a Navy trainer jet. On specific instruction of the President, the target planning for strategic warfare operations has been made an inter-Service task, with General Power as boss and an admiral as deputy, but with a staff largely from SAC. Admiral Burke is worried that the SAC will do a snow job and the first time Pete Aurand talked to me, he asked me whether I could set up a group to investigate the validity of methods used by SAC in determining nuclear warhead yield requirements for specific targets. I then said that this would be utterly improper for me to initiate and that only if the secretary of defense or the President specifically requested or ordered me to undertake such a study, would I do so. Pete told me now again that he had talked to Admiral Burke, who is getting more and more worried that SAC will develop demands for such large yields that enormous further expansion of SAC forces will be established as a national requirement.

11:30 AM–1:00 PM   Two staff people from BOB with their study of the government administrative setup for management of science and technology. Clearly an attempt to do one better on the PSAC paper on the subject. I played it coy and suggested that the conclusions would be different, depending on whether the concern was to balance the budget or to ensure such growth of science and technology as is necessary for our future leadership. We talked about many topics, but I doubt whether I made much of a dent.

2:45–3:15 PM   Ruina, who is in charge of electronics in York's shop, telling of his hopeless struggle against the military, who try to kill research in favor of systems development. I asked him to write me, anonymously if he preferred, citing as many concrete examples as he could find.

Dr. J. P. Ruina, electrical engineer. I never got his memo.

3:30 PM   Went to the review and reception on the retirement of General Twining. Fine ceremony. Afterwards stood in line at the reception, and who should join me but Secretary Brucker, who suggested that since his wife was in Canada perhaps I would be willing to act as one for him. All sweetness and light. He was most distressed to learn I am leaving Washington in January, etc. Afterward a brief chat with Dick Morse, who said that according to Secretary Brucker, I am the extreme example of un-American behavior in Washington.

Monday, October 3, 1960

This has been a lousy day!

Went to see Secretary Douglas and John Rubel about General Greer's plans for SAMOS and about our involvement in the military R&D budget. Douglas was not very explicit, but conceded that Greer may be spreading out too thinly and agreed to hold off endorsement of the program until the return of Charyk, which was what I requested. But on the subject of the Air Force being capable of successfully hiding development programs "in the black" he was insistent that it could be done—over my grave skepticism. He had to leave soon, so we never properly covered the budget issue, but Rubel and I discussed it in more detail. I am still in a fog as to what we are going to do.

At lunch in the White House Mess, I made the mistake of mentioning my October 14 speech to Freddy Fox, whereupon both he and Ann Wheaton urged me to include a laudatory reference to the President's birthday. Oh God! So I sent a draft to Freddy with a request to fit in such a reference.

The Reverend Frederic B. Fox, a speechwriter of the President.

Bob Gray, secretary to the cabinet, phoned to exult over the fine report of the Seaborg panel, assuring me that even though he himself was to the right of Senator Goldwater, he agreed with it. Then it developed he wants to put it on the cabinet agenda this Friday, which is terrible because it doesn't give me a chance to bring it to the President's attention privately and to adequately educate cabinet members. The problem which we haven't settled in our minds is whether Gray is anxious to get the Seaborg report published as a political move, or whether he is again lacking subjects for cabinet meetings and is casting about desperately. Friday will show.

Had a phone call from Teller, who told me in high indignation that Bethe's article in the *Atlantic Monthly* on the nuclear test ban is full of factual misstatements, but that Bethe refuses to join him, Teller, in a TV debate and that this is not gentlemanly behavior. I finally agreed to phone Bethe and suggest that he might reconsider but will not urge him to do so.

⟅⟆ Tuesday, October 4, 1960

A day divided between attendance at the PSAC-GAC panel on high-energy accelerators, and working on papers for the Friday cabinet briefing on the Seaborg report.

The meeting of the panel was instructive at times, but not throughout. Eugene Wigner rode his hobbyhorse, that work on high-energy accelerators takes scientific personnel away from more important tasks, and hung on tenaciously to this subject. He doesn't seem to listen to other people's arguments.

I told Bethe of Teller's phone call, and in the discussion with Hans became myself convinced that nothing good could come of the proposed TV debate. So I phoned Jerry Wiesner and withdrew my earlier advice to accept the invitation. Hans is flatly against involvement.

∼ *Wednesday, October 5, 1960*

Took Bob Bacher to lunch at the Mess. He thinks that the ten-year projection of the high-energy accelerators may overestimate the costs because of the success of the CERN machine, which completely revolutionizes our ideas of how to build really high-energy accelerators, so that the Argonne machine begins to look like a boondoggle. Talked about choices for my successor, but without any conclusive thoughts.

∼ *Thursday, October 6, 1960*

9:00–11:00 AM   NSC meeting. Before the first item, the presentation on limited warfare, began, the President turned to Gates and said he had been talking to George and he would like George to go to Omaha, to Offutt Air Force Base, and look at how General Power and the joint group are developing the strategic target list yield requirements for the emergency integrated war plan and report to him the findings. I saw a most pained expression on Tom's face and he said: "Well, Mr. President, Power and his deputy are coming to brief me shortly on what they are doing." A grin spread over the President's face. He said: "Well, and I want George to go there and report to me." Then, turning to Lemnitzer, he said: "And I want the Joint Chiefs to keep a close watch on what is being done, too." This puts me in the midst of the worst hornets' nest, of course. After the meeting I took Tom aside and explained what actually happened. Several weeks ago I went to see the President about something else and out of a clear sky he told me about the new unified strategic target list and the Single Integrated Operational Plan —in principle a really new departure in our military planning since it puts some restriction on SAC—and then implied that he would like me to get involved. I responded that I thought it would be very unfortunate if I did, and gave as reasons that this is a military operational rather than technical problem, and one involving intense inter-Service conflicts, etc. That was all that was said at the time. Then, as I have already noted, Pete Aurand came to me a couple of times egging me on to get involved and I said I wouldn't unless I got direct orders from the President. Obviously he was the guy that started the thing, because he whispered to me before the meeting that he saw the President yesterday and suggested that an impartial evaluation of procedures at Offutt Base would be helpful. I said to Gates that if he could convince the President that my involvement would not be useful, this would only make me happy.

The briefing on limited warfare took five concrete examples, scenarios as they were called, and came to conclusions which weren't very surprising. My own impression was that the conclusions were as follows: If the other side is obliging enough not to do anything inconvenient to us, we will achieve our objectives, i.e., containment and restoration of the status quo, by conventional forces. Otherwise, we have to start using nuclear weapons. If the other side refrains from retaliating in kind, we still can achieve our objectives. If they respond, nothing can be estimated about the outcome, and we must be prepared to engage in all-out nuclear war. In every case, the scenarios supposed that we would use nuclear weapons only after some initial period and only in limited quantities, e.g., in air-to-air battle and against strictly tactical targets.

Lengthy discussion followed with the President saying that he thought the whole thing was very unrealistic and that we were unfortunately so committed to nuclear weapons that the only practical move would be to start using them from the beginning without any distinction between them and conventional weapons and also, assuming there was direct Russian involvement, mount an all-out strike on the Soviet Union. This remark made rather pointless the things which I was going to say, so I kept still. Dillon made a long speech emphasizing that our "first use" of nuclear weapons would cost us our Allies and align the entire world against us. The outcome of this long presentation was an instruction to the JCS to study in more detail the consequences of both sides using nuclear weapons in the cases considered and report back. Then followed a briefing by Dulles. He took up Laos, the Congo, Algeria, and Cuba, and, if one took the verbiage out, the sum total of it was depressing, not to say frightening. In plain words, we are in full retreat, or shall we say the opposing side is in rapid advance on all fronts. What a mess!

In the afternoon I went to Secretary Perkins's office to listen to the Minuteman briefing. A lot of visitors present such as Charlie Lauritsen, Trevor Gardner, and others. According to the briefers, Minuteman is pretty much on schedule and most of the technical difficulties overcome, although the job is not finished by any means. Ended with typical Air Force presentation proving that the cost-effectiveness of Minuteman is several times better than that of Polaris. Wish one could get Air Force and Navy boys in the same room and have them argue out their figures.

Courtland D. Perkins, aeronautical engineer, assistant secretary of the Air Force for R&D.

∾ Friday, October 7, 1960

7:30 AM Breakfast with Jerry Persons, who showed me about half a dozen places in the Seaborg report which, as he put it, would give aid and comfort to the enemy (the Democrats), if taken out of context. I agreed since they are of that nature, but pointed out that these could be eliminated by simple editorial changes. He expressed annoyance about Robert Gray pressing this report on the cabinet.

8:15 AM Went with Persons to see the President. Jerry explained that I wasn't the one who put that item on the cabinet agenda and the President said he had no time to read the report and was unhappy about its being discussed under these circumstances. He then complained that we urged an increase in faculty salaries through government grants, but I reassured him, reading a quote from the report, that that was explicitly not what we recommended. He then said that the problem of patent policy worried him and he couldn't find that mentioned in the report as he leafed through it.

8:30 AM Went to cabinet meeting and the President, as soon as he sat down, announced that the item would not be considered. Obviously his mind was occupied with the problem since he forgot to have the usual prayer for the first time in all the cabinet meetings I have attended.

McCone spoke of his participation in the International Atomic Energy Authority session. Long presentation studded with accounts of our successes over the Soviet objections. Interesting sidelight is that Emelyanov told McCone privately that he is losing his influence and is likely to lose his job. This would be unfortunate because he is one of the few people it has been possible to deal with sensibly. Then followed a long account of UN activities by Wadsworth, who naturally was quite optimistic about our accomplishments.

Monday, October 10, 1960

Was called by Hagerty and Persons and asked to prepare a list of recent American scientific accomplishments for tonight's TV political speech of the President. Did it by 3:00 PM and spent an hour in the President's office while he was being prepared. Nothing very much to record except that I regret the extent to which the President believes that the Democratic candidates either talk nonsense or lie.

Tuesday, October 11, 1960

7:45–9:45 AM Pleasant breakfast with the President, Killian, and York. Began with personal chitchat and then swerved to partisan politics, the President delivering himself of a long discourse on how incompetent Kennedy is compared to Nixon, that even the more thoughtful Democrats are horrified by his selection, and that Johnson is the most tricky and unreliable politician in Congress. These remarks were interrupted only to ask "Who is that fellow Seymour Harris?" (alas poor Harvard) whose newspaper article was shown to the President yesterday and it is a pretty strong one because Seymour predicts that free society as we know it is bound to disappear and be replaced by planned society and socialism. I am really surprised at Seymour—what an assertion to make during a campaign, even though he isn't working directly for Kennedy! But he could never keep his trap shut. The President, after disposing of Harris, also talked at length about his concern about balancing the budget, pointing out that $12 billion worth of short-term U.S. obligations are held abroad and that in 1959, when the budget was heavily unbalanced, there was a tremendous run on our gold. He predicts that another run of catastrophic proportions will take place if the Democrats unbalance the budget and nearly deplete our gold reserves. At the end I asked the President about my trip to Omaha and he sweetly said that he thought I would like to do it, to which I replied I would do it only if ordered by him, as I considered it like a kamikaze dive, so he grinned and said okay, he will give me the orders. We then talked about internal difficulties in the Air Force and the Army, but without any decisive actions being agreed upon, which is not surprising in view of the short time remaining.

> Seymour Harris, a well-known liberal professor of economics at Harvard University, and a good friend of mine.

10:30 AM–12 M   Several gentlemen from the Brookings Institution questioned me about advice they should give to the incoming administration. I emphasized the importance of an early selection of a science advisor to the President, to give him indispensable advice on the technical qualifications of the dozen or more presidential appointees that should have technical competence. I believe that this point, as well as the point about the need of having assistant secretaries for R&D in the departments which don't already have them now, went across well.

3:00–5:00 PM   Attended briefing of Gates on Minuteman. A good one, without excessive promises and factual. When Gates was told that Minuteman could be equipped with a lighter warhead but one having twice the originally planned yield, he said he thought that was impossible because we weren't testing nuclear weapons. He was assured that this is certain to work, etc. He said he didn't believe it because McCone said it couldn't be done, so I slyly suggested that McCone's pessimism might have a political underpinning. I hope a seed of distrust in McCone has been planted in Gates's mind.

⤳ *Wednesday, October 12, 1960*

Visited the Pentagon to hear the preliminary briefing of their [Net] Comparative Evaluation Group on AICBM (ABM). Nothing much wrong with the study, but it certainly doesn't come to definite conclusions and probably can't. In subsequent discussion, Dillon emphasized the psychological effect of the announcement that the problem has been solved by one of the opposing parties and urged greater intensification of the Nike-Zeus project. Other comments were of more detailed nature.

Thursday, October 13, 1960

A special NSC meeting on the comparative evaluation of AICBM. The analysis was objective and, while not very conclusive, indicated the situation fairly well. Dillon again argued the danger of the psychological impact if the Soviets develop one ahead of us and so urged a strong program. The President emphasized the importance of a convincing demonstration that we have a capability but refused to accept the need of large deployment.

Then followed the regular NSC meeting, in which my presentation on international scientific activities was the last item. As in the week before, we never got to it.

Spent a pleasant afternoon at the Quantico Marine Base, being shown a demonstration of the prowess of the Marine Corps with modern weapons. Well staged, and I put it on thick, assuring them that I deeply believed in the importance of conventional warfare, so that we parted the greatest of friends.

Friday, October 14, 1960

Left at noon for Rochester. The plane was one and a half hours late, because of engine trouble, and I barely made Rochester in time for the panel meeting. The latter went well and since we were allotted only ten minutes each, I cut out a good third of my speech, including the last paragraph regarding President Eisenhower's birthday, about which I had misgivings anyhow. Questioning from the audience, 1200 undergraduates, was mostly directed at Ambassador F.E. Willis and dealt with the problems of Red China. Clearly this is a burning issue among the undergraduates and a source of misgivings about our foreign policy. Murrow gave his fairly typical prediction of a grim future for the U.S. and so my speech sounded very cheerful by comparison.

After cocktails and dinner, to the convocation to receive a degree
and to listen to Jim Baxter's eulogy of the Eisenhower administration's military and foreign policies. There is a staunch Republican! How wonderful to have that simplicity of faith.

> Edward R. Murrow, radio commentator; James P. Baxter, pres-
> ident of Williams College, and an old friend.

ᕫᕫᕫ *Saturday, October 15, 1960*

Had a discussion with Mesthene about his report on strengthening the Western World's science, which I felt needed a major
rewrite because it was more a sermon on all our sins than a
balanced analysis of problems. Spent most of the afternoon with
the AICBM panel, listening to appalling proposals by Convair
and Ramo-Wooldridge of satellite-based AICBM systems. Was
feeling so lousy by then that I was very discourteous (and don't
regret it either) to Convair's chief engineer. What an ass! His
whole introduction was devoted to how he was shuddering at the
thought that the Soviets will be the first to have a system like the
one Convair is proposing and so I said that while I had no
objections to his shuddering, I couldn't see what he was shuddering about.

ᕫᕫᕫ *Monday, October 17, 1960*

Chaired the PSAC meeting, and during the executive session
discussing nominees for PSAC had a long argument with Rabi.
He wants PSAC to devote their time to thinking about broad
problems, rather than to be members of a team with a special
assistant, working on more immediate problems of government. I
also listened to complaints from a few other members that our
agenda are too down to earth. Agreed to do something about it.

2:00–4:00 PM On Gates's invitation attended his briefing by
General Power and Admiral Parker on the Strategic Target System planning activity. Very impressive in one respect, namely

that for Gates's benefit there is a complete unification of the Services; everything is sweetness and light and no unresolved inter-Service problems. Went back to PSAC meeting with a feeling about what had been presented that something wasn't quite right but was unable to identify it.

PSAC then had a discussion of the latest version of the report on the organization of science in government, during which I lost my temper slightly because after all these months of effort, Rabi and Wiesner reopened a whole series of issues that appeared resolved months ago. I asked Rabi out of the meeting and rather forced him, because of his complaints in the morning, to assume responsibility for chairing the near session of the PSAC (Sunday, December 17) at which his "broad issues" will be discussed exclusively and warned him that he has to prepare those issues that he thinks PSAC should cover.

Spent the evening with the basic research panel and got home about 10:00 PM; took a seconal, went to bed and discovered I couldn't sleep, but kept thinking about the briefing by Power. Sometime later I suddenly realized that by their special interpretation of the President's directive on the probability of target damage that must be achieved, SAC has arranged things so that after a surprise attack our retaliatory strike would appear to be weak, according to their calculations. Naturally, this would become a powerful argument for insisting on massive increases in SAC forces. Smart boys! About 1:00 AM I gave up trying to go to sleep, turned on the light, and wrote up my analysis for George Rathjens to work out mathematically next day.

⟶ Tuesday, October 18, 1960

Had a long Pentagon briefing on command and control. Very interesting; it included valid criticism of the enormous electronic systems that the Air Force is developing; also a good analysis of the nature of the real problems involved, but the briefing was too long.

Phoned Gates and told him about my analysis of the Power presentation. He seemed to be quite indifferent and casual, but an hour or so later his aide, Gen. George Brown, called and said that Gates is worried about the implications for required force levels and wouldn't I send somebody from my staff to Omaha to get more information on the mathematics involved. I agreed and asked George Rathjens to go. In the meantime he made numerical calculations from which it appears that I was somewhat exaggerating the effect, but that it unquestionably exists and may amount to a fictitious proof that we need twice as many strategic delivery vehicles as otherwise.

∽ Monday, October 24, 1960

Interesting phone conversation with Admiral Burke about my visit to Omaha. He emphasized how he was afraid that SAC will use their study to build up additional force requirements for itself. I expressed my lack of enthusiasm for the job, but he said that the announcement that the President has asked me to do the job had already had an amazing effect on SAC. The bitter fights which SAC was having with Army and Navy personnel about methods and assumptions subsided overnight. He also told me that I must be careful with Admiral Parker, as he doesn't share Burke's concern.

Today I got the text of the President's directive. It's really quite a document.

> After instructing me about the information he wanted me to obtain from the Strategic Target System planning staff who were developing the first Single Integrated Operational Plan (SIOP), and authorizing me to take along associates of my choosing, the President's directive stated: "I am sending a copy of this note to the Secretary of Defense so that he may assure that you and your associates are provided all information and other assistance you deem necessary by the Commander-in-Chief SAC and Director, Strategic Target Planning Staff, as well as other elements of the Defense Department, at Omaha and elsewhere, as you consider appropriate for your technical study."
>
> I recall that with a grin he said that this paper gave me about as much authority as that of the secretary of defense and that I need not think my trip will be a kamikaze dive.

ᴄᴏ∿ *Tuesday, October 25, 1960*

Had a visit from Dr. Co Tui, who is going to Manila as a science advisor to Dr. Garcia. Discussed Philippine science problems with him; he has much the same ideas as I have—the need to strengthen the University of the Philippines and the need not to emphasize scientifically spectacular projects, but to put maximum effort into useful although less flamboyant work. Also taped for him a message to Garcia which was entirely noncontroversial.

1:00–4:00 ᴘᴍ FCST meeting, which was uneventful except for the criticism of Mesthene's draft of the summarizing paper on strengthening the Free World's science and technology. BOB is alarmed about the criticism of this administration implied in the report and Staats wants to withdraw the paper altogether, but the rest of the agencies felt that rewording is all that is necessary. Later I took the paper to Gordon Gray and asked him to read it overnight to determine if in a general way this is what he needs for NSC.

A navy captain from the JCS staff about the problems of countermeasures to protect our missile forces, something that I started them on quite some time ago. Considering the time and effort that went into the study, I am afraid the result could be described as a mountain bearing a mouse, but a mouse is better than nothing.

At lunch in the Mess sat next to Homer Gruenther, who said he was very busy with the campaign and so I pumped him. The Republicans feel that Kennedy is ahead and that they have an uphill fight. They were opposed to TV debates regardless of how effective the vice-president is, because each such debate builds up the image of Kennedy, who is less known to the voters. Homer intimated that much pressure is being exerted on the President to go campaigning to save Nixon, and that what he is planning to do, i.e., one speech in New York and perhaps one in Philadelphia, just isn't enough.

Homer H. Gruenther, special assistant to the President.

∽ *Wednesday, October 26, 1960*

Divided my time between the briefing of the Space Sciences Panel by NASA on the man-in-space ten-year plan and cost estimates and listening to the Limited War Panel. The NASA staff has obviously produced a realistic and detailed plan as well as cost estimate. Clearly the job of our ad hoc panel to present the costs of our space activities to the President and to the attention of the next administration will be comparatively simple. The data are there and only comments on their realism, etc., will be needed. The cost themselves are rather staggering: about $7 billion just to send men around the moon; more than $30 billion to land them on the moon, and a still much bigger sum to land men on Mars. These huge costs come about because the thrust of rockets now estimated necessary to carry out these missions is far greater than what PSAC estimated two and a half years ago in its public report on space. For instance, for the landing on the moon, a minimum of 9 million pounds' rocket thrust is estimated. In executive session of the panel, we talked about these things and I emphasized the need to spell out in our report what cannot be done in space without man. My opinion is that that area is relatively small and that, therefore, building bigger vehicles than Saturn B has to be thought of as mainly a political rather than a scientific enterprise. The panel seemed to agree with me.

The time with the Limited War Panel was not very productive. That panel has never congealed into an effective group and most of the discussion was diffuse and generalized. As has become his custom, John Foster delivered a speech on the promise of the pure thermonuclear weapons. Robert Bowie talked about the need of drastic political action cleaning up the Pentagon, etc.

> The President instructed me to make a study of the future costs of the NASA man-in-space programs and I asked our Space Sciences Panel, chaired by Donald Hornig, to do the job. The panel substantially agreed with NASA's cost and time estimates and when we presented them, I believe at a NSC meeting, the President was shocked and even talked about complete termination of man-in-space programs. I learned secondhand that the president-elect was also shown our report before inauguration and had then a negative reaction to the moon-landing proposition. The following spring after the Bay of Pigs fiasco and Yuri

Gagarin's Soviet orbital flight, President Kennedy announced the start of the Apollo Project. He stated over TV that NASA assured him that it would cost about $14 billion. Not a unique case of elastic cost estimates to fit the agency's wishes. The real costs of Apollo, including certain necessary auxiliary activities that went under other names (e.g., Gemini), were close to NASA's and PSAC's estimates in the fall of 1960.

⟳ Thursday, October 27, and Friday, October 28, 1960

Thursday and Friday spent with Zuckerman and his group from the U.K. Thursday we talked about the carrier task forces and tactical aircraft. My own thinking about the carrier problem has been influenced by this discussion, which was led by Harvey Brooks, and I am now somewhat dubious about our recommendation of a year ago to build smaller carriers. We heard strong arguments in favor of all-nuclear carrier task forces. The discussion of tactical aircraft was less useful. It was asserted that the newer antiaircraft weapons of the army are so effective that low-level flying will be suicidal unless carried out the way the future British aircraft will do, that is, by automatically following every land contour and therefore remaining hidden from such weapons. That evening I gave a dinner to the group, which was informal and pleasant except for an outburst by Rabi against Harold Brown in response to the latter's remark implying that he saw himself as a hostage in our midst. This was somewhat painful and required some soothing by me and Zuckerman. Rabi seems to dislike Brown intensely, as had already come out during the PSAC discussion of nominations for membership. He feels Brown has a closed mind, which I do not agree with.

Friday was devoted to the problem of battlefield surveillance. The WSEG briefer painted a very dismal picture of our efforts. A great deal of effort and money is going into the development of sophisticated electronic hardware that will probably be of no use, and no effort into better interpretation of aerial photographs, etc. The net result is that our position is not different from that of 1945. The discussion brought out that under conditions of nuclear warfare, the information that will be delivered to commanders will be inadequate and hence will inevitably lead to completely blind area-bombing.

The day ended with a dinner at Ambassador Lord Hood's home, a pleasant affair without any harsh exchanges. Before meetings began on Friday, I talked privately to Brown to smooth over Rabi's outburst and also explain my own position. I asked Brown to state to me whether his aim was preservation of peace or destruction of the Soviet Union, and he emphatically chose the first objective. I said I had that feeling about him and so asked him to join our meetings, whereas I didn't ask Teller, who is considerably more bloodthirsty. Brown agreed that Teller isn't uniquely concerned with the preservation of peace.

Monday, October 31, 1960

8:30–11:00 AM NSC meeting; the most extraordinary I remember attending. There were several items on the agenda, the first being a joint presentation by DOD and State on the military assistance program (MAP). It was introduced by Gates and then by Dillon; Jack Irwin gave the main presentation and was followed by Lemnitzer. It was a five-year outline, requiring $11 billion, which, as the presentation emphasized, was about $2 billion less than recommended by the Draper committee and less than half the $25 billion estimated to meet the JCS strategic objectives. The briefers stressed that it was in conflict with the NSC recommendation in October according to which Belgium, the Netherlands, Italy, Japan, and Portugal were deemed to be sufficiently wealthy so that we could taper off our military assistance to them.

After the briefing, the President said that the proposed level of expenditures would further hurt our balance of payments, which is even more critical than balancing the budget. The politicians running for office now promise everything to everybody, but whoever gets elected will have to face the hard facts of life. On the other hand MAP must continue, as the only alternative is isolation and the eventual victory of dictatorships. He said that we must be tougher with our Allies to improve our balance of payments. Thereupon Secretary Anderson delivered an emotional diatribe. He was chiefly concerned with the loss of our gold reserve and hence a loss of confidence in the U.S. He noted that balance of payments in the third quarter is so bad that it would imply a $5.6 billion rate of loss annually. His remarks were sprinkled with statements like these: We are in trouble right now, not next year and not six months from now. The Allies must pay for what they get. I cannot any longer carry this terrible load and somebody else must assume responsibility for the dollar. The gist was that we should drastically reduce all our foreign aid and should cut out MAP to the nations I mentioned previously. He was so carried away that he said we should stop aid to France, Germany, and England, and was reminded that we haven't been giving them a cent of MAP now for several years. Then he warned about the threatening deficit in the budget and said that the boys abroad will discover it very soon and will pull out their investments of $17 billion in the U.S. and there goes all of our gold. The President suggested with a grin that Secretary Anderson has been scaring him now for several years and has him so frightened now that going to war may seem to be a lesser evil than going off the gold standard. Then, grinning at McCone, he said that perhaps we should start selling our plutonium and U 235 all over the world to get back the gold.

Stans said emotionally that the expected 1961 budget surplus has already been eroded to nothing by increased military expenditures and decreasing profits from sale of copper. He expects a deficit in 1961 and that is intolerable. He pointed out that if we were to capitalize all federal commitments such as veterans' aid, military retirement, etc., the total federal debt would be three-quarters trillion dollars. Irwin, Dillon, and Herter went to the defense of the MAP plan. They emphasized that our standard of living is higher than that in other countries and so the arguments that we are contributing more and that our national debt per capita is higher are not convincing; the only way to induce others to increase their military budgets is by continuing our contributions. The discussion was ended by Gates who remarked that the overseas deployment of our forces is a greater burden on our balance of payments, and that it and MAP are intimately connected, and MAP in his opinion is a far better way to spend money. The decision was not to accept the plan, but to submit it for consideration by the Bureau of the Budget, since otherwise, as Stans pointed out, it could only challenge details and not basic planning.

Wednesday, November 2, 1960

Spent morning in York's office discussing various military R&D budget items. York, back to his usual form, very active in the conversation. On the whole no strong disagreements as I cautiously went about presenting our ideas on various items, except on the AICBM panel's recommendation for very limited production of Nike Zeus. York is violently opposed because he is convinced that once any production is started, it will not be possible to control it and we will have a fantastically large operation with very little return.

> Fortunately, York's position prevailed in forming the 1962 budget, and only R&D on ABM was authorized. Our panel, as I recall, was extremely critical of the Nike-Zeus system as totally incapable of defending against a large missile attack. It may be that personal friendships with the people of the Bell Telephone Laboratories, some of whom were involved in the panel briefings as the representatives of the contractor, helped in giving Nike Zeus the benefit of the doubt in the panel's report.

Thursday, November 3,
Friday, November 4,
and Saturday, November 5, 1960

Rathjens, Scoville, and I were received at the SAC headquarters in the Offutt AF Base politely, but not as VIPs, i.e., only a colonel met us at the airport Wednesday evening.

On Thursday, we made a ceremonial call on General Power and had a polite conversation and then proceeded with briefings. Shortly we were joined by Power, who immediately objected to Pete and George taking notes. Whereupon ensued a protracted and at times tense argument between him and me in which Power kept saying that he was responsible for the lives of his men and he must be given explicit orders to let us take notes. I won my point that we must be allowed to take notes, without showing him my presidential directive that I kept in the pocket as a last reserve. I did agree that we wouldn't take notes on operational details, i.e., specific sorties, flight profile plans, etc.

Actually, of the whole time we were there, we spent only half

ʿan hour being briefed on these details of the operational plan, during which time we didn't take notes. The rest dealt with methodology and other technical issues that we were instructed to study. My general impression is a rather troubled one in that I can see that this integrated targeting plan could be used to generate military requirements for tremendous additional forces, especially when the new administration takes over and the new people will not know the background, i.e., how these requirements were generated.

The evening of the first day we had a dinner, Dutch treat, with Butch Griswold, which was a pleasant occasion. On Friday night General Power gave a party for all the operational planning staff and their wives, some 500 people. General Bob Smith picked us up at the BOQ and took us to the club for the party, but then we were left to our own devices for a couple of hours. Talked to a lot of people, including a Colonel Gullion, who is representing CINCEUR [Commander in Chief, Europe], and who after a few drinks complained bitterly that they were being pushed around and simply told what they had to do.

On Saturday morning Pete and I spent an hour with Smith, being shown their intelligence operations layout. Really impressive and obviously expense is not a factor here. In the meantime, George was getting additional information that I asked him to get and he got quite a lot.

✺ Monday, November 7, 1960

Attended an NSC meeting, which included a briefing by Rubel on space activities of DOD. It certainly was a sanitized version and so presented that there was not much opportunity to bring out its incompleteness. A piece of bad news is that apparently the revolution in El Salvador has led to the infiltration of many Commies into key positions in the government. Are we having another Cuba so soon?

~~~ *Wednesday, November 9, 1960*

With Gates and Douglas for about 45 minutes talking about our Omaha trip. They were quite concerned about some of the things I told, expressed a desire to have another meeting next day, and proposed that all the Service Chiefs be present. I demurred, pointing out that I should first make a report to the President, whereupon I would be glad to meet with the Chiefs. So we compromised by agreeing to have Lemnitzer in.

~~~ *Thursday, November 10, 1960*

10:30 AM–4:15 PM   I chaired the seminar on Technical Assistance, in which we tried to elicit ideas from some fifteen people present on what the PSAC panel on Technical Assistance to Underdeveloped Countries could do to be helpful. The meeting was on the whole successful and those present not only felt that a new start is needed because of lack of basic philosophy and poor planning of our activities, but also that PSAC could be helpful. However, it was clear that the study of purely technological problems is not the answer and that there must be a close integration of social and technical planning. My conclusion is that we should have a panel and so I told Walsh McDermott, but also that the panel must have strong representation from social scientists and educators to be effective. I suspect that McDermott, as panel chairman, and my successor will have a hard time getting important contributions from the panel, but it is worth trying since the problems are so difficult, are unsolved, and need solution.

4:30–6:00 PM   With Gates, Douglas, their military aides, Lemnitzer, and Rubel. Reviewed again our findings at Omaha, emphasizing that the SIOP is derived on the basis of military judgment only and not from machine calculations as the planners would like us to believe, but is about as sound as could be hoped under these conditions. However, the chosen procedures may lead to less than optimum plans when they are applied to the small residual forces that may be available to us after a devastating surprise attack. I emphasized that the damage criteria in the directives to the planners are such as to lead to unnecessary and undesirable overkill. This has an evil effect on planning for small forces and also will be used by General Power as a justification for urging greater force requirements. I think it made a fairly strong impression on those present, but whether anything useful will come of it, I don't know. However, Gates and Douglas both thought that a revision of damage criteria should be ordered.

Friday, November 11, 1960

Spent all day with the PSAC panel chairmen working on our summary paper on military R&D budget. Fairly productive meeting and we seem to have gotten a reasonably good paper together. Jerry Wiesner was around part of the time. He is acting very coy and spent two hours having lunch with some Kennedy bigwigs, but is mum about the results.

Saturday, November 12, 1960

Paul Nitze phoned and asked if I would be interested in talking to him. I agreed and he came for about an hour. Some of the discussion dealt with the position of special assistant for science and technology. I was unhappy to learn that my successor will not be appointed until after inauguration day and emphasized that the president-elect needs a man to advise him on technical qualifications for quite a large number of important, if not top-level, presidential appointees and hence requires him early. I spoke also of the need of assistant secretaries for R&D to make the Federal Council into an effective body. Paul questioned me about my ideas of who would be a good successor after first asking me if I was available to continue. From the nature of his questions, I got the disturbing feeling that they are by no means settled on Jerry. He emphasized the need of a man with a "distinguished'' scientific reputation when I mentioned Jerry as my first choice.

Paul H. Nitze, economist and former government official was a member of the transition team (headed by Clark Clifford) of the president-elect.

∽ Monday, November 14, 1960

Breakfast with Persons and Goodpaster, at which I persuaded them that they should support the Seaborg report's being made public after minor editorial corrections, which Seaborg and I made later in the day. Also talked about the early appointment of the special assistant by the new administration and Persons agreed to mention the subject to Clark Clifford, whom he is to see today. He got from me a list of present and past members of PSAC for that meeting. I suggested that our paper on government organization for science should be transmitted to the next administration, and it was agreed that Goodpaster would look it over prior to transmittal. Hope his reaction will be as favorable as was that of Elmer Staats.

The rest of the day chaired the PSAC meeting, at which I had again a little run-in with Rabi, who was indignant that we were spending all of our time on ''unimportant details.'' He gave an account of his conversation with Emelyanov by whom, I think, he was completely taken in. Emelyanov claims that the USSR policy is, that if only one nuclear weapon explodes on USSR soil, they will immediately launch an all-out attack on the U.S. without determining whether it was an accident or intentional. I think it was a deliberate plant. All afternoon was spent on our military budget paper in the presence of York and Rubel, who were quite helpful. Alvin Weinberg created considerable difficulties by arguing persistently and not always to the point. During the morning when we were discussing possible future members, he was also very stubborn, insisting on Wigner and condemning several others. I am far from convinced that his net contribution to PSAC is constructive. Amusingly, it was Rabi who pushed him very hard for membership but now denies having done so and is angry at Weinberg.

Tuesday, November 15, 1960

With only half the membership remaining, we had a good session discussing broader issues related to Nike Zeus, the man in space, the Armand report to NATO, and the value of fallout shelters. Meeting broke up around 4 :00 PM.

THE WHITE HOUSE

WASHINGTON

January 6, 1961

Dear George:

I accept, as I must, your resignation as my Special
Assistant for Science and Technology, effective
January 20, 1961. I do so with real regret, for
the association we have enjoyed has meant a great
deal to me. You have served not only with the utmost
professional distinction but with a spirit of constructive helpfulness and outstanding dedication.

For your public-spirited service to your country and
your unfailing assistance to me, I am everlastingly
grateful. I especially appreciate your offer to continue to be available to me for future assistance.

As you return to the life of teaching and research --
to which I know you are so deeply dedicated -- you take
with you my very best wishes for a future as rewarding
and productive, both for yourself and for our country,
as has been the period just ending.   For my part, I
shall continue to prize the opportunity we have shared,
in company with your colleagues of the Science Advisory
Committee and its panels, to work together for the good of
the nation.

With warm regard,

As ever,

Dwight Eisenhower

The Honorable George B. Kistiakowsky
Special Assistant to the President
for Science and Technology
The White House

Epilogue

THE AUTUMN MONTHS of 1960 were still quite busy for me, but my diligence in recording the substance of what went on around me faded in the measure that the Eisenhower administration went into caretaker status. The election of Kennedy meant, of course, a quantum jump in that direction and so my diary ends in mid-November.

This leaves most of our activities dangling in mid-air and for those readers who may be experiencing a slight feeling of suspense about their fates here is a brief account of what happened, written from memory with the aid of some printed records.

I delivered to the President a detailed report (written jointly by Rathjens, Scoville, and myself) on our investigation of the target-planning activities at SAC headquarters for the first SIOP. I believe I did it a few days after describing our findings to the secretary of defense. The President was greatly interested and after discussing it he sent a copy to Gates who, I was told, gave it later to his successor, Robert McNamara. Secondhand I also heard that Eisenhower gave a copy to the president-elect.

In any case, the Kennedy administration ordered in the spring of 1961 a repeat operation by SAC and Navy with a directive which, I hope, was more specific than the one guiding the operation in the fall of 1960.

The National Reconnaissance Office continued in charge of

work on intelligence satellites and maintained effective progress in this highly important field.

Sometime in November the President called me in to return the draft of the Seaborg panel report which he had annotated. He asked me to make a few textual changes, whereupon he was ready to authorize its publication as a White House document. This was done and in an introduction the President endorsed the report's findings about the interrelations of graduate education, scientific research, and the federal government. The report may have contributed to a gradual change in the general policy of federal support of research in the institutions of higher learning that took place in the early sixties.

Our military R&D budget activities culminated with an extensive paper, copies of which went to the President, the secretary of defense, and to Herb York. I do not recall attending a meeting of the President with the BOB and Pentagon officials such as had taken place in the fall of 1959, or even hearing about one. The budget delivered to Congress in January 1961 was consonant with our recommendations; but since York and PSAC largely agreed, both can claim credit for moderating the wilder Service propositions.

On the civilian "science budget" I listened to the findings of our panels and used them orally to influence the Bureau of the Budget, not always successfully to be sure.

The SLAC project got going, the machine was built under the leadership of Pief Panofsky and has since contributed a great wealth of information on elementary particles of matter and their interactions.

The Material Sciences centers were established in several institutions of higher learning and have been active since.

The oversight of the oceanography program by the Federal Council of Science and Technology continued and was strengthened in the sixties. A successful international oceanographic exploration of the Indian Ocean did take place.

PSAC held regular meetings in December and January, but I recall few details except that the earlier one was held on a rather luxurious and comfortable private estate on the edge of the District of Columbia. Rabi, who was to chair the meeting, prepared no agenda and the discussion remained totally unstructured. It was enjoyable, as I recall, but resulted in no great thoughts for the next administration.

There are no entries on the Geneva test-ban negotiations in the diary after August because the Committee of Principals did not meet in the fall, and there were no significant new policy initiatives. In Geneva the meetings resumed in September and recessed in December without significant accomplishments, as the Soviet Union showed no interest in compromise. Its position continued to harden during the spring of 1961 when the United States and the United Kingdom offered substantial concessions to meet the USSR halfway, and Khrushchev even withdrew the offer to permit on-site inspections in the USSR. The negotiations reached a total impasse in the summer of 1961, and on August 30 the Soviet Union announced the resumption of nuclear weapons tests, exploding among others a monstrous 50-megaton bomb in the atmosphere. A week later the United States announced the start of its own testing.

I continued as a member of PSAC (chaired then by J. B. Wiesner) through 1963, and in 1962 was also appointed by President Kennedy to the membership of the General Advisory Committee of the newly created Arms Control and Disarmament Agency, but in neither capacity was I significantly involved with the nuclear test-ban issue. This involvement came through a Pugwash meeting.

In the late fall of 1962 Khrushchev wrote to Kennedy, offering "as a political compromise" two to three on-site inspections of suspicious seismic events in Russia per annum, an offer reaffirmed publicly by the Soviet Union in February 1963. The American position was then that at least seven annual inspections in the USSR must be authorized by the treaty, which still was to involve a threshold of 4.75 on the Richter scale (that is about equivalent to the shock produced by 20-kiloton weapons) and unilateral moratoria.

The secretary general of Pugwash invited small British, American, and Soviet groups of scientists to meet in London to discuss the progress and effectiveness of seismic identification of remote underground explosions. This we did on March 16 and 17, 1963. The Soviet group was headed by a distinguished physicist, Academician Artsimovich, who earlier played a major role in the Soviet H-bomb project. He later became very active in furthering cordial relations with Western scientists and is best known for his creation of the most successful experimental machine for controlled nuclear fusion, the tokamak, which has now been adopted by most laboratories in America and elsewhere that are working on this challenging problem.

During a break in the London meetings Artsimovich (whom I was meeting for the first time) took me aside, explained that he had regular personal contact with Khrushchev, and asked me to convey forthwith the message to the White House that Khrushchev would compromise on five annual on-site inspections in the USSR, provided the same numbers would apply to the United States and to the United Kingdom. Accompanied by Paul Doty, to whom I revealed the message, I flew to Washington.

In the reshuffle of the White House Office after the Kennedy inauguration the new special assistant for national security affairs, McGeorge Bundy, assumed direction of disarmament and assigned its monitoring to his deputy Carl Kaysen. To him we repaired on the eighteenth and I gave the message. It seemed to me to break the log jam in the test-ban negotiations, since the United States over a period of time had scaled down its demands for on-site inspections gradually from twenty to seven, so that the acceptance of five would represent a much smaller further concession than the doubling of his own number that Khrushchev now offered.

Unfortunately, to the best of my knowledge, this message generated no new initiative by Washington; for instance, on March 21 President Kennedy publicly stated about the test-ban on-site inspections that there was disagreement about the numbers, as if he had not been informed about the compromise offer.

In the summer of 1963 the test-ban negotiations, which had resumed in 1962, were terminated by the signing in Moscow of a partial test-ban treaty. This treaty had a welcome environmental effect by almost eliminating atmospheric tests and so their radioactive fallout, but exerted virtually no effect on the nuclear arms race, because it allowed unlimited underground testing, and this has since been extensively practiced by the superpowers. Thus almost all parties in Washington and Moscow were pleased with this substitute for the dragging negotiations and the Senate easily ratified the treaty.

I saw President Eisenhower twice, two days before the Kennedy inauguration. First, I was called to the cabinet room and there found assembled (without my knowledge) all my staff. The President came in and to my total surprise presented me a Medal of Freedom with a most generous citation that was read to us by Andy Goodpaster, who then received the medal himself. This recognition rather overwhelmed me, particularly since the President was very sparse with presentations of this medal, which he regarded as the highest award in peacetime.

The evening before I had been listening to Eisenhower's farewell address to the nation. I took no part in preparing it, but the contents were not a complete surprise because the President had talked to me more than once about his concern with what the speech called "the military-industrial complex." In speaking about the technological revolution he said that evening:

> In this revolution, research has become central; it also has become more formalized, complex and costly. A steadily increasing share is conducted for, by, or at the direction of, the Federal government. . . . Yet, in holding scientific research and discovery in respect, as we should, we must also be alert to the equal and opposite danger that public policy could itself become the captive of a scientific-technological elite.

The next day several of my scientist friends phoned me to ask whether Eisenhower was turning against science. That evening the Eisenhowers gave a farewell party for the White House staff. As I shook hands with the President he asked about the reaction to his talk. I told about the queries I had received, which seemed to make him quite upset, and when the formal reception was over he took me from the party to a private room, ordered tall scotches and first assured me that he was unhesitatingly for basic academic research and feared only the rising power of military science, and that I should make this clear publicly. He talked then at length about his concerns and I can do no better telling about what he said than to quote from his farewell address:

> This conjunction of an immense military establishment and a large arms industry is new in the American experience. The total influence—economic, political, even spiritual—is felt in every city, every State House, every office of the Federal government. We recognize the imperative need for this development. Yet we must not fail to comprehend its grave implications. Our toil, resources and livelihood are all involved; so is the very structure of our society.

> In the councils of government, we must guard against the acquisition of unwarranted influence, whether sought or unsought, by the military-industrial complex. The potential for the disastrous rise of misplaced power exists and will persist.

> We must never let the weight of this combination endanger our liberties or democratic processes. We should take nothing for granted. Only an alert and knowledgeable citizenry can compel the proper meshing of the huge industrial and military machinery of defense with our peaceful methods and goals, so that security and liberty may prosper together.

Abbreviations, Acronyms, and Code Names

ABM   Antiballistic missiles system (also AICBM), at that time involving the Nike-Zeus missile system

ABMA   Army Ballistic Missiles Agency, headquartered in Redstone Arsenal at Huntsville, Alabama; Dr. Wernher von Braun headed the civilian staff

AEC   United States Atomic Energy Commission, chaired by John McCone

AFTAC   Air Force Technical Applications Center, responsible for seismic and other methods for detection of Soviet nuclear explosions

AICBM   See ABM

ALBM   Air-launched ballistic missile

ANP   Aircraft Nuclear Propulsion project, managed by the Air Force and later canceled, was started shortly after the end of World War II and involved an expenditure of a billion or more dollars

ARDC   Air Research and Development Command, headed by Gen. Bernard Schriever, in control of most but not all research and development in the Air Force

ARGUS   High-altitude rocket-borne nuclear test explosions in 1958

ARPA   Advanced Research Projects Agency, set up in 1958 or early 1959 in the Office of the Secretary of Defense

ASW   Antisubmarine warfare

Bevatron   First of the large postwar particle-accelerator machines for nuclear physics research, built at the University of California at Berkeley by Ernest Lawrence and his collaborators

BMD   Ballistic Missiles Division of the ARDC, which also managed the mushrooming space projects of the Air Force

BMEWS   Ballistic missile early warning system : two installations (the first in Thule, Greenland; a later one in Alaska) of very large radars capable of detecting Soviet ICBMs in mid-flight to the North American continent and thus giving about 15 minutes warning of a coming attack

BOB   Bureau of the Budget, founded in 1921 by an act of Congress and moved into the newly created Executive Office of the President by President Roosevelt in 1939

Bomarc B   Aerodynamic medium-range air defense missile of the Air Force

BOQ   Bachelor Officers Quarters, on military bases

CBW   Chemical and biological warfare

CEP   Circular error, probable : a measure of missile accuracy

CERN   European Center for Nuclear Research, near Geneva, Switzerland

CIA   Central Intelligence Agency

CINCCARIB   American military commander in chief of the Caribbean zone

CINCEUR   American military commander in chief in Europe

COC   Highly shock-resistant underground command center near Denver, Colorado, of the North American Air Defense Command

COMINT   Intelligence gathered by the interception of radio messages

CW   Chemical warfare

DDRE   Director of Defense Research and Engineering and his office, set up in 1958, Herbert York being the first director

DEW Line   Distant Early Warning air defense installations in northern Canada

Discoverer   Photographic intelligence-gathering satellite project, in development

DOD   Department of Defense

Dynasoar   Rocket-launched long-range space glider project

ELINT   Intelligence from intercepted telemetry, radar, and other electromagnetic signals

EOB   Executive Office Building, next to the White House, across the West Executive Avenue, where our offices were located on the second floor

FAA   Federal Aviation Administration

FCST   Federal Council for Science and Technology

FDA   Food and Drug Administration, in the Department of Health, Education, and Welfare

FY   Fiscal year

GAC   General Advisory Committee to the Atomic Energy Commission

GAO   General Accounting Office, an arm of Congress monitoring federal expenditures of the executive branch

GSE   Ground support equipment, e.g., of operational missiles

HEW   Department of Health, Education, and Welfare

Hound Dog   A subsonic air-to-ground missile with a nuclear warhead launched from B-52 bombers

ICA   International Cooperation Administration

ICBM   Intercontinental ballistic missile

IDA   Institute for Defense Analyses: a nonprofit corporation in Washington

IGY   International Geophysical Year

IOC   Initial operational capability of a new weapons system (e.g., Atlas)

IRBM   Intermediate range ballistic missile

JCS   Joint Chiefs of Staff: the uniformed heads of the four (including the Marines) Services and their chairman with his staff

LOX   Liquid oxygen

MAP   Military assistance program

MATS   Military Air Transport Command, since disbanded

MIDAS   Military satellite project to detect enemy ICBMs in early flight

Minuteman   Three-stage ICBM propelled by solid propellant rockets, launched from silos

MITRE   A nonprofit corporation split from the Lincoln Laboratory of the Massachusetts Institute of Technology, which was engaged in military contract work

NACA   National Advisory Committee on Aeronautics, precursor of NASA

NAS   National Academy of Sciences, chartered by Congress in 1863

NASA   National Aeronautics and Space Administration, created in 1958

NASCO   National Academy of Sciences, Committee on Oceanography

NCRS   National Committee on Radiation [exposure] Standards, under the wing of the National Bureau of Standards

NEG   Net (or Comparative) Evaluation Group in the Pentagon, headed by Vice-Admiral Sides

NIE   National intelligence estimates, prepared by USIB

NIH   National Institutes of Health

Nike Hercules   The most advanced air-defense rocket-propelled missile

NORAD   North American Air Defense Command, unifying American and Canadian air defense activities

NRAC   Naval Research Advisory Committee, associated with ONR

NRC   National Research Council of NAS

NSA   National Security Agency

NSC   National Security Council

OAS   Organization of American States

OCB   Operations Coordinating Board, charged with ensuring that NSC actions are carried out

OCDM   Office of Civilian Defense Mobilization in the Executive Office of the President

OEECD   Organization for European Economic Cooperation and Development

ONR   Office of Naval Research

Orion   A proposal for space-vehicle launching in which rapid successive explosions of small atomic bombs under the space vehicle would by recoil accelerate it into space

OSD   Office of the Secretary of Defense

Pershing   Mobile long-range nuclear-tipped surface-to-surface Army missile

Planning Board   Interagency group charged with preparing policy papers for NSC

Pluto   Nuclear-powered ram-jet engine for the SLAM vehicle, under development by the Livermore Nuclear Weapons Laboratory; later canceled

PSAC   President's Science Advisory Committee

Pugwash   Nongovernment international conferences of scientists concerned with disarmament, etc., started at the initiative of Albert Einstein and Bertrand Russell in 1957, by Cyrus Eaton; the first conference was held in Eaton's estate in

Pugwash, Nova Scotia, Canada

    R&D   Research and development

    RDTE   Research, development, test, and evaluation (of military systems)

    REA   Rural Electrification Administration

    Rover   Nuclear rocket engine under development by the Los Alamos Laboratory; later canceled

    SAC   Strategic Air Command of the United States Air Force

    SAGE   Sophisticated semiautomatic ground system for air defense in late stages of development and deployment

    SAMOS   A multitude of sophisticated intelligence-satellite projects

    SIOP   Single Integrated Operational Plan

    Skybolt   Air-to-ground long-range ballistic missile project

    SLAC   Stanford linear accelerator project: a machine estimated then to cost $20 million, to accelerate electrons to highest energies

    SLAM   Nuclear-powered supersonic low-altitude missile project; later canceled

    STL   Space Technology Laboratories, a division of the Ramo-Wooldridge Corporation and the technical manager of ICBM projects for the Air Force

    TCP   Technological Capabilities Panel, a large study of national security problems in 1954

    USDA   United States Department of Agriculture

    USIA   United States [foreign] Information Agency

    USIB   United States Intelligence Board

    VELA   Project operated by ARPA to improve seismic and other detection of nuclear warhead tests

    VOA   Voice of America

    WSEG   Weapon systems evaluation group, attached to the JCS

    ZI   Zone of Interior, meaning the contiguous United States

Index